William Pollock Fraser (1867–1943)
Courtesy University of Saskatchewan Archives

Essays on the Early History of Plant Pathology and Mycology in Canada

RALPH H. ESTEY

McGill-Queen's University Press
Montreal & Kingston • London • Buffalo

Legal deposit first quarter 1994
Bibliothèque nationale du Québec

Printed in Canada on acid-free paper

This book has been published with the help of a grant from the
Social Science Federation of Canada, using funds provided by the Social
Sciences and Humanities Research Council of Canada.

Canadian Cataloguing in Publication Data

Estey, Ralph H. (Ralph Howard), 1916–
Essays on the early history of plant pathology and mycology in
Canada
Includes index.
ISBN 0-7735-1135-0
1. Plant diseases – Research – Canada – History. 2. Mycology –
Research – Canada – History. 3. Plant diseases – Study and teaching
– Canada – History. I. Title.
SB732.54.C3E88 1994   632'.3'072071   C93-090581-4

This book was typeset by Typo Litho composition inc.
in 10/12 Baskerville.

Contents

Preface

Plant pathology and mycology are recognized as important branches of botany, yet a fully referenced history of them, as they apply to Canada, has not been written. It was my intention to fill in, at least partially, that void in the history of science in Canada.

In reviewing nineteenth-century literature pertaining to agriculture and horticulture, I recognized the popularity of the essay as a vehicle for presenting ideas and information. It was so appealing that I chose the essay format for the presentation of this history of the genesis of plant pathology and mycology in Canada. In doing this, my objective has been to make each essay complete in itself, with its own references and endnotes. Thus the overall history is by nature somewhat episodic, with an inevitable unevenness between essays. There is also some unavoidable duplication, as, for example, when the work of one person is so encompassing that it has to appear in two or more essays.

It was my intention to ferret out of the early literature as many of the obscure accounts of plant diseases as possible, and the early efforts to understand them, plus the earliest accounts of the collection and identification of fungi. I did not mean to provide a complete listing of any individual's publications, but rather to draw attention to pioneering work in the general fields of mycology and plant pathology. Given such introductory information, additional references to their work may then be more easily found.

A work like this proliferates, like disease in an infected plant. For this reason it is restricted, almost entirely, to published accounts of work done in Canada. Regrettably, it leaves out the important contributions of technicians, plant-disease inspectors, supervisors of crop protection, mushroom growers, and many others whose type of work virtually precluded publications. It is also restricted in the number of crops with which it deals. For example, the tobacco crop, and several vegetable crops, have been omitted, because I did not consider the early work on their diseases important enough to devote a special chapter to it. Thus the reader will search in vain for P.G. Newell and R.H. Stover, who did some early research on tobacco diseases, or for Roger Desmarteau, Lucien Laporte, Louis Coulombe, Roger Paquin, and Lionel Cinq-Mars, who were involved in such a variety of plant diseases and disease consultation services in Québec that they had little opportunity to do much research on any one. They, and many others like them in other provinces, played important roles in the developing science of plant pathology in Canada.

With a few exceptions, this history ends around 1950. It ends there partly because significant contributions to the development of plant pathology by entomologists and horticulturists had virtually ended by that time, but mostly because many veterans of the Second World War had just completed studies for advanced degrees in universities and were entering the general fields of plant pathology and mycology. Their recent training in chemistry, physiology, ecology, genetics, microbiology, statistics, etc., prepared them for the introduction of a new phase in the history of both fields that was more professional, more biochemically oriented, and otherwise different from the earlier phases of both plant pathology and mycology.

Acknowledgments

The initial series of research visits to libraries and archives in every province of Canada was made financially possible by a grant from the Social Sciences and Humanities Research Council of Canada. The author is very grateful for that, and for the unfailing courtesy of the curators, librarians and others, in the many institutions that were visited. Such courtesy made the many hours of research a pleasant task.

Dr J. Ginns, Curator, National Mycological Herbarium of Canada, reviewed an early draft of the essay "A History of Early Mycology in Canada" and made several helpful suggestions, and his colleague Dr S. Redhead, Program Leader of the Economic Fungi Program, Biosystematics Research Centre, Ottawa, provided a number of useful references for that essay.

H.E. Gruen, Professor of Biology, University of Saskatchewan, read an early version of the section dealing with the history of mycology in that province, and Alex Wilson, Curator of Botany, Nova Scotia Museum, Halifax, read the section for that province. Their constructive criticism is appreciated.

Introduction

Plant pathology (phytopathology) is the branch of botany that deals with all aspects of diseases and disorders in plants, especially their causes, symptoms, prevention, control, and cure.

In many respects it is a uniquely difficult science to chronicle. Because it is not a fundamental science, it has been built on, or with parts of, other sciences; some might say it has parasitized other sciences. Thus, it has no beginning, and no precise boundaries for its scope have yet been drawn. The very expression "plant pathology" is of relatively recent origin, and a generally acceptable all-inclusive definition of those words had not evolved until near the end of the period covered by this history.[1] Nevertheless, in retrospect, one can detect aspects of the recognition of plant diseases, identification of their causes, and attempts at their control, in some of the earliest records of agriculture, horticulture, and forestry in Canada. This is possible because the increasing awareness of the importance of plant diseases that accompanied the steady development of those fields of endeavour made disease control a component of each, long before a distinctive science of plant pathology had evolved.

At first, the study of diseased plants was largely the work of entomologists, horticulturists, and botanists whose approach was almost entirely descriptive. Then, following a recognition, first by Europeans, of the parasitism of fungi, the study of disease in plants (in contrast to that of diseased plants) passed slowly into the realm of mycology, the science that deals with mushrooms, mildews, moulds, and other fungi. Even the early study of bacterial diseases, virus diseases, and various

Charter members Canadian Phytopathological Society. Left to right: Arthur W. Henry (1896–1989); William C. Broadfoot (b. 1899); Cecil E. Yarwood (1907–81); Ibra L. Conners (1894–1989).

non-parasitic diseases or disorders was often referred to as either "applied botany" or "applied mycology." This was largely a European phenomenon that did not last long in Canada.

The close association of mycology with plant diseases, in the latter part of the nineteenth century and the early decades of the twentieth, could easily lead to the notion that plant pathology developed solely as a branch of mycology. The fact is, however, that the genesis of plant pathology took place, in Europe, prior to the middle of the nineteenth century, during a period in which mycology played little part. A somewhat similar pattern of development is discernible in early Canada – that is, in the geographical area that is now Canada, although the premycology phase was relatively brief.

Mycology went through a lengthy descriptive, taxonomic phase during which Europeans and Americans set up classification schemes and otherwise provided a groundwork for studies, in Canada, of the life histories and the parasitic behaviour of many fungi. These were necessary prerequisites to the formulation of measures for their control.[2] This later phase was rapidly taken over, in the early decades of the twentieth century, by specialists who became known as plant pathologists. They did this to such an extent that the lines of demarcation between mycologists and plant pathologists, in relation to plant diseases, virtually disappeared by the end of the 1950s.

Early in the twentieth century, plant pathologists in Canada built a collective identity which culminated in the formation of the Canadian Phytopathological Society in 1929. The history of that society,[3] and its predecessor, the still very active Quebec Society for the Protection of Plants,[4] have been dealt with elsewhere. Professional mycologists have never had a distinctive society in Canada.

Note: Throughout these essays an attempt has been made to provide the year of birth and death for each individual, as for example, Melville DuPorte (1891–1981). However, when only the date of birth is known it is given as, for example Firmin Letourneau (b.1891). There are many instances in which neither the date of birth nor death are known.

1 Entomologists and the Genesis of Plant Pathology in Canada

The very early agriculturists had more plant pest problems than do modern farmers. They had to compete with wild grazing animals, mice and other small animals, birds, insects, and plant diseases. Because they had no concept of plant diseases, they tried to protect their crops from the pests that could be scared away or forcibly removed, and the most numerous and often the most troublesome of them were the insects.

The fact that insects were sometimes devastating to ancient farmers is reflected in the biblical stories of the locust plagues of Egypt. They were also a nuisance to Roman farmers, who are believed to have been the first in the Western world to use "chemicals" for pest control. To substantiate this claim, reference is usually made to Cato, a Roman agriculturist in the second century BC who wrote that if bitumen, sulphur, and amurca were boiled together in a copper kettle, cooled, and applied to the trunk and branches of vines, the mixture would keep caterpillars away.[1]

The history of crop protection closely follows the history of the increasing knowledge of the organisms that are visually troublesome. Early farmers could see insects chewing the leaves of plants and they learned to recognize insect damage without actually seeing the pests in action. It naturally follows that insects occupied a prominent place in the minds of early agriculturists, and an elementary study of insects preceded, by many decades, the study of less visible organisms that damage plants or produce diseases. In their work with insects in relation to plant pest problems, some of the early amateur entomologists made noteworthy contributions to what eventually became the science of plant pathology. Modern plant pathologists have acknowledged this in various ways. For example, the American Phytopathological Society honoured the Danish entomologist Johann Christian Fabricius (1745–1808) by republishing, in English, his 1774 work, *Attempt at a Dissertation on the Diseases of Plants*, as Phytopathological Classic no. 1. Similarly, the 1835–1836 work *Del Mal del Segno*, by the Italian amateur entomologist Agostino Bassi (1773–1856), has been published as Classic no. 10.

Agriculture, in the geographical area that is now Canada, developed slowly, because men of wealth and influence were, for a long time, more interested in the fur trade, shipping, lumbering, and the military than in the growing of crops. It was early in the nineteenth century before agriculture, including horticulture, began to replace the fur trade as a leading Canadian industry. The increasing importance of agriculture led to the organization of societies dedicated to its improvement, and it is in the minutes of the meetings of those societies, and in their publications, that one finds some of the earliest Canadian references to the ravages of insects and disease, and of suggestions for their control. But government officials in the early departments or bureaus of agriculture were so preoccupied with problems associated with immigration, land settlement, and public health that they paid little heed to the pleas of individual farmers, or those of the early agricultural societies, for help in pest control.

By the middle of the nineteenth century, insects and fungi were causing so much crop damage that they could no longer be ignored by the authorities. In 1856 the Bureau of Agriculture and Statistics, of the then Province of Canada, offered prizes of forty, twenty-five, and fifteen pounds sterling for the three best essays on "The origin, nature and habits – and the history of the progress from time to time and the course of the progress, of the weevil, Hessian Fly, midge and such other insects as have made ravages on the wheat crops of Canada; and on such diseases as the wheat crops have been subjected to, and on the best means of evading or guarding against them." The essays considered to be worthy of an award were judged by a committee named by the Board of Agriculture for Upper and Lower Canada.[2]

In response to that widely advertised appeal, twenty-two essays were submitted to the bureau, and the first prize was won by Henry Youle Hind (1823–1908), professor of chemistry at Trinity College, Toronto. His *Essay on the Insects and Diseases Injurious to Wheat Crops* was printed and distributed to farmers.[3] The second prize was won by the Rev. George Hill, rector of Markham, Ontario. The offering of those monetary awards, including the free publication of the prize essays, was one of the first plant pest services sponsored by any government in Canada.

Neither Hind nor Hill was a plant pathologist, botanist, or entomologist, in the modern sense of these terms, but the publicity associated with the call for essays stimulated several entomologists to enter the contest, and one of the most noteworthy of them was Abbé Léon Provancher (1820–92), of St-Joachim, Quebec. His entry, *Essai sur les insectes et les maladies qui affectent le blé* (Essay on the Insects and Diseases Affecting Wheat), won third prize. The French version of that essay was widely publicized, even though Provancher had submitted it to the judges under the name of Emilien Dupont, a man who at one time had been his beadle.[4] Provancher commented briefly on diseases due to environmental factors and then divided the remainder of his thirty-eight-page essay almost equally between insect and fungal parasites, with comments on the control of each.

When Provancher wrote that prize-winning essay in 1857, he was as much a botanist as an entomologist. His work was certainly botanical when he wrote and published *Traité élémentaire de Botanique* the following year. In that small book, Provancher commented on plant diseases and showed that he was abreast of the latest information by declaring that rust and smut of grain were species of fungi.[5] In 1862 he published *Verger Canadien* (Canadian Orchard), which was soon followed by *Le Potager Canadien* (The Canadian Vegetable Garden), both of which contained comments on fungal-induced diseases as well as damage due to insects. Provancher's botanical work culminated in the great, but to him disappointing, *Flore Canadienne*, in 1862. It was a let down to him because it was not well received by the scientific community – partly because he had used nearly four hundred illustrations from Gray's Manual of Botany without prior permission, and partly because it was considered to be amateurish in many respects. His disenchantment probably deepened when he did not get the position of botanist at Laval University that he had expected would follow the publication of his books. Those disappointments may have been major factors in turning Provancher away from botany. Whatever the reason may have been, he was much more of an entomologist than a botanist for the rest of his life.[6]

The publicity associated with the announcements of the winners of awards, and the publication of prize-winning essays over a period of two or three years, were among the factors that stimulated the Rev. C.J.S. Bethune of Toronto, an entomologist who published on both insects and fungus diseases,[7] and William Saunders, a pharmacist-horticulturalist-entomologist of London, Ontario, who had published on apple-scab control,[8] to consider the establishment of a society of entomologists. In 1862 they compiled, and Bethune published in the *Canadian Naturalist*, the names of all the people they could think of who might be interested in entomology. Provancher's name was on that list, but he did not join the Entomological Society of Canada that was established the following year.

Although considered by many to be Canada's oldest national and specialized society, the Entomological Society of Canada was neither national nor very specialized during its formative years. Its members were almost entirely drawn from Ontario and Quebec, and it encompassed sections that dealt with botany, ornithology, geology, and microscopy, in addition to entomology. It was mostly in the botany and microscopy sections, established in 1890, that members dealt with fungi and plant diseases, and the most outstanding of those who did so was John Dearness (1852–1954), one of Canada's foremost naturalists and a very early member of the Entomological Society.

Dearness frequently gave the main presentation of the evening, and he often dwelt particularly on fungi and their relation to plant diseases. This fact is recorded in the minutes of the London, Ontario, meetings of the Microscopical Section of the society for January 1891. They show that Dearness gave lectures and demonstrations on plant diseases, especially black-knot of cherry and various powdery mildews, together with methods for their control. In the Botanical Section, of which he was elected chairman in 1890, Dearness prepared displays of fungi and gave descriptive lectures on several that induce plant diseases.[9] Thirty-five years later he was still interested in plant diseases and made a locality list of "some fungus foes of the flower garden."[10]

Dearness was so active in the general field of entomology that he was elected president of the Entomological Society of Ontario in 1897. This led to his appointment to the San Jose Scale Commission, by the government of Ontario, in 1899. An infestation of that pernicious scale insect, known to have killed many fruit trees in the United States, had been discovered in orchards of the Chatham area of Ontario a few years earlier. Dearness became so dynamic in matters phytopathological that in 1927 he was elected president of the Canadian Division of the American Phytopathological Society, and when the Canadian Phytopathological Society was formed, in 1929, he was one of its charter members. He was made an honorary member in 1940. Thus, Dearness was acknowledged as both an active entomologist and a plant pathologist for more than half a century. The Montreal Botanical Garden named its plant pathology laboratory after him in 1941, when he was in his ninetieth year and still active. Dearness lived to be 102 years of age.[11]

Another early member of the Entomological Society of Canada was James Fletcher (1852–1908), the first entomologist to be employed by the Dominion Department of Agriculture, and the one who did more than anyone else to lay the foundations for plant pathology in Canada.[12] For nearly three years, he worked for the government without pay as an honorary entomologist. However, because he responded to so many inquiries concerning economic botany (his term for plant pathology), he was appointed to the joint position of Dominion botanist and entomologist in 1887, the position he held with distinction until he died, 8 November 1908. American entomologist Arnold Mallis considered Fletcher to have been one of Canada's greatest economic entomologists and applied botanists. Mallis suggested it was largely because of Fletcher's influence that the government voted relatively large sums for entomological and agricultural research at that time.[13] It was Fletcher, and American entomologist L.O. Howard, who drafted the original constitution of the American Association of Economic Entomologists, an association of which Fletcher became president in 1891.[14]

None of the early Canadian naturalists or biologists who worked in the general fields of either economic entomology or plant pathology, as we know them today, would have thought of referring to themselves as plant pathologists, because that designation had not come into common use. Instead, they may have thought of their plant disease work as applied mycology or as economic botany. Nevertheless, they were doing some pioneering plant disease work during the genesis of phytopathology in Canada.

When the botanical and entomological work of the Dominion Department of Agriculture was separated, following the death of James Fletcher, a new officer, C. Gordon Hewitt (1885–1920), was appointed Dominion entomologist. He soon became head of a newly created Division of Entomology. Although Hewitt was seldom directly involved in plant pathology, he was the person responsible for the administration of the Destructive Insect and Pest Act of 1910, and chairman of its advisory board. Thus he was indirectly involved with plant diseases and the various means for keeping new ones out of Canada.

Soon after the formation of the Division of Entomology, within the Dominion Department of Agriculture, provincial departments of agriculture began to appoint entomologists. In Quebec, the first appointee was the Rev. Canon Victor A. Huard (1853–1929), who became provincial entomologist in January 1913. Huard was a charter member of the Quebec Society for the Protection of Plants, which made him an honorary member in 1916.[15]

Canon Huard's plant pathological work was similar to Hewitt's, although somewhat more direct. While Huard was the provincial entomologist, he acted as a federal inspector of plants and had to "take the measures opportune according to the condition of such plants from the phytopathological point of view, for the admission or destruction of such vegetables."[16] This he did under the authority of the Destructive Insect and Pest Act for the Quebec region, and was thus directly concerned with the exclusion of unwanted insect pests and diseases. Huard gave many lectures on plant protection, including both insects and diseases, and published at least one paper that dealt with both topics.[17] Poor health forced him to resign in 1916.

Canon Huard was succeeded, later that year, by entomologist Georges Maheux (1889–1977). Maheux was somewhat more of a plant pathologist than Huard had been because, in addition to being Provincial Entomologist and professor of Entomology at Laval University, he taught forest pathology. Maheux became president of the Quebec Society for the Protection of Plants in 1930; chief of the Plant Protection Service of the Quebec Department of Agriculture, which included a vegetable pathology section and a weed section, in 1937; president of the Entomological Society of Ontario in 1938; director of the Information and Research Service within the Quebec Department of Agriculture in 1941; and president of the Quebec Agricultural Research Council in 1947.[18]

Maheux was a remarkable individual. For many years he attended meetings and conferences dressed in striped pants, long coat, and a high, starched collar. Thus his appearance alone commanded attention. In addition, he usually had something to say that held the attention of his audience. Because he was more of an administrator than an active worker in the field of either entomology or plant pathology, most of his published papers refer to the work of others, or they have some historical connotation.[19]

Virtually all of the early reports and bulletins pertaining to plant diseases published by the Quebec Department of Agriculture were prepared by the provincial entomologist. Those published prior to 1923 are listed in the report of the minister of agriculture for 1924, the year in which "Vegetable Pathology" was first listed under a Botanical Section in the report of the provincial entomologist.

The most noteworthy of the early entomologist-plant-pathologists in the colleges of Quebec was William Lochhead (1864–1927) at Macdonald College of McGill University. He founded and became a long-time president of the Quebec Society for the Protection of Plants (QSPP), the oldest such society in the world to be continuously functioning under its original name. Lochhead was primarily an entomologist, but he had taught some botany and published on plant diseases while at the Ontario Agricultural College, before going to Macdonald College to become professor of biology in 1907.[20] Lochhead's concept of a plant disease was more encompassing than that of the majority of

Edgar H. Strickland (1889–1962)
Courtesy Agriculture Canada
Research Station, Lethbridge

William Lochhead (1864–1927)

William H. Brittain (1889–1971)
Courtesy McGill University Archives

James Fletcher (1852–1908)

North American plant pathologists, because he considered plant galls and other abnormalities induced by insects to be diseases. At least one modern plant pathologist has suggested that entomologists who study insects that produce localized malformations or systemic phytotoxemia are in reality plant pathologists rather than applied entomologists.[21] Lochhead would have agreed.

The *Canadian Horticulturist* for 1906 has five brief articles on diseases of vegetables written by Lochhead. He had four plant pathological papers published in the first annual report of the QSPP in 1909; four in the report for 1910; three in 1911; two in 1912; and there are several in more recent years.

Another internationally known entomologist at Macdonald College who did some research and wrote articles on plant diseases, including one on insect transmission of disease, was Dr Melville DuPorte (1891–1981).[22] DuPorte also researched the possibility of an amoeboid organism causing a plant disease, but he could not convince himself – he was his own most exacting critic – that his inconclusive findings were worthy of being published.[23]

At the Oka Agricultural Institute, La Trappe, Quebec, Firmin Letourneau (b.1891), who, in 1915, became the first French-Canadian professor of entomology, published a paper on fire blight in the eighth annual report of the QSPP, the society of which he was a director for several years.[24] In the report of the Quebec minister of agriculture for 1916, Letourneau is listed as professor of entomology and vegetable pathology, an acknowledgment of his dual role.

One of Letourneau's colleagues at the institute, who also did some plant disease work, was the Rev. Father Leopold (1884–1947). Although there were times when Father Leopold must have considered himself to be more of a horticulturist than an entomologist, he is included here because he was president of the Entomological Society of Ontario, 1925–7, and at the same time an officer of the Quebec Society for the Protection of Plants. He published at least two papers on fungicides and insecticides in the annual reports of the latter society.[25] Leopold's academic position became that of professor of vegetable pathology and entomology in 1926,[26] the year before the University of Montreal awarded him an honorary DSA degree.

In the School of Agriculture at Ste-Anne-de-la-Pocatière, Georges Bouchard (b.1888), whose title in 1916 was professor of botany, entomology and vegetable pathology, appears to have been teaching as much plant pathology and botany as entomology. However, there is little, if any, record of work, other than teaching, that he may have done in connection with plant diseases. Bouchard left the teaching profession in 1919 to enter federal politics, and eventually became assistant deputy minister of agriculture.[27]

In 1911, members of the Pomological Society of Quebec requested the federal government to appoint a qualified person to help them deal with orchard pest problems in the southwestern part of their province. The government responded in July 1912 by engaging entomologist Charles E. Petch (b.1888) to provide them with the requested assistance. Petch, who owned and operated a fruit farm at Hemmingford, Quebec, was provided with a small, newly constructed laboratory at Covey Hill.[28] Almost from the beginning, and until about 1930, Petch was required to investigate plant disease problems, such as apple scab and fire blight, in addition to his studies of the local insect problems. He was also engaged in plant inspection work until the Plant Protection Division was established within Canada's Department of Agriculture in 1943.[29]

There were no professional plant pathologists in any Canadian college or university prior to 1920, when Guy R. Bisby (1889–1958) became professor of plant pathology in the Manitoba Agricultural College. In contrast, several colleges of agriculture had an entomologist on their staff. Consequently, it was normal practice for several of those entomologists to include plant diseases and methods of controlling them in their lectures. Thus it was not at all surprising when Tennyson Jarvis, lecturer in entomology and zoology at the Ontario Agricultural College (OAC), "provided plant pathology services to the Botany Department" in 1906, and until he resigned in 1914.[30] For a few years, Jarvis was interested in insect-induced galls on plants. He published articles on this topic in the *Report of the Entomological Society of Ontario* for 1906 and each of the two subsequent years. Several entomologists in southern Ontario around that time collected plant galls and studied the insects that were in them. One of the most widely known of these was a Toronto dentist, William Brodie (1831–1909), DDS, an ardent naturalist and prominent member of the Toronto Natural History Society. Brodie was known for his very large collection of galls, many of which were sent to him by others who knew of his hobby. For example, five hundred galls on goldenrod were sent to him from Manitoba in November 1892.[31] Brodie's direct contribution to plant pathology was minuscule. However, his talks on galls, and his exhibitions of them, focused the attention of several biologists on these and other forms of teratology in plants at a time when they were considered, by many entomologists, to be manifestations of disease.

Joseph H. Faull (1870–1961), while a lecturer in botany, University of Toronto, knew of Brodie's work with galls and became so interested in that topic that he persuaded his graduate student Absalom Cosens, MA, to do research on insect galls as the major topic for the PH D degree that he was awarded in 1913. Cosens, who made reference to

Brodie's earlier work, was the first Canadian to study the gallproducing stimulus and the resulting cell formations, which he photographed, in plant galls.[32]

Jarvis was not the only man at the OAC who at various times did the work of both an entomologist and a plant pathologist. Lawson Caesar (1870–1952), who graduated from the OAC in 1908, was appointed as an assistant in the Department of Entomology and Zoology, "with special duties of investigating insect injuries and fungous diseases of fruit trees and other plants."[33] In the following year, a new position was created, known as "Specialist in Plant Diseases," and it was filled by Caesar. In November of 1912, he became Ontario's first provincial entomologist, "with general charge also of plant diseases."[34]

Caesar's major contribution to both entomology and plant pathology was in connection with the provincial orchard inspection and spray services, over which he had administrative responsibility. His program for the control and eradication of "peach yellows" and "little peach" in the Niagara district of Ontario was so successful that those diseases virtually disappeared from that area before he retired, in 1940, from his dual position of professor of economic entomology at the Ontario Agricultural College and provincial entomologist.[35]

Although M.B. Waite had provided experimental proof that insects could transmit a plant disease in 1891,[36] Caesar at the OAC and DuPorte at Macdonald College (as mentioned earlier) were among the few Canadians writing about the role of insects in relation to the transmission of plant diseases in the first quarter of the twentieth century.[37] Early in his career Caesar saw a need for close collaboration between plant pathologists and entomologists, and made a public statement to that effect in his talk to members of the Canadian Division of the American Phytopathological Society, 10 December 1920. Therein he listed factors for the success of a young plant pathologist. Working in cooperation with entomologists, or securing a working knowledge of insects and insect control, was near the top of his list.

British Columbia, not unlike the federal government, had an honorary entomologist, the Rev. G.W. Taylor, as early as 1888.[38] Although Taylor, and his successors, Ernest Hutcherson (the first paid entomologist and inspector of fruits and pests, who resigned in 1894), R.M. Palmer (who held the position until 1906), and Thomas Cunningham (who was the "Government Entomologist" and inspector of fruit pests for ten years), must have had to cope with a number of plant disease problems in the course of their work, there is little published data to support such an assumption. An exception to this generality is Cunningham's 1907 bulletin on "Orchard Cleansing: Remedies for Insect Pests and Diseases."[39] Cunningham is known to have purchased a spray-pump for insect control on apple trees and, like so many entomologists in the first decades of the twentieth century, he included a fungicide to control the apple scab disease.[40]

In 1912 the government of British Columbia hired entomologist William H. Brittain (1889–1971), a BSA graduate of Macdonald College, and made him provincial plant pathologist – the first such appointment by any Canadian province. Brittain, who had been a student of William Lochhead, believed, as Lochhead did, that insects can induce plant diseases.[41] One of his first field assignments was an investigation of the "fruit pit" disease of apples. Brittain had little time to work on that project because he resigned his position in British Columbia to become provincial entomologist for the province of Nova Scotia and professor of zoology at the Nova Scotia Agricultural College, Truro, on 15 October 1913.

There was a time when the British Columbia Department of Agriculture had no fewer then three branches involved in both entomological and plant pathological work at the same time. These were 1/the Horticultural Branch, the head of which was the provincial horticulturist and inspector of fruit pests, who assisted with the work on fire blight of apples and wrote the text of bulletin no. 4, "Insects Injurious to Orchards," and no. 7, "Fungous Diseases of Orchards and Gardens;[42] 2/the Fruit Inspection Branch, sometimes referred to in official publications as the Inspection of Imported Fruit and Nursery Stock Branch, which, among other duties, was responsible for enforcing the provisions of the Federal Injurious Insect and Pest Act;[43] and 3/the Plant Pathology and Entomology Branch, whose head was variously known as the provincial plant pathologist, the provincial entomologist, or the provincial plant pathologist and entomologist.[44] During the period of Brittain's tenure, both he and W.H. Lyne, head of the Fruit Inspection Branch, were entomologists who, largely for geographical reasons, were each doing work as entomologists and plant pathologists in different parts of the province.

Brittain's successor as provincial plant pathologist in British Columbia was John W. Eastham (1880–1968), who had been assistant botanist, Central Experimental Farm, Ottawa. Eastham soon established himself in Vancouver, some distance from his senior assistant, entomologist Max H. Ruhmann (1880–1943), whose laboratory was in Vernon, where he had been Brittain's assistant. Because of this distance, Ruhmann "attended all inquiries concerning plant diseases received at the Vernon office,"[45] in addition to his many responsibilities as an entomologist. In other words, he was another example of an entomologist doing his own work and that of a plant pathologist. Ruhmann coped with those dual responsibilities until the two jobs were separated in 1918, after which he became assistant entomologist. Eastham, who had studied plant pathology at Cornell University, was one of the few plant pathologists in Canada who also did entomological work. When he wrote the bulletin on diseases and pests of cultivated plants, in 1916, he referred to himself as provincial plant pathologist and entomologist,[46] although, officially, he was provincial plant pathologist from 1914 until he retired in 1948.

Edward R. Buckell, a graduate of Caius College, Cambridge, England, who had been casually employed as an entomologist for the British Columbia government, resigned in 1922 to became a full-time employee of the Dominion Entomological Branch, and was stationed at Vernon. In 1929 he headed a team of entomologists whose objective was to discover the true relationship between aphids and a disease of apple trees known to plant pathologists as perennial canker. The entomologists introduced an aphid parasite which, together with a chemical spray, virtually eliminated the canker problem in the orchards of the Okanagan district.[47]

There were very few entomologists in the prairie provinces prior to 1940 and those few were so fully occupied with insect problems that they had little time to devote to plant diseases. One who did was Dr W.H. Coard, professor of agriculture and lecturer in entomology at the North-Western Agricultural College, Regina, Saskatchewan. Another was Edgar H. Strickland (1889–1962), the federal entomologist at Lethbridge, Alberta. In 1903, Dr Coard, who had been with the Dominion Department of Agriculture, offered a three-month course the entomological section of which dealt with "noxious insects and insecticides, fungi and fungicides."[48] Strickland, a graduate of Wye Agricultural College in Britain, had been sent to Harvard University, for the M SC that he obtained there in 1913, and to Lethbridge for experience, by the British Colonial Office. War intervened, and he went overseas with the Canadian Army and was wounded in 1918. Before he became professor of entomology at the University of Alberta in 1922, Strickland was briefly employed by the Dominion entomologist to study insect infestations of wheat being shipped from Canada, and he was involved with the first officially reported nematode problem associated with grain in Canada.[49]

By the time provincial entomologists were appointed in the prairie provinces, plant disease specialists had become available to assist farmers with their disease problems. Consequently, there was little incentive for the prairie entomologists to become involved with them.

As mentioned earlier, W.H. Brittain became provincial entomologist for Nova Scotia in 1913. He succeeded Robert Matheson, a graduate of Cornell University, who had been appointed in October 1912.

Matheson did little, if any, plant pathological work before he resigned at the end of his first year.[50] In contrast to this, Brittain was soon involved in plant disease problems, including fire-blight inspection of apples and diseases of potatoes, and he stayed on the job for thirteen years. In 1914, he initiated a scheme for the disease-inspection of potatoes destined for Bermuda, which had placed an embargo on potatoes from Nova Scotia.[51] Those activities were in addition to his teaching responsibilities at the Nova Scotia Agricultural College and the "normal" duties of a provincial entomologist. Brittain earned a PH D from Cornell University in 1921 while on leave from his job in Nova Scotia. His involvement with plant disease problems was almost nil after he became professor of entomology at Macdonald College in 1926, and dean of the Faculty of Agriculture and vice-principal, McGill University, in 1934.

At Annapolis Royal, Nova Scotia, George Saunders (b.1884), entomologist in charge of insecticidal investigations, worked with Arthur Kelsall (b.1892) in formulating combinations of insecticidal-fungicidal dusts that were used in the orchards of many countries for the control of insects and diseases.[52] Saunders, a native of Nova Scotia who had a BSA degree from the University of Toronto, moved to Kentucky in 1922.

Kelsall, who was promoted to Saunders's position, had migrated to Nova Scotia with his parents from Yorkshire, England, in 1907. He graduated from the Nova Scotia Agricultural College, and then obtained a BSA degree in entomology at Macdonald College of McGill University, before accepting a position at Annapolis Royal. Shortly after the Dominion Laboratory of Plant Pathology was opened at Kentville, with plant pathologist John F. Hockey (1895–1980), BSA, as officer in charge, Kelsall and Hockey became joint authors of Dominion Department of Agriculture pamphlet no. 65, *Nova Scotia apple spray and dust calendars*, published in 1926. Such cooperation between the entomologists at Annapolis Royal and the plant pathologists at Kentville was so obviously beneficial to orchardists in the Annapolis Valley that the Nova Scotia government assembled a staff of entomologists to do field work. Their reports were coordinated with, and included in, those of the plant pathologists who issued bulletins, to press and radio, on insect and disease control.

In New Brunswick, Raymond P. Gorham (1885–1946), BSA, began work as an entomologist with the horticulture department of that province in 1911. One could rightly refer to Gorham as a plant protectionist because he wrote bulletins on insect and fungal diseases of plants and gave lectures to young farmers and schoolteachers on insect pests and plant diseases each year until he was appointed to the Division of Entomology of the Dominion Department of Agriculture in 1919.[53]

In the early days of forest tree disease investigations in Canada, entomologists were often the first to notice a new disease and the first to make an organized study of it. There were very few forest pathologists, in relation to the number of forest entomologists, working in Canada prior to the Second World War. It was entomologist R.E. Balch, officer in charge of the Dominion Entomology Laboratory, Fredericton, who first noted that the birch trees in the central and southern areas of the province were presenting an increasingly unhealthy appearance. He kept this unusual condition under observation for several years and, with J.S. Prebble, reported that there was a particularly striking expansion in the area of this condition in 1938, and again in 1939.[54]

The first research on this birch tree problem was begun in 1943, by R.F. Morris, another entomologist at the Fredericton laboratory, who established a series of permanent plots for a long-term study of the disease, variously known as birch dieback, birch decadence, or birch decline. Morris got some welcome assistance from L.S. Hawboldt, provincial forest entomologist in Nova Scotia, who worked in his own province and in New Brunswick.[55] A few years earlier, entomologist Jean Burnham, also from the Fredericton laboratory, and an authority on aphids, revived an interest in the transmission of viruses and other plant disease organisms by insects. She did this so successfully that the Dominion botanist commented, with tongue in cheek, on the welcome entrance of an entomologist into the field of plant pathology.[56] From about 1943, when plant pathologist J.F. Hockey began to assist Hawboldt in studies of birch dieback, forest tree disease research began to be dominated by plant pathologists.

When nematodes were found to be a threat to the sugar beet growers in the township of Sarnia, Ontario, in 1939–40, the Dominion Department of Agriculture called upon entomologist Alexander D. Baker (1894–1974), BSA, M SC, to investigate the problem, and Baker looked for people with training in entomology to assist him.[57] Apparently the government of Ontario assumed that the provincial entomologist would look after provincial interests with regard to the nematode problem, because the regulations made under the Plant Disease Act, known as "The Sugar Beet Nematode Regulations," stipulated that "The Provincial Entomologist shall furnish assistance to all inspectors appointed under this Act."[58] Nevertheless, from about the mid 1940s it became increasingly unlikely that an entomologist alone would have the leading role in the study or research of a plant disease in Canada. Trained plant pathologists had begun to assume leadership in such studies, although cooperative work and sometimes joint leadership with entomologists continued. The federal government encouraged this cooperation in the realm of forest pathology when it recognized the bond between entomology and forest pathology by creating the Division of Forest Biology within the framework of Science Service of the Canada Department of Agriculture, 1 January 1951.

2 Contributions of Horticulturists to the Early Development of Plant Pathology in Canada

The amateur gardeners and gentlemen farmers of Europe, especially those who had glasshouses or orangeries, as they were called until about 1725, were among the first to take positive steps toward an understanding of plant diseases, their causes, and how to control them. Those horticulturists were writing about their successes and failures in the battle against mildews, blights, and rots of various kinds in the *Gardeners' Chronicle* in England, and in similar periodicals on the continent of Europe, beginning late in the seventeenth century. Thus it is easy to understand why horticulturists in Canada, with their numerous contacts in these countries, were among the first in North America to take an interest in and to do research into the causes and the means of controlling plant diseases.

When a plant pathologist makes a list of plant diseases, he will almost invariably include those that result from exposure to low-temperature. Therefore, any method or device that is used to protect plants from frost or low-temperature injury is as much an aspect of plant pathology as is the use of a fungicide to protect plants from diseases induced by fungi. It is in this general area of plant pathology that horticulturists have excelled and in which they have been real pioneers. Glasshouses, or the so-called greenhouses, are the domains of horticulturists, and a major function of glasshouses, hotbeds, and cold frames is the protection of plants from the deleterious effects of low temperature.

Hotbeds, the heat for which was commonly generated by decomposing animal manure, were used to germinate seeds and to produce seedlings early in the spring, by people who did not have glasshouses, long before the outside soil was warm enough to support young plants. The hotbed was, basically, a rectangular wooden frame containing manure and soil and usually covered by movable panels of glass. To enable people to build them on the prairies, where lumber and glass were both scarce and expensive in the early days, horticulturist Robert McNeil, who owned a nursery in Saskatchewan, publicized a method of constructing the frames for hotbeds or cold frames out of poplar poles. Once constructed, he suggested that the frames could be covered with white cotton cloth dipped in linseed oil to make it semi-transparent and somewhat waterproof, thus obviating the necessity of using expensive glass.[1]

In Canada, as elsewhere, most plants are grown outside of such protective structures, and here again horticulturists were among the first to devise methods of protecting plants from frost. The Nova Scotian orchardist S.B. Johnston, writing in the *Nova Scotia Journal of Agriculture* in 1869, described how he had protected his peach trees from frost by building fires "of old logs or anything that will make a heavy smoke" on the windward side of his trees.[2] Orchardists had long known that a very light covering will often protect fruit from the effects of frost, whether it has been harvested or is still on the tree. P.H. Gosse, writing in his book *The Canadian Naturalist* in 1840, described how apples on a tree had been preserved from frost injury by "a linen cloth thrown loosely over them."[3]

Horticulturist George B. McGill, writing in the first number of the 1878 issue of the *Transactions and Reports of the Fruit Growers of Nova Scotia*, published the first comprehensive classification of plant diseases by anyone in Canada. He stated that a careful consideration of the different kinds of disease would probably warrant the following classification: "1) Those produced by external conditions of life, such as redundancy or deficiency of the ingredients of the soil, of light, of heat, air and moisture. 2) Those produced by poisonous agencies, as poisonous gasses [sic] miasmata in the air, or by poison in the soil. 3) Those arising from the growth of parasitic plants, such as the various fungi. 4) Such as are caused by mechanical wounds, and by the attacks of insects." McGill gave examples to clarify his classification scheme and cited black knot of cherry and plum, and mildew, as examples of diseases due to parasitic plants. It is noteworthy that he followed the lead of many European biologists in considering insect damage to be a diseased condition.

The Honourable Charles Ramage Prescott (1772–1859), a wealthy Halifax merchant who more or less retired from business life at the age of forty and became a renowned horticulturist, was an enthusiastic promoter of scientific agriculture. He tested different varieties of grain for their freedom from smut, and turnips for their general suitability, in Kings County, Nova Scotia. However, his principal horticultural accomplishment was in pomology through the importation of foreign varieties, grafting, and testing for winter-hardiness and fruit quality. Sometime between 1830 and 1835 he introduced the Gravenstein apple. Prescott was president of the short-lived (1825–8) Horticultural Society of Kings County, and when the Nova Scotia Horticultural Society was formed in 1836, he was named its vice-president. Prescott's horticultural accomplishments are briefly outlined in the *Dictionary of Canadian Biography,* volume 8.

In 1891, W.O. Creighton, who referred to himself as a horticultural graduate of the Provincial Agricultural School, published *Fruit Growing for Profit,* in Halifax, Nova Scotia. In that book, one chapter was devoted to the prevention of diseases. He admitted that some of the recommended remedies "are only partial."

George Moore's book, *Fruit Culture in the Province of Quebec, and in particular the Eastern part of the Province,* published in 1892, had one paragraph on apple scab and no other reference to diseases. He did, however, comment on problems associated with frost and the short growing season.

The procurement of cold-hardy plants, by selection or otherwise, was a major preoccupation of virtually all Canadian horticulturists for several decades, and that objective continues to be pursued by a few of them. Fruit growers were among the first to select and to breed plants that would be resistant to or could avoid periods of cold weather, and, though not the first in Canada, one of the most noteworthy was Charles Gibb (1864–1900) of Abbotsford, Quebec. In 1882, in company with a horticulturist of the Iowa State Agricultural College, Gibb went to Russia in search of fruits that might be expected to endure the extremes of temperature to which they would be exposed on his Quebec farm. He not only made the trip at his own expense but also distributed some of the seeds and trees, free of charge, to fruit growers in Quebec and Ontario, to the Central Experimental Farm in Ottawa, and to the Botanic Garden of Montreal. Thus, some of his material became incorporated into the breeding stock of many horticulturists who were trying to develop cold-hardy plants at that time, and for many years later.

Charles Gibb was one on the early advocates of experimental fruit stations under the control of the government. He was supported in this by the Montreal Horticultural Society, through its journal, the *Canadian Horticultural Magazine,* which it began publishing in 1897.[4] In 1887 Gibb went alone to Russia, Norway, Sweden, and Denmark with the object of finding more hardy fruit trees and ornamental plants. Other trips were made to various countries, including India, China, and Japan, where he continued his search for hardy horticultural material. He was on his way home from one such trip when he became ill and died in Egypt, 8 March 1900.[5]

Another Quebec orchardist who made a significant contribution toward the selection and development of winter-hardy plants was Jean Charles Chapais (1850–1926) of St-Denis. The results of his tests on large and small fruits in his "Northern Orchard" were incorporated into the annual reports of the Dominion horticulturist for 1895, and again for 1899.

When William Saunders (1836–1914) became director of the Dominion Experimental Farms, one of the first things he did was to import seed of hardy apples, including the Siberian crab apple, from the Imperial Botanic Garden, St Petersburg, Russia, for the purpose of developing hardy sorts of apples for the colder areas of Canada.[6] Saunders not only planted seeds and seedlings at the Central Experimental Farm for the purpose of selecting those that were cold-hardy, he also did a great deal of hybridizing and made the first ever interspecific cross of apples.[7] One of his goals was the development of fruit trees and shelter-belt trees and shrubs suitable for the harsh climate of the western prairies. As director of all federal experimental farms, he saw to it that horticulture had a prominent place at each of them, particularly those on the prairies where there had been no real tradition of having trees and flowers around the farmhouses.

Fortunately for Saunders, his horticultural enthusiasm coincided with a reforming zeal that was evident throughout society. That zeal was seen in the actions of suffragists, prohibitionists, evangelists, and others. Nature study and gardening became popular throughout the country, and provided the impetus for a great movement to improve the environment of Canadian homes, especially those on the horticulturally barren prairies. Saunders's early directives to the superintendents of the experimental farms are reflected in their annual reports. All of them, for the first few years, told of how many of each sort of tree, shrub, and other horticultural plants they had planted, and how many of each kind survived the winter.

Some of the farms did not employ horticulturists as such, but they had gardeners doing horticultural work. For example, at the Scott Research Station, H.C. Love, gardener from 1912 until 1921, and his successor, James Allaway, gardener from 1921 until 1943, continued the work of the first superintendent, R.E. Everest, who, following Saunders's instructions, ensured that horticulture would have a prominent place on the station by setting out a number of varieties of apples and small fruits. Each of the two gardeners enlarged the horticultural work by adding new strains or varieties, including currants, gooseberries, strawberries, vegetables, and flowering plants.[8] At first, their plant pathological work, in addition to routine spraying for the control of diseases and insects, consisted almost entirely of testing the plants or seeds brought to the station for winter-hardiness or adaptation to local conditions. Later, they did some basic plant breeding, followed by selection for hardiness, freedom from disease, and other desirable qualities. The crabapple variety "Rescue" was one of the selections made at the Scott Station.[9]

The horticultural situation at the Indian Head Station was similar to that at Scott. The first superintendent there, Angus MacKay, soon expressed his views in this regard when he stated, "it was deemed very important that something be done in fruit and forest tree culture without delay."[10] That his words were put into action is shown by his report of twenty-three thousand trees and other plants being set out during May and June 1888. In his annual report, MacKay referred to a horticulturist's house at the station, through he did not name anyone as having that title. Most, if not all, of the horticultural work was supervised by MacKay himself until George Lang was hired in 1890. Lang was always listed as a gardener, but when he was replaced by John Walker, in 1924, the latter was given the title of assistant to the superintendent for horticulture.[11]

Titles of those who did horticultural and pest-control work at the Swift Current Research Station, which opened in 1920, had little more meaning than those at Indian Head. For example, H.J. Kemp, who was employed there from 1923 until 1944, was simultaneously assistant superintendent, horticulturist, cereal crops evaluator, and builder of plot equipment.[12]

Among S.A. Bradford's early projects, as superintendent of the Brandon Farm, was the planting of trees. By 1891, more than one hundred thousand trees, shrubs, and bushes, including apples, crab apples, pears, plums, cherries, currants, grapes, blackberries, and gooseberries, had been planted to determine their hardiness and suitability for Manitoba conditions.[13] Because of the severe climatic conditions at Brandon, however, relatively few of the fruit trees planted there survived longer than three or four years. That problem led to the establishment of the Morden Experimental Farm in 1916. It was in the Morden district that A.P. Stevenson, who died in 1922, did "more to prove that apples could be grown in Manitoba than any other man ... and inspired many to test their fruits there."[14] Although the testing of vegetable varieties and ornamentals continued at Brandon, the lead in horticultural work on the prairies, by the federal Department of Agriculture, was taken by Morden from that time onward.

In the early years of the Dominion Department of Agriculture, it was not uncommon for private orchardists to cooperate with the experimental stations in fungicidal and plant-disease experiments. The results of cooperative trials in the orchards of William Craig, Jr and J.M. Fisk of Abbotsford, Quebec, are included in the report of the dominion horticulturist for 1891.

Although the horticulturist Rev. Father Leopold (1884–1947), who managed the orchards at the Oka Agricultural Institute, La Trappe, Quebec, was not officially a cooperator, he became an authority on insecticides and fungicides for the control of orchard pests and published the results of his experiments for the benefit of others.[15] In Nova Scotia, William Saxby Blair (1873–1967), a diploma graduate of the Nova Scotia School of Horticulture who was the horticulturist at the Dominion Experimental Farm in Nappan from 1896 until 1905, found several orchardists in that province willing to cooperate with him in spray testing. In some instances he went to the orchards, did the spraying, and recorded the results.[16] Blair, along with almost every horticulturist of that period, conducted experiments on the use of Bordeaux mixture on orchard trees and on various other horticultural plants. The results of his trials with Bordeaux for the prevention of potato rot appeared in his reports for 1892 and 1893, and his efforts to control the black knot disease of plum trees is in that of 1900.

After an interval as the first head of a Horticultural Department at Macdonald College, Ste-Anne-de-Bellevue, Quebec, 1905 to 1912, Blair returned to Nova Scotia to become superintendent of the experimental station at Kentville, where his interest in experiments with dust and liquid formulations of Bordeaux mixtures continued until at least 1918,[17] and probably until he retired in 1938. He was awarded an honorary D SC by Acadia University in 1930.

William Blair's son, Donald Saxby Blair, BSA, M SC, who was born in 1909 when his father was on the staff of Macdonald College, followed in his father's footsteps by becoming a horticulturist. After doing some pollination studies and other horticultural research in Nova Scotia, Donald joined the staff of the Horticulture Division, Central Experimental Farm, Ottawa. It was while there that he published in 1935 a paper on winter injury to apple trees in eastern Canada.[18]

William Blair's successor as head of the Department of Horticulture, Macdonald College, was T. Gordon Bunting, BSA, who had been on the horticultural staff of the Central Experimental Farm, Ottawa. It was largely because he had Archibald H. Walker, a well-known florist, on his staff, and four modern greenhouses, that Bunting's department became a focal point for the florists of Montreal and vicinity. Their society made a practice of meeting there around Christmas and Easter to discuss their mutual problems, including the prevention of wilt and diseases. Bunting died, and Walker retired, in 1945.[19]

Throughout the 1920s there was very close cooperation between the horticulturists and plant pathologists at the college, with respect to the culture of disease-free flowering plants. At that time, around 250 varieties of iris were being grown, more or less cooperatively, by the horticulture and plant pathology departments, with the horticultural staff carrying out the disease-control recommendations of the plant pathologists. An iris disease that was particularly troublesome was a common leaf spot, and plant pathology graduate student Champlain Perrault (1903–88) made it a major research topic for the master's degree that he was awarded in 1927. That was the year in which John E. Machacek (1902–70), who was then a student and laboratory assistant in the plant pathology department, published a paper on a penicillium rot of gladiolus in the annual report of the Quebec Society for the Protection of Plants. Machacek had reports of research on diseases of horticultural crops in each of the three succeeding annual reports of that society. Another graduate student in plant pathology who studied a disease of a horticultural plant was Arthur J. Hicks. His study of a disease of narcissi became the topic for the master's degree that he obtained in 1931. His work must have appealed to Frank L. Drayton (b.1892) of the Dominion Department of Agriculture, who had also been studying diseases of horticultural plants, because Hicks became Drayton's assistant for several years before serving with distinction in the Canadian Army during the Second World War.

While still a member of the staff of the horticulture department, Macdonald College, Frederick M. Clement (b.1884) expressed the general concern about the different forms of winter injury to fruit trees in a lecture in 1913 to members of the Quebec Society for the Protection of Plants.[20] Clement left Macdonald College to become director of the experimental station, Vineland, Ontario, in 1914, and he eventually became dean of the Faculty of Agriculture, University of British Columbia.

Canadian agriculturists tend to think of William Saunders as a grain breeder, which he was, but he was first and foremost a horticulturist. In 1873 the Fruit Growers' Association of Ontario awarded him a prize for his essay on the cultivation of the plum, in which he described the black knot disease and how to control it.[21] When members of the association visited the Saunders fruit farm that year, they could have seen 2,500 apple trees, 2,300 pear trees, 700 plum trees, 330 cherry trees, 100 peach trees and 500 vines, in addition to the many currants, raspberries, gooseberries, etc., on a farm of about one hundred acres immediately opposite the insane asylum on Dundas Street, London.[22] In the minutes and proceedings of various horticultural societies throughout Canada over the next ten or twelve years, there are references to Saunders and his occasional comments on plant diseases. One early example is his talk on proposed remedies for apple scab that was published in the *Canadian Horticulturist* in 1884.[23] With the progressive expansion of commercial apple orchards, the apple scab disease became as serious a source of loss to the fruit growers as the late blight disease was to potato growers or the rust disease to wheat growers. Therefore the published account of Saunders's comments on that disease would have been avidly read by progressive horticulturists.

In a very real sense, the first gardener-agriculturist to select good, sound, mature seed for the succeeding crop was, unknowingly, selecting the propagules of plants that had either escaped or successfully resisted the organisms and environmental factors that induce diseases of seedlings and immature plants. This type of selection against disease is nearly as old as agriculture itself.

In Canada, as elsewhere, the improvement of plants has gone through two phases. The first, selecting and retaining the best seeds from the mixture of natural hybrids, was being practised by the Hurons and other native people long before European settlers arrived, as it had been in Europe itself. Settlers continued to select the best plants for seed stock, and that practice has continued to this day, although it is being replaced by products of the second phase.

The second phase is the deliberate crossing of selected plants with the object of improving some aspect of their growth, their beauty, their yield, or their freedom from disease, and horticulturists in Canada were following this practice as early as 1864. In that year, Peter Dempsey, of Albury, Ontario, one of the more forceful and enthusiastic of the horticulturists in southern Ontario at that time, grew several varieties of grapes in pots, under glass, and arranged them so that he could retard or advance their time of flowering by raising or lowering the temperature. By this method he induced some of the exotic varieties to be in bloom at about the same time as his native plants. Then, he wrote, "I removed the stamens of about one third of the blossoms from about two bunches of each plant, the remaining were removed entirely, I fertilized the Hartford Pacific with pollen from the Black Hamburg ... and vice versa. From the seeds ... I grew about fifty plants ... many of them would now resist the mildew."[24] From a plant pathological standpoint, it is noteworthy that Dempsey was selecting plants resistant to mildew more than a century ago.

Dempsey, who served as president of the Ontario Fruit Growers' Association in 1880–82, was not only a noteworthy plant breeder but also an innovative user of chemicals for the control of plant diseases. In 1884, he told a meeting of Ontario fruit growers that he "used to take sulphur and place it in a pot and throw a stone of quicklime in and boil it up a little, and then ... strain the liquor and place it in vessels ...

put a little of this in the water that we syringed our vines with under glass, and as long as we used it we never saw any mildew on the vines."[25] This may be a record of the earliest use of boiled lime sulphur as a fungicide in Canada.

One year later a horticulturist referred to only as Mr Beadle – probably Delos W. Beadle (1823–1905), editor of the *Canadian Horticulturist* and author of *The Canadian Fruit, Flower and Kitchen Garden* – was advocating the use of sulphur dust to control mildew, while his colleague, Mr Morton, recommended caution in the use of a French preparation composed of sulphate of copper and limewater to control mildew of grape vines.[26] That is the first Canadian reference to what became known as the Bordeaux mixture, invented, and publicized in 1885, by P.M.A. Millardet in France.[27]

In scanning the reports of the Ontario Fruit Growers' Association for the last fifteen years of the nineteenth century, one gets the impression that virtually all horticulturists of that period were experimenters, and that they were trying many different methods for the control of plant pests, including those that caused disease. When that association met at Peterborough in December 1893, the members appointed a committee to devise a scheme for experimental horticulture, and the experiments they outlined included several pertaining to the control of diseases of fruit trees and vines. The resulting Fruit Experimental Stations of Ontario, as they were called, provided annual reports of their experiments for many years, beginning with 1894. Those stations were the forerunners of the provincial Horticultural Experiment Station and Products Laboratory at Vineland Station, the most important work of which has been plant breeding.[28]

Plant breeders, or those who cross-pollinated plants, whether of the same species or not, were, for many years, referred to as "hybridizers." Horticulturist Charles Arnold (1818–83), of Paris, Ontario, distinguished himself as "Canada's great hybridist" by producing new varieties of apples, corn, grapes, peas, raspberries, strawberries, and wheat, some of which were selected because they were free or reasonably free of disease. Arnold's best-known production was the American Wonder pea, which he sold to an American company for the then handsome sum of $2,000.00.[29] Hybridization of plants became so popular with progressive horticulturists during the last quarter of the nineteenth century that departments of agriculture were offering prizes for good essays on that topic. Two such prize essays, both under the general heading of *Hybridization, and its Canadian Results* – one by D.W. Beadle and one by P.E. Bucke – were published by the commissioner of agriculture and arts for the province of Ontario, in his annual report for 1877.

Harold R. McLarty (1891–1988)

John Frederick D. Hockey
(1895–1980)

Before the first decade of the twentieth century had passed, the theory and method of hybridizing had become well known and a small army of amateur and professional plant breeders were at work in Canada. However, not all horticulturists were keen hybridizers, and a few were opposed to it because they considered it to be an unnatural means of obtaining new or improved varieties. One gets the impression that Dorothy Perkins (*nom de plume* of Adele Austin) may have been one of the latter. In her *Canadian gardening book*, published in 1918, she referred to flowers "that are pure and innocent, for they have not been hybridized and crossed by ardent horticulturists who seek to outstrip nature."

George Lawson (1827–95), PH D, LL D, professor of chemistry at Dalhousie College, Halifax, Nova Scotia, to whom horticulture may have been little more than a hobby in 1884, did noble work for the cause of plant breeding by repeatedly describing the process and otherwise advocating it as a method by which members of the Fruit Growers' Association of Nova Scotia could produce new and disease-free varieties.[30] Lawson, and the other members of that association, were very progressively minded and willing to make sacrifices in time and money for the benefit of future horticulturists in their province. It was almost entirely through their efforts that the first school of horticulture in Canada was established, on the campus of Acadia University, Wolfville,

Nova Scotia, in 1893. That school opened for students 9 January 1894, under the direction of E.E. Faville, professor of horticulture.[31]

During his first year there, Faville examined fungal-induced growths on plants, observed the formation of spores, and lectured on the best methods of spraying to control plant diseases and insect pests.[32] He resigned after three years at the school and his successor, F.C. Sears, who soon became involved in the investigation of a disease affecting apple trees, locally known as a collar rot, was even more interested in plant diseases.[33] Sears stayed at the school until 1905, when it was absorbed by the Nova Scotia Agricultural College, in which he was the horticulturist until he resigned in 1907. Percy J. Shaw, who succeeded Sears, wrote about the "Hard Knot Disease of Pears" in his 1908 report to the principal, and his 1921–2 report included a statement about his experiments to control potato scab.[34]

In commenting on Nova Scotian horticulturists who contributed to the development of plant pathology, it is noteworthy that fruit grower W. Ferguson was using a solution of vitriol (copper sulphate) to cleanse wounds made during the excision of black knot from cherry trees in 1874,[35] several years before Millardet's discovery of the Bordeaux mixture in France. This becomes all the more remarkable when one recalls Millardet's claim to have been the first person to advocate the use of a copper compound for the control of a plant disease.[36]

In New Brunswick, the early horticulturists did not publish reports of any research on plant diseases that they may have undertaken, other than the ubiquitous spraying tests and demonstrations. However, Arthur G. Turney (1885–1960), a 1909 graduate of the Ontario Agricultural College who was provincial horticulturist from 1910 until 1955, gave lectures on diseases and pests of orchards and gardens to various groups, including schoolteachers. Turney was frequently so occupied with administrative duties, including the organization of the New Brunswick Fruit Growers' Association, that his assistant for several years, Raymond P. Gorham, often gave those lectures.[37]

Gorham, a 1911 graduate in horticulture from Macdonald College, was assistant provincial horticulturist from 1911 until 1917 when he became an instructor in the provincial Normal School. During that time he was the author of two bulletins, *Chief insecticides and fungicides for orchard and garden crops* and *Powdery scab of the potato*. Both bulletins lack dates, but they were available to farmers and orchardists in 1915, and Gorham distributed them among his audience whenever he lectured on those or related topics. Gorham joined the Entomological Branch of the Dominion Department of Agriculture in 1919 and took charge of field-crop and garden insect investigations at Fredericton. After that he did little, if any, plant pathology work.

Francis P. Sharp (1823–1903) was a noteworthy plant breeder-horticulturist who has often been referred to as the "Father of Fruit Culture in Northern New Brunswick" because of the number of well-liked varieties he produced. Sharp gave advice freely about his methods of disease control, but he was mistaken when he stated that the yellows disease of peach and club root of cabbage were different expressions of black knot of plum, and that all of them came from the soil.[38]

A search through the agricultural and horticultural literature of both Prince Edward Island and Newfoundland failed to find anything pertaining to plant diseases published by the horticulturists there, prior to the end of the Second World War.

There were several horticultural societies in Ontario before 1860,[39] but horticulture, as a course of study, had a slow start at the Ontario Agricultural College (OAC). Only occasional lectures on that subject were being given in 1881, when J. Playfair McMurrich (1859–1939), BA, MA, was professor of biology and horticulture.[40] McMurrich did, however, include the topic of plant diseases in his lectures. The annual report of the OAC for 1880 shows that he asked his students to describe the life history of the rust fungus *Puccinia graminis* on the examination paper in systematic and economic botany that year. In addition, McMurrich may have been the first professor in a Canadian educational institution to do research on a plant disease: he studied black knot of plum trees in 1882.[41]

Fifty years later, John H.L. Truscott (b.1905), BSA, MS C, PH D, of the horticulture department, was studying post-harvest diseases of garden produce, especially storage rot of celery. That work, and the waxing of fruits and vegetables for storage and marketing – a type of post-harvest disease prevention – was new to that department in 1936,[42] but it soon became a popular subject for research. Over the next two years, Truscott, who had been a charter member (student) of the Canadian Phytopathological Society, supervised the research of three candidates for master's degrees, each of whom studied some aspect of the storage of pears or tomatoes. His semi-popular articles on frozen foods, locker storage, and the processing of fruit and vegetables appeared in the *Canadian Refrigeration Journal* over a period of several years, particularly during the early 1940s. In 1945, when Truscott was assistant professor and chief of horticulture research, he wrote Ontario Department of Agriculture bulletin no. 445, *Temperature control in relation to the storage of food.*

In the years between 1884 and 1936 the horticulturists at the OAC created a number of new varieties of ornamental and other plants, some of which were selected for their resistance to disease but which were not advertised as such.

In western Canada, nurserymen-horticulturists had long been selling barberry bushes (*Berberis vulgaris*) for use as hedge fences, and for their ornamental value. However, when they became convinced that those bushes were host plants for the grain rust fungus they were among the early advocates of their eradication. During the first conference of the Manitoba Horticultural and Forestry Association, following the damaging rust outbreak of 1916, it was resolved that the association petition the provincial legislature to enact legislation prohibiting the sale and planting of the common barberry and the destruction of those bushes already in existence. As a result of their petition, and a similar one by the Manitoba Branch of the Canadian Seed Growers' Association,[43] the Manitoba Noxious Weeds Act was amended in 1917 so as to include the barberry.

When the Manitoba Agricultural College accepted its first students, in the autumn of 1906, Frederick W. Brodrick (b.1879), BSA, was there as a lecturer in horticulture and forestry. Until 1937 he was the only academic member of what became the horticulture and forestry department. During much of that time he was also secretary-treasurer of the Western Horticultural Society and had little time for plant disease research, except for the usual spraying trials of his period.

The extension horticulturists in Manitoba – J.R. Almey (1921–29), John Walker (1929–37), and C. Ray Ure (1939–43) – became responsible for the inspection of nurseries soon after the Act to Prevent the Introduction and Spreading of Insect Pests and Fungus Diseases came into effect, 1 September 1927. They also conducted some experiments for the control of plant diseases and insects, in cooperation with Brodrick in the horticulture department of the Agricultural College.[44] Brodrick was replaced in 1937 by John Walker (b.1893), B SC, MS, who had been an extension horticulturist. After Walker transferred to the college there was little incentive for him to become involved with plant diseases, or otherwise to advance the cause of plant pathology in Manitoba, because, from 1920, the college had a plant pathologist on its staff.

The departments of agriculture of Saskatchewan and Alberta did not appoint provincial horticulturists until sometime after the Second World War. However, George Harcourt (1863–1940), who had been deputy minister of agriculture, was appointed head of a new Department of Horticulture at the University of Alberta in 1915, where he remained in charge until he retired in 1934, and Cecil F. Patterson (1893–1961), BSA, M SC, PH D, established a similar department at the University of Saskatchewan in 1921. Neither of those men published anything in the recognized journals of science, prior to 1949, that could be referred to as plant pathology. However, when Patterson introduced the Perfection squash he had virtually eliminated the button or turban from Buttercup, because that structure was believed to encourage the invasion of fungi and rotting organisms. Disease prevention had thus been an objective.

It was not uncommon for horticulturists to act as inspectors of orchards, gardens, nurseries, etc., working under some provincial or federal legislation for the control of plant diseases and insect pests. This was particularly true in British Columbia, which had its own inspectors of fruit pests as early as 1892. Those provincial inspectors eventually assumed the task of enforcing the federal Destructive Insect and Pest Act of 1910 and continued to do so until 1933 when federal officials took over.[45]

The Horticultural Society and Fruit Growers' Association of British Columbia had been so concerned about the possible entrance and spread of foreign insects and diseases in their province that it persuaded the provincial government to bring in the act that created a provincial Board of Horticulture in 1892. Both the association and the board provided strong support for the horticulturists who were appointed as inspectors.[46] Thomas Cunningham, the first provincial inspector of fruit pests, was the author of a bulletin titled *Horticulture, orchard cleansing, remedies for insect pests and diseases*, in which he provided information about both fungicides and insecticides. That bulletin underwent its third revision in 1908.

When Cunningham died in 1916, R.M. Winslow was given the dual position of provincial horticulturist and inspector of fruit pests. He, somewhat like his predecessor, was responsible for the preparation and publication of a series of horticulture circulars, including one in cooperation with the provincial plant pathologist, W.H. Brittain, that dealt with diseases in orchards and gardens. Winslow and Brittain also gave several lectures and demonstrations on fire blight and its control.[47] Winslow's successors, M.S. Middleton and W.H. Robertson, also held those dual positions.[48]

Several horticulturists became inspectors at the district level, where they were referred to as "district men"; some were, or soon became, active in the control of plant diseases. For example, Edward W. White (b.1887), BSA, district horticulturist and inspector of fruit pests for Vancouver Island and the Gulf Islands, lectured on "Apple-tree Anthracnose or Black-spot Canker Control," and because the essence of his lecture was based largely on his own work, which included demonstration spraying, it was incorporated into the annual report of the minister of agriculture for 1921. That was not exceptional, because in the 1920s all reports of the district horticulturists included a section on insect pests and diseases, and were often included in the annual reports of the minister, like the one by H.H. Evans, who, in 1929, was in charge of research on club-root of cabbage.

Assistant horticulturist P.E. French was the author of Horticulture Branch circular no. 10, which dealt with commercial potato culture. In that circular, published in 1910, he described early blight, late blight, and scab, and gave the accepted methods for the control of those diseases.

Harold R. McLarty (1891–1988), BA, MA, PH D, officer in charge of the Dominion Laboratory of Plant Pathology at Summerland, made major contributions to British Columbia horticulture. He was the author of Dominion Department of Agriculture circular no. 15, *Diseases of plums and their control*, and he played the leading role in the discovery of a control for "corky core" of apples. Together with his assistants, George Ewart Woolliams (1901–92) and J. Charles Roger (d.1935), and a group of federal and provincial entomologists, plant pathologists, and horticulturists, McLarty formed an Okanagan Horticultural Club. For a while, all the recommendations from local governmental officials were made "on the authority of that Club." When it was first learned that boron would prevent "drought spot" or "corky core" in fruit trees, members of the club were called in for a discussion on the best way to present the then incomplete but very promising results of research to local growers. The consensus of the club was to recommend to growers that they apply the boron via holes bored into the trunks of the trees, and to explain that further tests on other forms of boron application were being carried out.[49]

For several years after the Central Experimental Farm (CEF) was established at Ottawa, James Fletcher (1852–1908), Dominion botanist and entomologist, had assumed virtually all responsibility for answering growers' questions about plant diseases. However, by 1893 the three-in-one job of botanist, entomologist, and plant pathologist had become so onerous that the work on plant diseases was transferred to John Craig (1863–1912), the first horticulturist in the Department of Agriculture to be paid for his plant pathological work.[50]

Craig, who had been horticulturist to the CEF since 15 January 1890, received his early training on the farm of Charles Gibb in Québec, and subsequently with Professor J.L. Budd of the State Agricultural College in Iowa.[51] There is no evidence to indicate that W.W. Hilborn (1849–1921), Craig's predecessor as horticulturist to the CEF, carried on any plant pathological experiments during his brief term of office. From an early age, Hilborn was interested in wild and cultivated raspberries and the possibility of crossing them. This he eventually did and produced what became quite well known as the Hilborn variety. After only two years as horticulturist in Ottawa, Hilborn resigned and purchased a farm for fruit growing. It was while operating that farm, near Leamington, that he cooperated with Craig in several experiments, including one on how to prevent peach leaf curl.[52]

Craig's series of spraying experiments, both at Ottawa and in orchards at Abbotsford, Quebec, led to the publication of the first federally sponsored spray calendar, which is in CEF bulletin no. 23, published in 1895. Some of Craig's earlier experiments with fungicides are noted in the CEF report for 1890, which also has an account of his work on winter injury of strawberries. In this latter report, he placed ninety-seven varieties of strawberries in ten groups, based on percentage of winter injury. The following year he tested the effect of mulching as a means of preventing rot in tomatoes, and he tried various fungicides to prevent grape mildew and gooseberry mildew.[53] In fact, there are brief accounts of his work in relation to plant diseases and his experiments with fungicides, in all of the CEF annual reports from 1890 through 1897. Craig was also interested in and wrote about the cold storage of fruits and the effects of frost on vegetable tissues, and he tried to determine the relative amount of winter injury sustained by peaches and plums throughout Ontario. Craig's ideas and reports were just beginning to make a significant impact on fruit growers and other horticulturists throughout the country when he resigned his position as Dominion horticulturist, in 1897, to become professor of horticulture at Cornell University. He died at Ithaca, New York in 1912.

Another Canadian horticulturist to be paid by the federal government for his plant pathological work was William T. Macoun (1869–1933), son of the famed botanist Professor John Macoun. At the age of nineteen, in 1888, young Macoun was appointed to the staff of the CEF as William Saunders's assistant and foreman of forestry. Ten years later, following the resignation of John Craig, Macoun became horticulturist to the CEF and curator of the arboretum and botanical garden.

Having acted as his father's botanical assistant prior to joining the staff of the CEF, William Macoun, without benefit of a university education, was an accomplished botanist, and he learned the techniques of cross-breeding while working as Saunders's assistant. From 1892 until 1898, Macoun had charge of the different varieties of wheat produced at the farm, under Saunders's direction. It was during this period that Marquis wheat had its beginning. Young Macoun's knowledge of plant breeding, combined with his love of plants, led him to devote his personal attention to the breeding and selection of horticultural plants, especially apples, soon after his appointment as horticulturist.

One of the earliest accounts of Macoun's work with plant diseases appeared in the CEF Report for 1899, in which he commented on Baldwin spot of apples. In the 1901 report, he wrote about asparagus rust and several diseases and pests of apples, and by 1904 his work included experiments on the control of blight and rot of potatoes.[54] In 1913 he prepared a "Spraying Calendar" that was widely publicized as a free poster from the Department of Agriculture.

When the organization of the Dominion Department of Agriculture underwent a major change following the death of James Fletcher and the appointment of Hans Güssow (1879–1961), most of the plant pathological work became the responsibility of the latter. Although Macoun was no longer officially responsible for the plant-disease work after 1910, he continued a number of experiments for the control of a wide range of plant diseases, and included the subject in his bulletins and other publications. For example, in the bulletin he prepared in 1916, under the simple title of *The apple in Canada*, there are comments on various kinds of frost injury, russeting of fruit, and water core.

In 1923 Macoun became the first recipient of the Carter Gold Medal for achievement in the advancement of horticulture. This was the first of several awards given to this worthy horticulturist, who also made his mark as one of the early Canadian plant pathologists. He became president of the American Society of Horticulturists in 1912, and was awarded a D SC by Acadia University in 1929. In the following year, Macoun became president of the Canadian Society of Technical Agriculturists (now the Agricultural Institute of Canada).[55]

Macoun's successor was Malcolm Bancroft Davis (1890–1979), who had obtained a BSA from Macdonald College in 1912, the year in which he became manager of Sunnyside Farms, Limited, in Bridgetown, Nova Scotia. In December of 1913, he was appointed assistant to the Dominion horticulturist but did not begin the job – essentially that of pomologist – until early in 1914. Within a year he published a paper on the use of fire pots, or heaters, for the protection of orchard trees against frost,[56] and in 1916 the results of his experiments, involving the use of six styles of heaters, convinced him that the best heaters could provide adequate and reliable frost protection.[57] After serving overseas in the Canadian Artillery, Davis was appointed chief assistant to the Dominion horticulturist in 1919 and, following the resignation of William Macoun, became, in effect, Dominion horticulturist, although his official title was chief, Division of Horticulture, a position which he held until he retired in 1955.

Fear of frost damage to his orchard prompted one ingenious horticulturist in Ontario to devise in 1914 a unique frost alarm system. By installing a specially designed thermometer in the orchard, and attaching it to an electric bell in his bedroom, the owner could warn himself of impending freezing conditions in time to light fires or heaters.[58] The winter of 1917–18 was one of the most disastrous to orchards in Ontario and Quebec that had ever been recorded. Thousands of trees were killed. However, that disaster provided research opportunities for salvaging trees that survived but which had one or more dead or severely damaged branches. In 1919, some experimental work in tree surgery was undertaken by horticulturists in Ottawa in which different compositions of cementing materials were used for filling cavities.[59]

In July 1927, Hinson Hill (b.1902), BSA, M SC McGill, joined Davis's staff and devoted his time to research on deficiency diseases of horticultural crops, both before and after earning a PH D at the University of London in 1935. This was, albeit indirectly, an extension of Davis's innovative research in plant nutrition.[60]

Davis was well aware of the need to extend the post-harvest life of fruits and vegetables, and thus prolong the period of their use or saleability. In this regard he agitated for, and was largely accountable for, the construction of the first experimental cold storage unit in Canada.[61]

In 1931 a low-temperature laboratory for studying controlled environmental conditions for the storage of perishable fruits and vegetables was built by the Division of Horticulture, in Ottawa. It consisted of six chambers of about thirty cubic meters each. One was devoted to freezing experiments while the others were used to observe the storage behaviour of various fruits and vegetables. That facility was expanded to twenty-one chambers in 1936, including an additional two for freezing experiments, all of which were managed by William R. Phillips (b.1906), BSA, MA, who was assisted by Peter A. Poapst (b.1921), B SC (Agr), under the general supervision of Davis.[62].

Davis was also responsible for some early research on the preservation of fruits and vegetables by freezing, and for some of the first research on gas storage of apples in North America. While the cold-storage studies were being carried on, he was surveying the various methods employed for gas or controlled-atmosphere storage, before deciding on the one to be used in the gas storage rooms of the Division of Horticulture.[63]

In his research on the freezing and dehydration of fruits and vegetables, Davis had the competent assistance of Mary MacArthur, B SC, AM, PH D, and Cecil C. Eidt (b.1898), BSA MacArthur, whose early specialty was plant histology and cytology, became well known for her leadership of the Canadian work on dehydration, which included fundamental research on methods for determining the inactivation of enzymes in plant tissues prior to dehydration. She had a large dehydration tunnel built in Ottawa in 1942, in which several hundred experiments were carried out, and before the end of the Second World War she published a paper on the freezing of commercially packaged asparagus, strawberries, and corn.[64] Incidentally, Mary MacArthur eventually became the first woman "Fellow" of the Agricultural Institute of Canada.

In British Columbia, D.V. Fisher, J.E. Britton, and H.J. O'Reilly conducted cold-storage experiments with peaches in 1940 and 1941 at the Summerland Experimental Station.[65] That was approximately when Robert S. Willison (b.1898) was doing valuable research, in Ontario, on the fungal wastage in peaches during storage and transit.[66] Willison, who had BA and MA degrees from McMaster University, also did some noteworthy research on wound-gum on peaches and grapes for the PH D that he obtained from the University of Toronto in 1929. Willison's major contributions to horticulture during the 1940s, in relation to plant pathology, were in the realm of virus diseases of fruits, and peach canker.[67]

To a plant pathologist, the preservation of harvested fruits and vegetables comes under the heading of post-harvest pathology, so in that sense Davis and his assistants and co-workers across the country were working in an area of plant pathology that had been largely neglected, and which has never received nearly as much attention as it deserves. Retaining the quality of fruits and vegetables for as long as possible after being harvested has been an objective of horticulturists, especially fruit growers, for almost as far back as there is a history of gardening. Thus it is not unexpected to see recipes for storing grapes in charcoal, or peas in dry sand or infusorial earth, or instructions on how to construct pits for the storage of fruits and vegetables, in the early reports of Fruit Growers' and Horticultural Associations in Canada.

A Canadian horticulturist who attained international recognition in the general field of controlled-atmosphere storage of apples is Charles A. Eaves (b.1908). Eaves, a native of Liverpool, England, began his horticultural career at the Central Experimental Farm, Ottawa, in 1929. After earning a BSA from McGill University in 1932, for work done at Macdonald College, he was awarded a scholarship by the Imperial Order Daughters of the Empire that enabled him to study at the Low Temperature Research Station of Cambridge University, in England. Eaves wrote a history of the early development of gas storage research, including some of the preliminary trials at the Central Experimental Farm, Ottawa.[68] Shortly after his return to Canada in 1940, Eaves was employed as an assistant in low temperature research, Dominion Experimental Station, Kentville, Nova Scotia, and that is where he earned an enviable reputation as the inventor of a safe, inexpensive, and effective method for the reduction of carbon dioxide in apple stores.[69]

Although Malcolm Davis had not been a pioneer in the prevention of post-harvest diseases or the deterioration of fruits and vegetables in storage or transit, he and the others mentioned above were among the earliest in Canada to put that work on a modern scientific basis. This essay ends with Davis and his colleagues in post-harvest disease prevention because any plant pathological work by horticulturists – and there has been a significant amount – more recent than theirs is too modern to be considered as a contribution to the early development of plant pathology in Canada.

ADDENDUM

There have been a number of plant pathologists and many others who, although not really horticulturists, have been doing horticultural work in relation to plant pathology. The dividing line between a horticulturist doing the work of a plant pathologist and a plant pathologist doing horticultural work is sometimes very blurred. Individuals may consider themselves to be horticulturists even though, for bureaucratic reasons, they are listed under some other classification. Examples of this are the plant pathologists and their students at Macdonald College doing research on diseases of horticultural plants. Garven H. Berkeley (b.1894), Robert S. Willison, George O. Madden (1890–1944), and Gerald C. Chamberlain, at St Catharines, Ontario, were classed as plant pathologists while they were doing some excellent disease research on raspberries,[70] whereas George H. Harris (b.1898), BSA, MS, PH D, doing plant pathological work on a raspberry problem in British Columbia,[71] was a horticulturist.

Garven H. Berkeley, BA, MA, PH D, deserves special mention for his outstanding work on tomato diseases over a period of more than twenty-five years. He was the author of Dominion Department of Agriculture bulletin no. 51, in 1925, titled *Tomato diseases*, and publication no. 552, *Prevention of virus diseases of greenhouse grown tomatoes* in 1937.

Plant pathologists A.A. Hildebrand (1896–1969) and L.W. Koch were doing horticultural research, both individually and jointly, when they studied diseases of strawberries in the Niagara Peninsula.[72] Koch also studied diseases of fruit trees and raspberries,[73] to mention only two more of his contributions to horticulture. The work of J.F. Hockey (1895–1980) and G.C. Chamberlain on apple scab, of R.S. Wilson on cherry diseases, of L.V. Busch on tomato late blight, and of C.D. McKeen on vegetable seedling diseases – all of which were recorded in the 1948 *Proceedings of the Canadian Phytopathological Society* – are of a horticultural nature, although performed by plant pathologists.

Similarly, Randal E. Fitzpatrick (1907–80), Frances C. Mellor (b.1916), and Maurice F. Welsh (b.1916) were doing horticultural work when they studied diseases of apple trees in British Columbia.[74]

Fitzpatrick had studied a disease of peach trees for his PH D thesis, University of Toronto, in 1933. After ten years as a plant pathologist at the Summerland research station, he became head of the Dominion Laboratory of Plant Pathology when it was established in Vancouver in 1946.

When Lloyd T. Richardson (b.1913), BA, MA, PH D, was doing research on tomato diseases, he elicited the cooperation of Harry Katznelson (1912–1965),[75] a bacteriologist who had found a bacillus that produced a thermostable substance which inhibited the growth of seventy-seven out of eighty-one fungi.[76] Katznelson was neither a horticulturist nor a plant pathologist, but, like many chemists, physiologists, and others too numerous to be mentioned in this brief history, he made a positive contribution to the development of both horticulture and plant pathology in Canada.

3 Potato Diseases and the Beginning of Plant Pathology in Canada

Plant pathologists know that it was a potato disease, now called late blight, that stimulated much of the early development of plant pathology in Britain and northern Europe in the middle of the nineteenth century, but it is not generally so well known that diseases of the potato also provided the greatest stimuli to the early development of plant pathology in Canada. By the beginning of the nineteenth century the potato had displaced grain as the major source of carbohydrates in the diets of the peoples of northern Europe and their descendants in North America. For this reason, any failure or significant decrease in the yield of potatoes was quickly reflected in the eating habits and in the cost of living of almost everyone in those areas.[1]

For various undetermined reasons, the potato was considered to be an unreliable crop throughout eastern Canada long before late blight, the disease that was largely responsible for a famine in Ireland, made its appearance there. This is evidenced by comments such as those of the editor of the *Times* of Halifax, Nova Scotia, who, in the issue of 23 September 1834, referred to the partial failure of the potato crop for the previous two or three years, and of the editor of the *Standard* of St Andrews, New Brunswick, who in the issue of 3 November 1838 referred to "the extensive failure of the potato crop for several years past."

One such failure on Prince Edward Island is documented by an unnamed correspondent to the *Royal Gazette* of Charlottetown, whose letter referring to "the failure which has taken place in the potato crop" appeared on 26 July 1836. Grain as well as potatoes were in such short supply through the fall and winter of 1836–37 that the Prince Edward Island House of Assembly found it necessary, on 17 April 1837, to pass a bill called "An Act to Prohibit the Exportation of Grain Meal and Potatoes ...": it was feared that there would be insufficient potatoes and grain to seed a new crop. Nova Scotia had found it expedient to pass a similar act on 31 March 1837. The misery in New Brunswick, resulting from the crop failures there is reflected in the appeals of hungry people from various parts of that province that are recorded in the *Journal of the House of Assembly*, 10–14 July 1837. Apparently there had been a complete or near-complete failure of crops throughout the Maritime provinces in 1836, because the *Yarmouth Herald*, of Yarmouth, Nova Scotia, on 5 May 1837, made a derisive reference to the "Embargo Act" that had been passed by the legislature of that province and suggested that relief in the form of seed and bread should be provided in Nova Scotia as was being done in both New Brunswick and Prince Edward Island.

With such well-publicized failures of the potato crop in mind, it is understandable that some growers, grasping at straws, would tend to believe stories about "wonder potatoes" – potatoes that would not succumb to the dreaded blight and give higher yields than any other variety. Just such a "miracle" variety appeared on the scene around 1837. It was known as the Rohan Potato, and so many people were trying to purchase a few tubers that the editor of the *Commercial News and General Advertiser* of Saint John, in the issue of 13 November 1839, made reference to the "Rohan Potato mania" that had broken out in Fredericton. In Ontario, the fame of the Rohan increased when the 14 November 1839, issue of the Brockville *Recorder* reported a farmer as stating: "while the blight has, long since, killed almost every kind of potato growing, which was planted as early as these Rohans, they have continued to flourish." Yet, through stories about the Rohan Potato were common in the farm papers of eastern Canada for several years prior to the appearance of the late blight disease that is induced by the fungus *Phytophthora infestans*, apparently it was as susceptible as most other varieties to that fungus, because it was seldom mentioned after 1844.

The year in which the late blight disease of potatoes first made its appearance in Canada may never be known, but references to a blight of the potato crop, such as the one mentioned above for the area around Brockville, were not uncommon in the 1830s and early 1840s. Unfortunately for the historian, the term "blight" had a much broader meaning in those days than it does now. For example, in 1835, William Evans, secretary of the District and County of Montreal Agricultural Society, wrote in his book on agriculture in Canada that there were four generally accepted forms of blight. One was thought to originate in cold and frosty winds, one in "a sort of sultry and pestilential vapour," one from "want of due nourishment," and one from "the propagation of a sort of small and parasitical fungus."[2] That latter statement, made by a resident of Montreal in 1835, must have been considered a bit radical, because the fungal nature of plant diseases was not generally accepted by naturalists for at least another decade. However, the blight that was mentioned in farm papers earlier than about 1844 is unlikely to have been due to *Phytophthora infestans*, the incitant of "late blight."

Except for a report that "potato dry rot" was common in Nova Scotia in 1840,[3] there were very few references to potato diseases in the newspapers of eastern Canada from that year until the autumn of 1844 when, on 16 September, the editor of the *Quebec Gazette* commented on "the great alarm about the potato crop." The *Morning Courier* of Montreal referred to the prevalence of "the potato disease" in its issues of 5 and 8 October, and the St Catharines, Ontario, *Journal* of 20 September 1844 had a comment on the failure of the potato crop there. Without citing further examples, it is sufficient to state that almost every newspaper published in what is now Ontario, Québec, New Brunswick, and Nova Scotia in the autumn of 1844 referred to the potato "disease" or the potato "blight" in local crops.

In contrast to those reports, the editors of papers published in Prince Edward Island were commenting on the remarkable season of fine weather and the excellence of the potato crop, as the *Royal Gazette* of Charlottetown did, on 29 October 1844. Apparently late blight had little or no effect on the potato crop of the Island that year. However, there is no doubt about its effect there and elsewhere in the Maritime provinces and Newfoundland in 1845. The *Royal Gazette* of Charlottetown, on 23 September, carried this comment with reference to the potato: "The present season is the first within the memory of the oldest inhabitant, in which such a disease as that which now appears to prevail, has been known to affect that valuable esculent, although Canada and New England suffered in the same way last year." The *Times and General Commercial Gazette* of St John's, Newfoundland, had a note about the excellent quality of local potatoes in its 13 November 1844 issue, but on 15 November 1845, its editor was remarking that "the potato rot must be stayed in its progress or it will be of incalculable injury to the country." The *Gleaner*, of Newcastle, New Brunswick, was also commenting on the extent of the potato rot throughout eastern Canada in its issue of 20 September 1845. Thus it is reasonably certain that late blight, as it is now known, was rampant throughout eastern Canada, except for Prince Edward Island and Newfoundland, in 1844, and in those islands as well in 1845.

Farming in Newfoundland did not become a significant industry, in the true sense of that word, until the Second World War and later. Before that it was, for the majority of producers, a gardening operation to provide vegetables and meats for home use, with the small surplus being sold locally for money to supplement income from fishing, and to a lesser extent, from logging. Doubtless, potato late blight had a devastating effect on individual growers who were dependent upon that crop to help feed their family, but it had relatively little effect upon the economy of Newfoundland in the late 1840s.

Crop failures, or periods of scarcity, in the Maritime provinces were not unusual in the nineteenth century. The Journal of the House of Assembly of Prince Edward Island, in a record of the second session of 1848, refers to periods of "scarcity" in 1836, 1846, and 1847. In 1848 the "visitor" (inspector) of district schools on the Island remarked that he was "well aware that the pressure of the times, caused by the late partial failure of the crops, prevents many parents from sending their children to school."[4] The failure of the potato crop was so devastating that the House of Assembly resolved, on 7 February 1848, "That owing to the failure of the potato crop, and the subsequent scarcity of that article of food, it is expedient to introduce a Bill having for its object to prohibit the exportation of potatoes from the Island." Such a bill was introduced, and, according to the minutes of the House, received third reading five days later. Thus, for the second time in little more than a decade, Prince Edward Island was forced to forbid the sale of potatoes off the Island.

Potato blight became such a problem in eastern Canada, from 1845 onward, that the editor of the *Agricultural Journal and Transactions of the Lower Canada Agricultural Society* was advising his readers in 1848 that "It is a dangerous speculation, under the circumstance of our total ignorance of the cause of the disease, to cultivate potatoes extensively."[5] Between 1845 and 1851, late blight, locally known as the potato rot, was so destructive on Cape Breton Island that it "reduced the island to a state of near starvation."[6] It also caused such losses to potato crops in the region of Toronto that George Buckland (1805–85), who was then editor of the *Canadian Agriculturist*, considered the potato to have become "too precious a crop to be relied on as a staple article of food. It must descend to the rank of a garden luxury."[7] Henry Youle Hind (1823–1908), lecturer in chemistry and natural philosophy at the

Normal School in Upper Canada (now Ontario) and an authority on agriculture, was referring to "the universal prevalence of this malady" in 1850.[8]

One year later, the governor of Massachusetts offered a ten-thousand-dollar reward for the discovery of a "Sure and Practical Remedy for the Potato Rot."[9] That offer would be of little significance to plant pathology in Canada except for the fact that one of the published letters, sent in response to that offer, was submitted by John W. Dawson (1820–1899) of Pictou, Nova Scotia – the man who later became principal of McGill University, and its first professor of agriculture. Dawson's views on the cause and control of "potato rot" were later incorporated into his book, *First Lessons in Scientific Agriculture for Schools and Private Instruction*, published in Montreal by John Lovell in 1864.

The Board of Agriculture of Upper Canada was offering somewhat more modest awards for essays on the control of plant diseases at that time. In an essay for which he was awarded a prize of twenty pounds in 1852, John Harland of Guelph remarked that "potatoes, which used to be somewhat extensively and profitably cultivated, have of late years been almost a general failure, and few persons at the present time risk the planting of more than to produce sufficient for their own culinary purposes."[10] The annual report for 1862 of the Board of Agriculture of the province of New Brunswick has a quotation from a prize essay on potatoes and potato diseases, indicating that province was also offering awards to anyone who could solve the problem of the disease. Perhaps the general feeling about the potato late blight disease was best expressed by the Hon. Adam Fergusson when he stated that all other crop diseases "sink into comparative insignificance when brought into consideration with the awful and apalling [sic] visitation which has, for the second consecutive season in Canada, ravaged our potato crop."[11]

Widespread plant diseases that interfere with a nation's food supply, and epidemics of human or animal diseases, tend to stimulate research and the advancement of knowledge. However, the stimulus to plant pathology in Canada that came about as a direct result of the late blight disease of potatoes was negligible in comparison to the enormity of the problem. Many people considered it to be an act of God and therefore beyond human ability to rectify, except through prayer. Even though late blight was the disease that posed the greatest threat to potato growers for more than half a century following its first appearance in eastern Canada, most of the "research" on it, for several decades, consisted of innumerable tests on various formulations of the Bordeaux mixture, a means of chemical control that had been discovered and developed in Europe. In Canada there were tests to determine the best strength of materials to use, best time of application, best rate of application, and the optimum number of applications, to ensure a disease-free crop.

Other spray and dust materials were tested during that period, but not to the extent of the Bordeaux mixture and its variants.

The experiments on methods for the chemical control of late blight were often paralleled by those on physical and cultural methods, such as those advocated by Henry Y. Hind. Quoting a British source, he recommended that seed potatoes should be cut lengthwise so that some "eyes of the rose end be in every seed-set." He further advocated the spacing of drills at least a yard apart and bending the emerging stems to one side so that they could be covered with earth "like the roof of a house with the shaws [stems] growing out one side." This would prevent rain being conducted to the developing tubers. Hind also thought it a good idea to cut off the stems as near the earth as possible after the flowering is over, and then to "cover the incised stumps of the stems with at least an inch of earth."[12]

The editor of the *Islander* reported, in the issue of 21 August 1846, that a farmer in Prince Edward Island had the tops of his potato plants "mowed down when they became blighted" to allow new non-blighted stems to spring up. Apparently the editor did not follow up on that story to see if the new shoots became blighted: the subject was not mentioned in succeeding issues of his newspaper.

Hybridists (plant breeders) of that era were trying to develop cultivars of potatoes that would be resistant or even immune to the blight disease and still have desirable culinary qualities. One of the most noteworthy was Peter C. Dempsey, president of the Fruit Growers' Association of Ontario, who, beginning in 1861, developed the Dempsey potato, which was reputed to be blight-resistant.[13]

Before the fungal nature of late blight was generally known, there were many theories about its cause. One of the most interesting was that of W. Bustin, as recorded in the first volume of the *Nova Scotia Journal of Agriculture*. Bustin thought some peculiar electrical state of the air or earth might be the cause. His theory, "based on electrical science and successful experiment," was that the disease could be controlled by placing a pole in the centre of each small field of growing potatoes and having copper wires running from the top of the pole, in a tentlike fashion, over the potato plants to the earth.[14] Apparently the Nova Scotia government took little notice of Bustin's theory, because it seriously considered importing potatoes from the native home of the tubers in the hope of getting some that were immune to the blight. Instead, however, it decided to purchase varieties reputed to be blight-resistant that had been developed by the Rev. Chauncey E. Goodrich, in New York State, from tubers or true seed obtained from Peru. The Board of Agriculture purchased seventeen named varieties and distributed them "among 20 gentlemen throughout the Province," who were expected to report on the resulting crops. Several of the ensuing reports claimed that a few of those varieties, including Garnet Chili, which became popular, were so disease-resistant that they would save the country millions of dollars.[15]

In 1887, William Saunders, director of the Dominion Experimental Farms, obtained 245 varieties of potatoes, mostly from Europe, for comparative purposes at the Central Experimental Farm, Ottawa. Ranking varieties for their yield, culinary qualities, and resistance to blight was a favourite "test" of potatoes for many years. The Ontario Agricultural College (OAC), which had published a bulletin, "The Potato Rot Its Cause and Remedies," in 1886, was comparing the merits of 195 varieties in 1896.[16] Similar trials were conducted at several places in Canada around the end of the nineteenth century and the early decades of the twentieth. Every Experimental Farm Report, from 1895 until 1901, referred to a potato variety called "Brown's Rot Proof" in their variety trials. It was just one of the more than one thousand variety names listed in bulletins and reports during that period. They included such exotic names as Rose's Beauty of Beauties, Black Elephant, Lark's Eye, Member of Parliament, and Lizzie's Pride.

James Fletcher (1852–1908), Dominion entomologist and botanist, was conducting experiments at the Central Experimental Farm in 1891, to attack both the scab and the rot of potatoes at the same time.[17] One year later he told a standing committee of the House of Commons that, although the efficacy of the Bordeaux mixture for the control of late blight had been thoroughly proven, he regretted that "very few of our Canadian farmers have been persuaded to try it."[18] Fletcher must have known that the spraying of crops for disease control had not been mentioned in Canadian publications prior to about 1885, and that farmers were naturally reluctant to use such a relatively new method of disease control. That reluctance persisted until the First World War, when, as part of the effort to increase the food supply, governmental agencies, such as the Greater Production Service in Quebec, made it easier for farmers to buy the ingredients and to purchase or rent spraying equipment.[19]

Before a satisfactory control method became generally known, it is understandable that some gullible growers might, in 1872, pay as much as fifty dollars for a tuber that was reputed to be immune to the blight.[20] It is much more difficult to understand why anyone in 1903, when the efficacy of the Bordeaux mixture had been widely publicized, would pay $1,250.00 for a pound of the so-called "Eldorado Potato," as was alleged to have been done.[21]

The greatest and most significant stimulus to the early development of plant pathology in eastern Canada came not from late blight but from three diseases of potatoes that were relatively unknown in North

America. The first of those to come into national prominence was a foreign disease, one known variously as potato canker, black wart, or wart, which had never been found in Canada. Another was powdery scab, which, although present in Canada, had not come to the attention of anyone who recognized it as such. The third, a so-called degenerative disease, was much more obscure at that time.

Potato canker came to the notice of the Dominion botanist, Hans Güssow (1879–1961), when diseased tubers that had been grown on Red Island in Newfoundland and sent to the editor of a newspaper in Montreal were forwarded to him for identification in October 1909.[22] Güssow, who had seen potato canker in Europe, recognized the disease at once, and considered it to be more destructive to potatoes than blight "because there are no remedies known for the canker."[23] He took almost immediate action to prevent the admission of additional diseased tubers into Canada. To do this he cooperated with the Dominion entomologist, C. Gordon Hewitt, in urging the Canadian government to enact a law that would place an embargo on the importation of plants from countries in which certain diseases and insect pests were known to exist.

Newfoundland was a Crown colony at that time. Because there was very little plant disease research in Newfoundland before it became a province in 1949, the story of plant pathology there does not form a significant part of this history, most of which ends in the late 1940s. However, the government of that Crown colony had established a demonstration farm, where imported potatoes were being grown and selected, prior to 1940, for resistance to both the canker and late blight.[24]

The efforts of Güssow and Hewitt, plus those of H.W. Smith, professor of biology at the Nova Scotia Agricultural College,[25] and other like-minded people, resulted in the promulgation, by the federal government, of "An Act to Prevent the Introduction or Spreading of Insect Pests and Diseases Destructive to Vegetation" on 4 May 1910.[26] That act, best known by its short title, "The Destructive Insect and Pest Act," was modelled somewhat after a similar one in England. The Canadian version authorized the governor in council to "make any such regulations as are deemed expedient to prevent the introduction or admission into Canada, or of spreading therein, or the shipment beyond her borders, of any insect, pest or disease destructive to vegetation." One of the items promptly banned, by an Order in Council dated 27 February 1911, was the importation of potatoes from Newfoundland or from Europe, where potato canker was known to occur.

Güssow had a bulletin on potato canker prepared and published before the end of the year, which was a remarkable accomplishment. In various ways, he did all he could to ensure that growers became aware of the disease, and he urged them to be on the lookout for it.

Güssow's efforts to keep foreign potato diseases out of Canada were frustrated later in 1911, when the law restricting the importation of potatoes from Europe was relaxed to compensate for a shortage in Canada that year. Between 1 October 1911 and 31 March 1912, more than twenty thousand bushels of foreign potatoes were allowed into the country.[27] Güssow knew how disastrous it could be to the potato industry in Canada if some European potatoes infected with the canker were used as "seed," so he tried to prevent anyone from doing this by persuading the authorities to issue another Order in Council making it illegal to "receive or use, for seed purposes, any potato imported from Europe."[28] During the summer of 1912 tubers suspected of having canker were discovered in one or two areas of Ontario and Québec.[29] Those districts were placed under a quarantine that was not lifted until 1922, and the disease did not become established in Canada. After that it was found only once or twice in small gardens, including one in Halifax in 1941.[30]

To assist in the identification of the disease, Güssow had an illustrated poster, "Potato Canker Danger," issued in August 1912, with the request that if a potato was found that looked like the illustration it should be sent to the Division of Botany, Dominion Department of Agriculture, for identification. In response to that poster, many samples of diseased tubers were sent in, from Cape Breton in the east to Alberta in the west.

Among those from Quebec were some that Güssow identified as having powdery scab, a disease that had not been identified as such anywhere in North America. Güssow, being a scientist, albeit somewhat politically naïve, published in an American journal the fact that powdery scab had been found in Canada.[31] Potato growers in the United States saw this as a possible means of excluding Canadian tubers from their markets, so they lobbied their congressmen to place an embargo on the importation of potatoes from Canada. It was not long before American authorities found the disease in a shipment of potatoes from the province of New Brunswick. That was the excuse the Congress needed. In December 1913, after a public hearing, and in spite of objections from Canadian officials, it placed an embargo on the importation of potatoes from Canada and other countries where the disease was known to occur.[32]

That embargo was seen as a severe blow to Canadian potato growers, and it received a great deal of publicity in the farm papers. However, within a few weeks, potatoes originating from the state of Maine but being transported through New Brunswick were found to have the disease. As a consequence of that discovery, and of appeals from Güssow and other Canadian authorities, the United States Congress lifted the embargo on Canadian-grown potatoes, but with the important proviso that before such potatoes could enter the United States they would first have to be inspected by competent officials and declared to be free of the disease. Furthermore, the potatoes had to have been grown from clean seed and on land that had never been infected with powdery scab.[33]

That action by Congress proved to be a phenomenal boon to the development of plant pathology in Canada, because, to comply with its ruling, competent inspectors had to be recruited and trained to recognize potato diseases. A few inspectors were hurriedly trained, but before they had become efficient in their work one carload of potatoes containing tubers with powdery scab went to Boston, where the disease was discovered by United States officials. That incident resulted in the withdrawal of permits for the entry of potatoes into the United States for the remainder of the season.[34] Once again there was a market loss for potato growers in Canada.

Another event that tended to highlight the need for better inspection and more stringent controls over the quality of exported potatoes was the embargo placed on the importation of Canadian potatoes by Bermuda in 1914. That embargo was put on because farmers in Bermuda were getting low yields of poor-quality potatoes when they used seed-tubers from Nova Scotia. Representations initiated by W.H. Brittain, provincial entomologist for Nova Scotia, persuaded the Bermuda authorities to admit potatoes from that province, providing they were first inspected and certified to be free of disease.[35]

The publicity associated with the embargoes, plus increasing agitation by growers for better potato disease information, made it easier for the federal government to inaugurate a Seed Potato Certification Service in 1914, and to justify the creation of two new plant pathology laboratories in the Maritime provinces: one in Fredericton and one in Charlottetown. The latter location was chosen because Prince Edward Island was reputed to be the worst area for potato blight in Canada.[36]

In New Brunswick, Raymond P. Gorham (1885–1946), BSA, prepared a bulletin on powdery scab for that province's Department of Agriculture, which he gave to farmers attending his lectures on potato diseases.[37]

Plant disease control, including the inspection and certification of potatoes for disease, was under Güssow's jurisdiction, and he was very cognizant of the need for well-trained, knowledgeable inspectors and research workers. For that reason he appealed to young men in agricultural colleges and universities to specialize in plant pathology, where, he stated, "the future ... is most promising."[38]

Güssow also recruited the best people he could find to help solve the potato disease problems. One of these was Gilbert C. Cunningham, a native of Ontario and a graduate of the OAC who, for the five previous years, had been associate plant pathologist at the Agricultural Experiment Station in Vermont, USA, working on diseases of potatoes.[39] Cunningham became the first assistant in charge (all such appointees were considered to be Güssow's "assistants") of a new laboratory of plant pathology in Fredericton, on 1 July 1915, and was given control of the potato inspection services for New Brunswick and Quebec.

According to Güssow's 1914 report to the director of the experimental farms, eleven men had been employed to make field and cellar inspections of potatoes for powdery scab prior to Cunningham's appointment. They were placed under Cunningham's supervision, and he was required to recruit and train additional inspectors while setting up potato research projects at Ste-Anne-de-la-Pocatière and Lennoxville, Quebec, as well as ones in the Fredericton area.

Because he suspected that the alleged "degeneration" of potatoes in Bermuda was due to virus-infected tubers that had been shipped there, Cunningham searched for virus-free potato plants. In 1916 he found a few, and they became the nucleus for a variety of certified seed potatoes in the Maritime provinces.[40] By 1921 Cunningham had become more knowledgeable about potato virus diseases and was conducting experiments on potato mosaic, leaf roll, and related disease problems. He was also conducting spraying demonstrations, rather than additional spraying experiments. He considered potato spraying for insect and disease control to be beyond the experimental stage.[41]

Cunningham resigned from the federal service in 1923, but continued to promote the potato industry, both privately and as an employee of the New Brunswick Department of Agriculture, until the time of his death in 1949. People who were qualified to evaluate his work believe that Gilbert Cunningham did more than anyone to promote the production and export of high-quality New Brunswick potatoes.[42]

Cunningham was succeeded by Scott Foreman Clarkson, a McGill graduate with BSA and MSC degrees. Clarkson had been an employee of the Dominion Department of Agriculture in Fredericton, working mostly with apples and strawberries, from 1934 until he resigned to occupy the newly created position of director of the Plant Protection Service in the New Brunswick Department of Agriculture. From that time onward he became more and more involved with potato protection and promotion, fields in which he eventually earned for himself an enviable reputation.

The second plant pathologist to be hired by Güssow for work on potato diseases in the Maritimes was Paul A. Murphy (1887–1938), a graduate of Dublin University and the Royal College of Science for Ireland. Murphy, with a travelling scholarship awarded by the English Board of Agriculture, had extended his knowledge of plant pathology by one-year periods of study at the Imperial College, London, England, at universities in Germany, and at Cornell University in the United States.[43] In 1915, he was appointed assistant in charge of a field laboratory of plant pathology that had been built that year in Charlottetown. He was so pleased with his new laboratory that he boasted, "It is doubtful if there exists anywhere an institution as well fashioned and furnished for carrying on the work for which it was designed."[44] Within a year he had begun more than seventy-five field experiments, some of which dealt with clubroot of turnips and scab of apples, in addition to the ones on late blight and other diseases of potatoes.[45]

It was not uncommon for someone doing research on potato diseases also to study one or more turnip diseases at the same time. That was certainly true in the Maritime provinces, where, in the early decades of the twentieth century, clubroot and water-core or brown-heart diseases of turnips or rutabagas became such a problem that many farmers abandoned turnip growing until satisfactory methods for their control were discovered in the 1930s.

One of Murphy's major responsibilities was supervision of the newly formed Seed Potato Inspection Service for Prince Edward Island and Nova Scotia. Güssow had appointed him to head that work for those two provinces, as Cunningham was doing for New Brunswick and Quebec. During the summer of 1916, Murphy worked in Nova Scotia with provincial entomologist W.H. Brittain and E.J. Wortley, a representative of the Bermuda Department of Agriculture who wanted to satisfy himself that potatoes destined for his country were being carefully inspected by competent people. They were particularly concerned with the potato leaf roll disease, because it was suspected of being the cause of the serious decline in yield of potatoes in Bermuda that were grown from seed originating in Nova Scotia.[46]

The leaf roll disease was difficult to diagnose in the field because rolling of the leaves could be caused by other agencies, other such as water-logged soil, drought or extreme heat, and certain fungus diseases. The colour of the foliage of leaf-roll infected plants commonly deviated from that of normal plants, but that symptom varied with the potato variety, the environmental conditions, and the severity of the infection. When leaf-roll infected tubers were cut in cross sections they often exhibited a discolouration of the phloem tissues that was referred to as "net necrosis," but this was not a reliable diagnostic indicator either, because it too could be caused by other factors.

Attempts to clear the potatoes of leaf roll by selecting seed potatoes from healthy plants failed in both Bermuda and Nova Scotia. Murphy suspected that something, such as an insect, must be carrying the disease-inducing factor from infected to healthy plants while they were growing in the field. The report of the OAC for 1917 indicates that the investigators there were also implicating insects in the spread of leaf roll and mosaic. Proof that insects transmitted the viruses that caused those diseases came from research done on the mosaic diseases of tobacco and potato in Europe and the USA.

In 1921, P.J. Shaw, the provincial horticulturist for Nova Scotia, began conducting some experiments for the control of the potato scab disease by applying sulphur to the soil.[47] For several years that was a popular topic for study. The National Research Council (USA) had a special sulphur fellowship committee, and the Texas Gulf Sulphur Company donated fellowships for student research. One such fellowship was awarded to Catherine Graham Welch, a student of Professor G.H. Duff, University of Toronto. Her MA thesis, "Sulphur as a control of potato scab," was published in 1927.[48] Several chemists, including Walter A. DeLong (1894–1976), who, at that time, was assistant provincial chemist in Nova Scotia, studied the relationship between sulphur, soil acidity, and the incidence of potato scab.[49] Before the end of 1949, so much research had been done on potato scab, in Canada and elsewhere, that the Ontario Crop Improvement Association paid the University of Western Ontario one thousand dollars just to review the literature. This, and a note on scab research at the OAC and at the St Catharines Laboratory of Plant Pathology, is mentioned in the Ontario Minister of Agriculture Report for 1949.

During the winter of 1924, W.K. M'Culloch, who became head of the Seed Potato Inspection Service for Nova Scotia, following the resignation of Murphy in 1920, indexed tubers in a greenhouse at the Dominion Research Station, Kentville, by planting a single eye from each potato. The resulting virus-free tubers were then multiplied at the station and on selected farms. Thus he initiated the tuber-indexing method that became widely adopted as the basis of "foundation" seed-potato production in Canada.[50] To encourage young people of Nova Scotia to become interested in the production of high-quality potatoes, Junior Potato Clubs were organized in 1928. Two years later the members of those clubs were each given a bulletin titled *Diseases and insects of the potato, with methods for their control.*[51]

In the meantime, the publicity surrounding "new" potato diseases, the embargoes on potato exports, and the regulations pertaining to the inspection service, at a time when Canada was at war with Germany, had a devastating effect on the private lives of Hans Güssow and his family. It began with grumbling at meetings of farmers, and others, about "that German in Ottawa."[52] Even Canadian plant pathologists, who should have been better informed, authorized one of their committees to "wait upon the Minister of Agriculture to urge the appointment of a Dominion Botanist who can command the confidence of Plant Pathologists."[53] The situation, for Güssow, deteriorated to such an extent that his name was excluded from the annual reports of the minister of agriculture for the years 1918 to 1921 inclusive.

After the war, when his phenomenal contributions to plant pathology in Canada were looked at more dispassionately, Güssow was readily accepted as a charter member of the Canadian Phytopathological Society and chosen to be its first president, 1929–30. In recognition of his contributions to Canadian agriculture, Queen's University bestowed upon him an honorary LL D in 1931, the year in which he was elected Fellow of the Royal Society of Canada. Güssow had been a charter member of the American Phytopathological Society, and was elected its president for 1935–36. Needless to say, he had been reinstated as Dominion botanist, the position which he held with dignity and respect until he retired in 1944.[54]

During his tenure of office, Güssow was largely responsible for the establishment of laboratories of plant pathology from coast to coast, and he was the prime mover in the Certified Seed Potato Program that resulted in more than a threefold increase in disease-free potatoes in Canada during his working life. No man has made a greater contribution toward the development of plant pathology in Canada than Hans Theodor Güssow.

On Prince Edward Island, Murphy had been doing some pioneering work on the "running out" or "degeneration" disease that had been investigated by Brittain in Nova Scotia in 1914. It was later determined to be a virus disease, the first virus disease of plants to be recognized as such in Canada.[55] From around 1920, when the potatoes in the variety and cultural tests at the Central Experimental Farm, Ottawa, were found to be so badly infected with viruses that the tests were temporarily discontinued, virus diseases became a major problem for potato growers.[56]

Because of the breadth of Murphy's knowledge of potatoes, and for the sake of uniformity in inspections, in 1918 he was placed in charge of all potato certification work in eastern Canada.[57] His comprehensive study, titled *Investigation of potato diseases*, was published by the federal

Department of Agriculture, as bulletin no. 44, a year after he resigned his position to return to Ireland in 1920. Subsequent to Murphy's departure, the potato inspection and certification program for all of Canada was conducted from the Central Laboratory, Dominion Department of Agriculture, Ottawa. It was supervised by Geo Partridge, who had local supervisors and inspectors in the major potato-growing areas of the country.[58]

Murphy was succeeded in the Charlottetown Laboratory by John B. McCurry (b.1894), a BSA graduate of the Ontario Agricultural College who continued the potato research initiated by Murphy, and expanded it to put more emphasis on the chemical treatment of "seed" to prevent seed-piece decay and disease. McCurry was transferred to Ottawa in 1923, and S.G. Peppin was acting assistant in charge until Richard R. Hurst (1895–1961), a native of Woodstock, New Brunswick, took charge in 1925.[59] For Hurst, who had graduated from the OAC with a BSA, in 1922 and been employed as an assistant plant pathologist in Saskatoon from 1923, it was a return to Charlottetown, because he had worked there as a plant disease investigator, under Murphy, in 1918.

Beginning in 1927, and for several subsequent years, the annual reports of the Dominion botanist show that Hurst was involved in a variety of plant disease problems, including ones on hollyhocks and turnips, in addition to those of potatoes. The report for 1929 shows that while continuing a plant disease survey, he found several diseases that had not previously been reported on the Island, including Alternaria tuber rot of potatoes and Phoma rot in association with powdery scab of potato. It also shows that by using the tuber unit method, begun in 1927, he obtained potatoes of twenty varieties which, he thought, were free from virus diseases. Hurst continued the potato investigations that had been left to him by McCurry, and expanded them to include further studies on late blight, Rhizoctonia, Verticillium, and treatments of seed and soil for disease control.[60]

Hurst, whose title changed to that of officer in charge in 1927, was fortunate in having J. Lorne Howatt (1895–1959), a native of Charlottetown, as a student research assistant during the summers between 1925 and 1930, and George Ayers (b.1911), from 1930 to 1936. Both were destined to become esteemed plant pathologists. Howatt became employed in the Fredericton laboratory, after earning a BSA at Macdonald College in 1931. Ayers had a Macdonald College B SC(Agr) degree and, with a scholarship in horticulture from the Quebec government, returned to the college to study celery storage diseases for the M SC degree that he obtained in 1937. Almost immediately after graduation he became a full-time employee at Charlottetown and was soon working on several disease problems, including clubroot of turnips, in addition to those of potatoes, particularly the Verticillium wilt disease. He continued those diverse studies until he joined the Royal Canadian Air Force, early in the Second World War.[61] After the war he returned to the Charlottetown laboratory as an associate plant pathologist, and continued studies on plant diseases, especially those of turnips and potatoes.

As a wartime replacement for Ayers, Hurst persuaded Lorne C. Callbeck (1912–79), B SC, a professor of agriculture at Prince of Wales College, Charlottetown, to join his staff on a part-time basis while continuing his teaching at the college. In 1946, Callbeck reversed those roles by becoming a full-time member of the Science Service Laboratory while continuing to teach agriculture at the college on a part-time basis. He became well known for his research on chemical control of potato late blight. After the Dominion Department of Agriculture set up a potato fungicide committee, in 1949, to coordinate the work of testing fungicides for the control of late blight, new ones were screened at Summerside, and those materials that showed promise were tested in other provinces. Callbeck was also one of the first to conduct controlled experiments on the value of killing the foliage of potatoes, referred to as "top killing," as a means of disease control.[62]

In November, 1945, a seed potato inspector in Prince Edward Island found tubers with a disease or injury that he had not seen before. Samples sent to plant pathologists in Ottawa eventually got to A.D. Baker, head of the nematology unit, who sent specimens to Gerald Thorne, Division of Nematology, United States Department of Agriculture. Thorne was known to be an authority on nematode identifications, and had just recently described a new species from Idaho potatoes that he named *Ditylenchus destructor*, the "potato rot nematode." Thorne advised that the nematodes from Prince Edward Island were of the same species as those from Idaho.[63] R.R. Hurst of the Dominion Laboratory of Plant Pathology, Charlottetown, reported, in the 1947 *Proceedings of the Canadian Phytopathological Society*, that he found no varietal resistance to the nematode in any of the ten varieties of potatoes in his test series but that a soil fumigant was an effective means of control.

Finding the potato rot nematode in Prince Edward Island was considered to be such a serious menace to the potato industry that a potato-rot nematode research committee of Science Service, Dominion Department of Agriculture, was set up under Baker's direction. That nematode situation was studied, largely by V.E. Henderson, until sometime in the 1950s, and thus is too recent to be dealt with here.

After the resignation of Cunningham, research on potato diseases at the Fredericton laboratory was carried on by J. Frederick Hockey

Scott F. Clarkson

Jacques Champlain Perrault
(1903–88)

(1895–1980), who was just getting things organized when he was transferred to the Kentville, Nova Scotia, laboratory in 1924. Hockey's successor was Donald J. MacLeod (1894–1990), a native of Dunvegan, Ontario, who had BA and MA degrees, with emphasis on botany and bacteriology, from Queen's University.[64] MacLeod, as head of the plant pathology laboratory, automatically assumed supervisory responsibility for the potato inspectors. That responsibility continued until 1927 when Clarence Hurdman Godwin was appointed district plant disease inspector, thus enabling the plant pathologists to concentrate on more fundamental research.

The transition of the potato research from that of Cunningham to MacLeod was aided by James Keith Richardson (1899–1960), a native of Montreal who had been a plant disease investigator under Cunningham in 1920. After obtaining a BSA degree in 1921, and an M SC in 1923 from McGill University, he returned to Fredericton to work with Cunningham as an assistant plant pathologist, the position he was holding when MacLeod became officer in charge. Richardson was an able assistant to both officers in their potato research, until he transferred to the St Catharines Laboratory, in 1930, where he became well known for his ability to diagnose potato ring rot.[65] His place in Fredericton was taken by J. Lorne Howatt, who had recently obtained a BSA degree from Macdonald College of McGill University. In 1933, Howatt went on leave to Macdonald College, where he studied the mo-

Paul A. Murphy (1887–1938)
Courtesy Agriculture Canada
Research Station, Charlottetown.

saic diseases of tobacco for the M SC degree that he earned two years later.

The Fredericton laboratory was moved, in 1925, from the centre of the city to the Dominion Experimental Station, a few kilometres down river. There the plant pathologists had more ready access to greenhouse and field plot space, and closer cooperation with the superintendent, C.F. Bailey, and his staff. In 1933 the Fredericton station was designated as the site for the national potato breeding program. For the first few years the major objectives of that program, with the cooperation of horticulturists, entomologists, plant breeders, and plant pathologists, but largely under the leadership of potato breeder Lewis C. Young (1902–90), was the development of varieties resistant to mosaic and late blight. Other specific objectives were added later. Throughout that continuing program the plant pathology laboratory was responsible for testing the resistance of selected potato varieties to diseases.[66] By 1949 nearly 91,000 seedlings had been tested for resistance to late blight.

Some potato breeding was carried on in other provinces, and Canada cooperated with the United States Department of Agriculture in its national potato breeding program. That cooperation made possible the development of a potato variety, named "Canus" after the two countries, in 1949.

In Fredericton, Howatt assumed most of the responsibility for the fungal diseases, and MacLeod took a particular interest in the virus part of the testing program. In 1936 MacLeod obtained leave and studied virology under R.N. Salaman and Kenneth Smith in the virus research laboratory at Cambridge University, in England. Two years later he returned to Fredericton and continued his thesis research for the PH D that he was awarded in 1944.[67]

In the 1930s there was great interest in plant diseases due to deficiencies of certain minor elements in the soil or to the inability of plants to absorb them. When Frank Dunlop, of Millville, New Brunswick, saw that the leaves of his potatoes displayed an unhealthy, discoloured appearance, he took some of them to MacLeod, who suspected it was a manifestation of magnesium deficiency. He and his colleague Howatt had read about the effects of magnesium deficiency in the soils of other potato-growing areas. When Dunlop followed their suggestion by including some magnesium oxide in his fertilizer, the problem was solved.[68] That incident encouraged both MacLeod and Howatt to intensify their research on deficiency diseases in other crops, particularly the brown-heart disease of turnips, which was investigated in cooperation with R.R. Hurst of the Charlottetown laboratory.[69] Nevertheless, potato diseases, especially potato viruses, remained MacLeod's major interest throughout the war years and beyond. Incidentally, he sent material to René-O Lachance (1909–92) that was helpful in the study of boron-deficient plants that formed the basis of Lachance's PH D research at Macdonald College in 1940.

The plant pathologists at Fredericton and Charlottetown cooperated on virtually all of their respective potato disease projects, and were sometimes joint authors of published papers. They in turn had the cooperation of workers at other research stations, and that of a number of private potato growers. After a period at the beginning of government-sponsored agricultural research, when Ottawa was the focal point of potato investigations, Fredericton and Charlottetown became the major centres for potato research in Canada.

In 1939 the government of the province of New Brunswick, through "An Act Respecting the Potato Industry," authorized the Governor in Council to establish foundation seed stock areas, and certified seed stock areas, and to regulate virtually all aspects of potato growing, digging, storage, and transportation within those areas. Controlled or quarantine areas, as a supplementary measure for further seed improvement, were instituted through a proclamation in the *Royal Gazette*, on 24 July 1940. These consisted of eight foundation seed areas comprising 104 growers, and seven certified seed areas, comprising 258 growers. The controlled areas were a combination of isolation and control. Each area was a continuous block of adjacent farms on which only certified seed was allowed to be planted, reasonably isolated from other farms which were not under control.

Growers of foundation seed commonly planted their potatoes in "tuber units," meaning that at planting time each tuber was cut into four seed pieces which were planted in a unit with a prescribed space between each quarter and twice the prescribed space between units. The knife used to cut the seed pieces had to be disinfected before cutting each tuber. When a disease was subsequently found in one or more plants in a unit, all four plants of the unit had to be removed and destroyed.

The removal of diseased plants from a plot or field was referred to as "roguing." To be most effective, roguing was started as soon as disease could be detected in the young plants so that as many as possible of the virus infected ones could be removed before aphids appeared in the field. It had been learned, elsewhere, that aphids transmit virus diseases from infected to healthy potato plants. In 1941 the New Brunswick Department of Agriculture began to assist growers in the roguing of their seed plots, and in 1947 it conducted a potato "Roguing Course for Growers," according to its annual report for that year.

Although tuber-unit planting was not compulsory, it was, until the 1920s, one of the best ways of ensuring that a field of potatoes would pass the two compulsory field inspections and be classed as foundation material. Every grower interested in the production and maintenance of high-quality seed potatoes was encouraged to have at least a small seed plot that was tuber-unit planted.

Soon after the "tuber index" method of planting was described, in the first issue of *Phytopathology* for 1921, that method became popular for two or three decades. In essence, it consisted of removing from each tuber a small piece bearing a single eye and planting it in a greenhouse during the winter months. If the plant produced by that small seed piece developed symptoms of any virus disease, the tuber from which the piece was cut would be discarded. Only those tubers the seed pieces of which produced normal plants in the greenhouse were saved for planting in the field. Such indexed tubers were usually planted by tuber units in seed plots where the stock was multiplied for planting the following year. The practice of tuber indexing was based on the premise, later proven to be incorrect, that virus particles were uniformly distributed throughout an infected tuber so that every eye would produce an infected plant.

Another provincial program to assist in disease control was the "advance testing" during the winter months on a farm in Florida. That scheme, inaugurated in 1945, was, like the indexing method in green-houses, intended to make available advanced information on the presence of diseases in the seed stock of the growers who participated in the program.

Tolerance levels of disease and other certification standards for tubers were adopted in 1922 and standards for the growing plants, or field inspection, in 1926. By 1930 it became illegal to sell "seed potatoes" that had not been certified, and labelled as such, by an authorized inspector. As the quality of potatoes increased, the certification standards were periodically raised, and for a few years immediately before the Second World War, a field or plot of potatoes could not have more than 0.5 per cent of its units showing evidence of disease at the time of the first field inspection, and not more than 0.1 per cent on the second, to be classed as foundation material. Disease tolerances for certified seed were approximately double those of the foundation seed.

The objectives of the 1939 Potato Industry Act of New Brunswick, and those of somewhat similar acts that came into force in Prince Edward Island, and, later, in other provinces, were to further the production of better seed through the improvement of disease control. The effect of such acts was greatly to enhance the reputation of those provinces for the production of superior quality, disease-free seed potatoes.

A history of the certified seed potato industry in Canada, including a commentary on certification standards, was reviewed by W.J. Scannell to members of the Quebec Society for the Protection of Plants, and published in the 1949 report of that society.

In Quebec, Abbe Léon Provancher (1820–92) published an article on the diseases of potatoes in the very first issue of *Le Naturaliste Canadien*, the journal that he initiated in 1869. Because relatively fewer potatoes were being grown in Quebec for export to other countries, there was comparatively less government-sponsored potato disease research in that province during the first two decades of the twentieth century. The situation was quite different, however, at the agricultural colleges in that province. For example, Macdonald College of McGill University, in Ste-Anne-de-Bellevue, was testing twenty-nine varieties of potatoes for yield and disease resistance in 1911,[70] and William Lochhead (1864–1927) was advocating formalin as a treatment of potato seed-pieces for the control of scab and Rhizoctonia.[71] A few years later, B.T. Dickson (1886–1982) investigated a "black dot" disease of potatoes,[72] and J.G. Coulson (1893–1974), with field plots in Plaster Rock, New Brunswick, and at Macdonald College, studied the effects of gypsum on the common scab disease of potatoes.[73]

In 1931, Bernard Baribeau (1899–1974), of the Dominion seed potato certification service that had been established at Ste-Anne-de-la-Pocatière, Quebec, in 1922, found a disease which, at first, he called bacterial wilt, then bacterial blight.[74] Although Baribeau is commonly credited with having been the first person to discover that disease in North America, Paul Murphy may have seen it a few years earlier in Prince Edward Island when he found what was "identical with Appel's bacterial ring disease."[75] Regardless of who was the first to see the disease in Canada, the earliest scientific study of it in this country was done by Douglas B.O. Savile (b.1909), who not only isolated the causal bacterium, *Corynebacterium sepedonicum*, and proved its pathogenicity, but also developed a smear technique for the rapid identification of infected tissues, and showed that the pathogen was easily spread on knives used to cut the "seed."[76] Savile, who was born in Dublin, Ireland, graduated from Macdonald College with a BSA in 1933, and an M SC in 1934. He earned a PH D from Michigan in 1939, and then joined the Royal Canadian Air Force. He and his supervisor, Homere Noe Racicot (1893–1968), when researching the "new" bacterial disease of potatoes, suggested that the name "Bacterial Wilt and Rot" would be representative of both the vine and tuber symptoms,[77] but the disease soon became generally known as ring rot.

Racicot, born of Huguenot parents in Quebec, obtained a BA degree from McMaster University in 1921 and, after doing some graduate work at the University of Toronto, became plant pathologist and assistant in charge of the new Dominion Laboratory of Plant Pathology at Ste-Anne-de-la-Pocatière, Quebec, in 1923. He thought it was because of religious differences that he had problems with his staff, and that this was responsible for his transfer to Ottawa, in 1930, to be plant pathologist for Western Quebec.[78] In 1936 he was appointed head, Horticultural Crop Disease Section, Division of Botany and Plant Pathology, Ottawa.

Racicot was the author of two undated miscellaneous publications of the Dominion Department of Agriculture: *Bacterial ring rot of potatoes* and *Bacterial wilt and rot of potatoes*. He also compiled for publication extracts from the various presentations of participants in a symposium on bacterial ring rot of potatoes, sponsored by the Canadian Phytopathological Society and published in the *Proceedings* of that society in 1944. His paper, "Nematode diseases of potatoes in Canada," was abstracted in the 1946 *Proceedings*.

Racicot's successor at Ste-Anne-de-la-Pocatière was J. Champlain Perrault (1903–88), BSA, M SC, who soon became one of the leading plant pathologists in Quebec and an authority on potato diseases. He studied the rate of spread of virus diseases in the 1930s, and continued with that and other potato disease research through the years of the Second World War, and beyond.[79]

In Ontario, 440 varieties of potatoes were tested at the OAC between 1885 and 1940.[80] Comparative tests of potato varieties, spray materials, and seed treatments were conducted at the OAC for many years, and the college eventually became one of the testing sites for potato varieties developed at Fredericton. Specific potato diseases were under investigation at the OAC as early as 1886, when J. Hoyes Panton (1847–98), professor of chemistry, who had BA and MA degrees from the University of Toronto, was writing about the potato rot, its cause and remedies.[81] It was Panton who prepared the spraying calendar, complete with formulae for the Bordeaux mixture and several other fungicides, that was published as an unnumbered special bulletin by the Ontario Department of Agriculture in 1895.

Francis C. Harrison (1871–1952), a BSA graduate of the OAC class of 1892, who was an assistant to Panton and librarian for at least a year, described the blackleg disease of potatoes and named the causal organism *Bacillus solanisaprus* in 1906.[82] Ten years later that disease was featured in the Dominion Department of Agriculture Experimental farm circular no. 11, written by Paul Murphy and, in 1930, in pamphlet no. 105, new series, written by Donald J. MacLeod, who studied the disease between 1926 and 1930.

Charles A. Zavitz (1863–1942), BSA, professor of field husbandry and director of field experiments at the OAC, conducted various experiments with potatoes, including ones for the control of scab and of late blight, during the first decade of the twentieth century.[83] He was awarded an honorary D SC by his old alma mater, the University of Toronto, in 1916, and retired in June of 1927.

J. Eaton Howitt (1881–1966), BSA Toronto, professor of botany at the OAC, surveyed much of the potato-growing area of Ontario for potato diseases during the growing seasons of 1918 and 1919. This was part of an effort that was being made to improve the potato crop in southern Ontario. It included a system of inspection and certification for diseases in seed potatoes, and the testing for both disease and yield of seed originating in northern Ontario, southern Ontario, and New Brunswick. The seed pieces from those sources were grown side by side in eighty different plots in forty counties of Ontario. Howitt found that the seed grown in northern Ontario had the least leaf roll.[84]

In the following year, he and Roland Elisha Stone (1881–1939), B SC, M SC, PH D, associate professor of botany, reported that Verticillium wilt of potatoes, easily mistaken for Fusarium wilt, was quite common in that province.[85] Then, and for at least six years in the 1920s, Howitt, and William G. Evans (b.1885), BSA, MS, demonstrator in botany, experimented on methods for the control of Rhizoctonia, including the treatment of seed-pieces with corrosive sublimate.[86]

The widespread use of corrosive sublimate (mercuric chloride) was of such concern to Hans Güssow that he persuaded the Dominion chemist, Frank T. Shutt, to issue a warning about the potential hazards associated with its use.[87] Despite that warning in 1912, mercuric chloride was still being used in potato seed-piece and soil treatments in 1940 when Henri Genereux (b.1911) BA, BSA, was doing research for the master's degree that he obtained from McGill University.[88] Genereux was destined to become a highly respected plant pathologist, and an authority on potato diseases, while working as a plant pathologist for the Quebec Department of Agriculture 1940–48, followed by a more lengthy period with Canada Department of Agriculture. In 1948 he and Elzéar Campagna (1898–1987) had two papers on potato diseases in the annual report of the Quebec Society for the Protection of Plants.

Güssow was as interested in the Rhizoctonia disease as he was concerned over the methods for its control, so he persuaded Frank L. Drayton (1892–1970), who became assistant plant pathologist and bacteriologist shortly after being awarded the BSA degree by McGill University in 1914, to study the lesions induced by the causal fungus.[89] Drayton's research was set aside while he served overseas as an officer in the Canadian Army, 1915–19. When he returned to Canada, after having been severely wounded in action, Drayton was promoted to plant pathologist, the position he held until 1937. This includes a leave period to study plant pathology, mycology, and plant physiology at Cornell University, which awarded him a PH D in 1932.[90] Drayton's Cornell interlude took him away from potato diseases, and he became an authority on certain ascomycetous fungi and the diseases of ornamental plant bulbs. His title was changed to that of agricultural scientist in 1937.[91]

POTATO DISEASES IN WESTERN CANADA

The potato was not a major crop, and late blight was almost non-existent in the prairie provinces, before the Second World War. In 1917, Vincent W. Jackson (b.1876) BA, M SC, professor of botany and biology, Manitoba Agricultural College, reported the province to be comparatively free of the worst potato diseases, and that "blight seldom appears."[92] Neither G.R. Bisby nor I.L. Conners saw the disease there until 1927.[93] However, Bisby did find the soil infested with the fungus that incites Rhizoctonia, a potato disease that he and his colleagues worked on for three years.[94]

Paul Murphy visited Manitoba in 1918 and called attention to the seriousness of leaf roll and certain other little-known potato diseases,[95] and bacterial ring rot was reported there in 1938. However, relatively little potato disease research was carried out in that province prior to the establishment of a provincial potato committee in 1942. Shortly after that, a potato-breeding program was initiated at the University of Manitoba which eventually resulted in the development of a new variety named Manota, early in the 1950s.

Saskatchewan was not very different from Manitoba with regard to potato diseases. There was no late blight to study, so, in 1914, John Bracken (1883–1969), B SC, professor of field husbandry, University of Saskatchewan, began a series of tests to determine the relative value of formalin and corrosive sublimate for the control of other diseases.[96] When J.K. Finlayson spent three weeks in that province, in the fall of 1926, "looking over the potato crops," he found very little evidence of Rhizoctonia, but "scab was everywhere present."[97] Ring rot of potatoes was not reported in Saskatchewan until 1940.[98] The annual report of the Saskatchewan Department of Agriculture for that year tells that potato clubs were being organized and that certified seed had been obtained for their use. It was expected that by these means the quality of potatoes in that province would be enhanced. The considerable improvement that resulted is evidenced by the formation of a Saskatchewan Certified Seed Potato Growers' Association, in January 1945. The near failure of the potato crop that prompted the federal Relief Commission to purchase 296 carloads of potatoes from Manitoba for distribution in Saskatchewan in 1934 was due to drought.[99]

The potato disease situation in Alberta was quite different from that of the other prairie provinces. Potato diseases there had reached such an alarming state by 1920 that the Edmonton Potato Producers' Association began, that year, a cooperative test of eight standard varieties for yield, cooking quality, and disease resistance.[100] Members of the association prevailed upon the provincial Department of Agriculture for authoritative information about potato diseases, and the government responded by issuing circulars on potato diseases and potato seed treatment in 1921.

The most noteworthy early potato research in Alberta was conducted by Guthrie B. Sanford (1890–1977), BSA, MS, PH D, who began a study of the common scab disease of potatoes before he was transferred from Saskatchewan to Alberta in 1928. He studied that disease and its causal organism over a period of more than twenty-five years.[101] From his own experiences and observations, Sanford came to doubt that conclusive results could be obtained when commonly used treatments for scab were used under field conditions. To satisfy himself in this regard, he conducted a series of controlled tests under both field and greenhouse conditions. It was during his research on potato scab that Sanford began thinking about the possibility of biological control and became the first scientist in North America to suggest that the disease might be inhibited by the activity of saprophytic bacteria developing in decomposing green manure.[102]

Sanford and S.B. Clay watched the development of an apparently new disease in the potato fields of Alberta for a period of nearly ten years before publishing an account of it in 1941. Because they were unable to isolate or otherwise find bacteria or fungi in association with the disease, which they suggested should be called "purple dwarf," they assumed it to be of virus origin. The following year, Sanford wrote a brief account of another strange potato disease in Alberta.[103] His report of finding bacteria in normal potato plants is recorded in the 1947 *Proceedings of the Canadian Phytopathological Society.*

The unexpected appearance of brownish dead or dying phloem tissues, or "net necrosis," of potato tubers derived from certified seed stock became so common in Alberta that Sanford and agricultural assistant Jack G. Grimble (b.1917), B SC, M SC, decided to make a study of it. Although they failed to arrive at any definite conclusions about the cause of the condition, they published their observations in 1944.[104]

Several plant pathologists were studying the phloem tissues of diseased potatoes in the late 1930s and early 1940s, one of whom was D.B.O. Savile of the Division of Botany and Plant Pathology, Science Service, Ottawa. In 1939 he was examining some potatoes that had "rusty internal flecks," somewhat similar to a net necrosis, when he noticed that a high proportion of the starch grains in cells adjacent to the necrotic areas were distinctly abnormal in form. Thinking that these abnormalities could have some diagnostic value, he made a study of them in relation to several well-known potato diseases. As a diagnostic tool, the malformed starch grains proved to be disappointing, because they were found in diseased specimens of a variety of fungus, virus, and environmental disorders.[105]

Except for Sanford's work, very little research on potato diseases took place in Alberta until long after the ring rot disease was discovered there in 1937. That disease, found in the irrigation districts in the south of the province, spread rapidly through the commercial potato-growing areas until special control measures were implemented in 1940, when the annual survey revealed its presence on 235 farms.

In 1942 all districts where ring rot had been found in Alberta were placed under a modified quarantine by authority of the Agricultural Pests Act. In each successive year, regulations governing the planting, producing, and marketing of potatoes within established pest areas were gradually tightened to constitute a virtual quarantine, with two objects in view: a/to reduce the infection within affected areas, and b/to protect all other sections of Alberta and neighbouring provinces from the disease. Those special measures achieved a degree of control by the introduction of disease-free seed, vigorous enforcement of quarantine regulations, and an educational program respecting sanitation and other precautions against infection.

Regulations to prevent the contamination of healthy stocks of certified seed required that the "table stock" (potatoes not intended for use as seed) of each grower be inspected, together with the seed stock, and if ring rot was found in any field of the farm no seed stock from that farm could be certified. The regulations further required growers of certified seed to plant at least one-tenth of their potato-growing area in unit rows, unless foundation seed or seed approved by an inspector was used.[106] To meet a demand for high quality seed, the Field Crops Branch of the Alberta Department of Agriculture, in cooperation with the University of Alberta and the Seed Potato Inspection Branch of the Dominion Department of Agriculture, assisted in the production of disease-free foundation stock by the tuber-index method, for distribution to qualified seed producers.[107]

The use of "eye sets," (small pieces of tuber each containing a bud or "eye") became popular in the seed potato trade of the prairie provinces during the Second World War. They could be widely distributed at little cost, and seed houses could ship them by express or parcel post. That relatively new program prompted research on such factors as the best size of the eye set, most suitable package, effect of storage temperature on keeping quality, best treatment before and after the eyes were removed, etc. J.W. Marritt, district inspector, stationed at the Dominion Plant Inspection Office, Edmonton, reported on those studies in 1944.[108]

The provincial legislature of Alberta passed the Seed-Control Areas Act during its 1948 session. That act made it possible for not less than 60 per cent of the occupiers of land in a proposed seed-control area to petition the Lieutenant Governor-in-Council to have such an area established and to have the kinds and varieties of crops to be grown or prohibited in the area specified. That set the stage for further improvements in the quality of Alberta seed potatoes, and their freedom from diseases.

In British Columbia, Thomas A. Sharp began selecting from twenty-four varieties of potatoes grown on the Dominion Experimental Farm at Agassiz, in 1891,[109] where there were numerous "tests" of varieties and spraying materials over the next two or three decades. When assistant horticulturist P.E. French wrote circular no. 10, *Commercial potato production*, for the British Columbia Department of Agriculture in 1913, he remarked that the principal potato diseases in that province were early blight, late blight, and scab. He described those diseases and outlined control measures for each.

The story of potato disease problems in British Columbia has some parallels with the one for Nova Scotia. Potatoes shipped from British Columbia to the United States in 1915 were found to have powdery scab, and the embargo placed on further shipments prompted the introduction of an inspection program by the provincial Fruit Inspection Branch, later in that year.[110] Before the embargo was lifted, a representative of the United States Department of Agriculture went to British Columbia to ensure that the potatoes destined for his country were properly inspected. That is reminiscent of the representative of Bermuda who went to Nova Scotia for a similar purpose. British Columbia, also somewhat like Nova Scotia, organized boys' and girls' potato competitions and issued Bulletin No. 62 to guide them. In the early days, however, British Columbia had its own rules and regulations regarding import and export of plant materials. Potatoes were imported from China, Japan, and the United States, with scant heed for the federal laws pertaining to such importations.[111]

A more rigorous program of seed-potato inspection and certification was initiated late in 1920 by William Newton (1893–1973), chief soil and crop instructor. For various reasons the first inspections were not made until 1921, when they were carried out by his colleague Cecil Tice (1892–1944), a 1919 BSA graduate of the Ontario Agricultural College. From that date onward, Tice effectively directed the seed-potato certification service of British Columbia until it was transferred to the federal authorities in 1933. The annual reports of the British Columbia Department of Agriculture list Tice variously as soil and crop instructor, potato specialist, and, after Newton resigned in June 1923, as chief agronomist. His title changed to that of provincial agronomist in 1926, and he was appointed secretary of the newly formed British Columbia Certified Seed-potato Growers' Association in 1933. At the time of his death in 1944 he was listed as field crops commissioner. The transition to federal inspection authority was a smooth one, because almost from the beginning Tice had tried to enforce the regulations of the Dominion Department of Agriculture and, from 1924, he and his staff were assisted each year by two or more inspectors provided by the federal Division of Botany.[112]

Newton, who had a BSA degree from McGill and an M SC from the University of California, returned to the latter university for the PH D that he was awarded in 1923. After an interval of teaching and research in California, he came back to British Columbia in 1928 as a plant pathologist for the Dominion Department of Agriculture. He was stationed at the University of British Columbia until 1929, when he moved to Saanichton and became officer in charge of the Dominion laboratory of plant pathology until he retired in 1958. Newton retained an interest in potato diseases throughout the 1930s when he wrote about the rhizoctonia disease,[113] and, with H.I. Edwards, potato viruses and the production of antisera by inoculating a potato virus into chickens.[114]

The provincial plant pathologist, John W. Eastham (1880–1968), B SC., who had done some potato research while working as Güssow's assistant in Ottawa,[115] helped Tice set the standards for the potato inspection program in British Columbia. He also did some potato research on land that was placed at his disposal on the Dominion experimental farm at Agassiz. His projects there included ones on such virus or suspected virus diseases as spindle tuber, witches' broom, mosaic, and streak, and tests of seed disinfectants.[116]

Eastham's assistant in the Certified Seed Potato Inspection Service was Walter Jones (1890–1957), an agricultural graduate of the University of Aberysywyth, Wales, who had an M SC from the University of California. When Jones transferred to the Dominion Laboratory of Plant Pathology at Saanichton, he conducted research on several plant diseases, including one, in 1945, on a pink rot disease of potatoes. That research was initiated several years after he and district inspector H.S. MacLeod had written an article about dry rot of potato tubers.[117]

In the winter of 1920–21, Harold Ross McLarty (1891–1988), then a graduate student at McMaster University, was invited to apply for the position of plant pathologist and be in charge of a Dominion laboratory of plant pathology that was being constructed in Summerland, British Columbia. He was appointed to that position in 1921, the year in which he completed his studies at McMaster and was awarded an MA degree. McLarty, who took time off to study for the PH D that he earned at the University of Illinois in 1932, became well known for his discovery of a method for the control of the corky core disease of apples, but he was also a keen observer of other plant diseases, reporting on a "witches' broom" disease of potatoes in 1926.[118] McLarty's first full-time assistant, George E. Woolliams (b.1901), who had an MS degree from the University of Idaho, worked for several years on the development of disease-free strains of potatoes.[119]

In October, 1930, district inspector Haddon S. MacLeod called the attention of William R. Foster (b.1905), assistant plant pathologist and geneticist in the British Columbia Department of Agriculture, to what appeared to be a new kind of tuber stem-end rot.[120] Foster sent a specimen to Güssow in Ottawa, who considered the causal fungus to be new to science and, with Foster, described it in 1932.[121] Foster, who became provincial plant pathologist in 1948, was a native of Edmonton and had attended the University of Alberta, where, after obtaining a B SC degree in 1928, he became the first student to register for the M SC degree in plant pathology. His research for that degree, awarded in 1930, was supervised by Arthur W. Henry (1896–1989), who in 1936 wrote University of Alberta bulletin no. 46, "The potato crop in Alberta."

According to the annual report of the federal minister of agriculture for 1942, the bacterial disease of potatoes was found for the first time in British Columbia in a few tubers from the 1941 crop. However, there was no important research on it in that province until after the Second World War.

At the University of British Columbia, Norman S. Wright studied a species of Stemphylium as part of the early blight complex of solanaceous plants (which included potato and tomato) as the thesis topic for the MS degree that he was awarded in 1946. Wright, destined to become an outstanding plant pathologist in British Columbia, published on a stemphylium leaf spot of potatoes in 1947 and on witches' broom of potatoes in that province in 1949. Both papers were reported in the *Proceedings of the Canadian Phytopathological Society* for those years.[122]

POST-HARVEST DISEASES

The pathology of potatoes does not end when they are harvested. They are subject to a number of post-harvest disease problems, one of which is physiological and due to frost. The pioneer English settlers soon learned that the traditional British method of storing potatoes in clamps (elongated mounds of tubers covered with straw and earth) was generally unsuitable in most parts of Canada because of the very cold winters. In devising other methods of storing potatoes, many farmers found that they could avoid frost by storing their tubers in pits. Henry Y. Hind, who was exploring westward from the Red River to the south branch of the Saskatchewan River for the Canadian government, described some storage pits he had seen there in 1858. They had been constructed by "excavating chambers near the high bank of the Assiniboine river"; the owner had access to the chambers through a small movable cover over a hole in the roof of each. Hind remarked that with the use of such pits or chambers there was no difficulty in storing quantities of potatoes and turnips, free of frost, through the severe winters of that region.[123] Such information, plus that gleaned from others, must have been very useful to the officials who had to purchase and store a large quantity of potatoes for the several thousand vegetarian Doukhobors who arrived in western Canada during the winter of 1898–99. Safe, frost-proof storage was essential because a sufficient quantity of potatoes was needed to last over that winter and spring, plus a sizeable reserve for planting.[124]

In eastern Canada, the early French settlers did not grow potatoes until around the latter part of the eighteenth century or early in the nineteenth. When they did, they generally followed the practices of their English neighbours, the majority of whom, by that time, were using cellars under their houses as their major vegetable storage chambers. Those cellars usually had two entrances – a relatively large one, through which potatoes were brought in from the field or garden for storage and which was covered or "banked" in the fall to keep out the cold, and a second, smaller one, to which access was commonly gained through a trap door in the floor of the kitchen.

In 1852, William Berczy, a member of the Lower Canada Agricultural Society, referring to potato storage in cellars and the rot that was often found in them, told the special committee looking into the conditions of agriculture in Lower Canada, "Experience has proved to me, that after taking up potatoes, before cellaring them, it is necessary to dry them in barns or sheds for at least 15 days, taking care to shift them at certain intervals."[125]

In 1912, when officials of the federal Division of Botany inspected stored potatoes, especially those suspected of having powdery scab, they discovered that the losses due to various "rots" were considerable, and of far greater economic importance than had been generally realized. The experimental farms reports for 1913 tell of instances in which as many as 40 per cent of the stored tubers had become worthless.

In 1914, H.T. Güssow, realizing that something should be done to prevent losses resulting from the storage of potatoes in unsuitable root-houses and cellars, experimented with variants of the older method of pit storage. He thought the pit method would be a more economical way of preventing post-harvest losses than the cellar method, especially if special cellars or buildings had to be constructed for that purpose. In the early spring of that year, Güssow also explored the possibility of dehydrating surplus potatoes as a means of extending the period of their post-harvest usability.[126] He was a long way ahead of his time in that regard. Also in 1914, Professor W.P. Fraser at Macdonald College reviewed some of the potato storage problems that he had encountered, grouped them into four major types, and offered advice about how most of the problems could be overcome.[127]

To encourage farmers in British Columbia to take better care of harvested potatoes and other vegetables, Edwin Smith, who was in charge of storage and transportation investigations, produced a well-illustrated bulletin of drawings and photographs of several storage installations, some of which had been successfully used in the United States.[128]

Shortly after the First World War, several plant pathologists concentrated on studies of the particular causes of storage losses. One of these was H.N. Racicot, who in 1926, while officer in charge, Dominion Laboratory of Plant Pathology, Ste-Anne-de-la-Pocatière, Quebec, was interested in a spotting and shrinking disease of potatoes in storage.[129] Racicot maintained an interest in potato storage problems for many years. He suggested the topic, and collaborated with L.T. Richardson and W.R. Phillips in the study of low-temperature breakdown of potatoes in storage that they began in 1941.[130]

During the Second World War there was a revival of interest in post-harvest diseases and other problems of stored potatoes and vegetables, including a reassessment of the early method of pit storage. The Department of Horticulture at the OAC experimented with fifteen storage pits, representing wide variations in type, and compared their efficiency with similar air-cooled units.[131] In this connection, Edward H. Garrard (b.1905), BSA, MSA, professor and head, Department of Bacteriology, studied a storage rot of potatoes, during the last two years of World War Two, that was caused by fluorescent bacteria.[132] Beginning in 1945, and for the next five years, Dorothy F. Forward (b.1903), BA, MA, PH D, Fellow of the Ontario Research Foundation and plant physiologist in the Department of Botany, University of Toronto, studied the response of harvested potato tubers to a period of anaerobiosis.[133] At the Dominion Laboratory of Plant Pathology, Charlottetown, Prince Edward Island, George Ayers studied the fusarium storage rot of potatoes for at least three years, beginning in 1946.[134] That was about the time L.T. Richardson and W.R. Phillips, in Ottawa, were studying the susceptibility of several different varieties of stored potatoes to low-temperature breakdown.[135]

4 Grain and Forage Crop Diseases and the Early Development of Plant Pathology in Canada

Because the cereal grains have been our major source of carbohydrate food for as far back as we have any written history of agriculture, it is not surprising that some of the earliest records of plant diseases would refer to those affecting grain. The very early Sumerian references to a "samanu" disease causing barley to turn red[1] are not sufficiently descriptive to permit a modern plant pathologist to identify the disease; nevertheless, they indicate the importance the early farmers attached to disease and other irregularities in growing grain.

The first grain to be sown and harvested by white people in Canada was probably fall rye or winter wheat, sown by Jacques Cartier in September 1541, near the present Cap Rouge, Quebec. Because of some trouble with the natives, Cartier abandoned the place early in the following spring, and it is in the journal of Jean Alphonse, a pilot of Roberval's expedition, that reference is made to harvesting grain sown by Jacques Cartier.[2] One of the earliest Canadian records of a plant disease, albeit a physiological one, also pertains to grain. A superior of the Ursuline Convent in Quebec, in a letter to her son, 20 August 1663, mentioned that the crops had all yellowed because of the drought.[3]

A thorough search of earlier records may reveal other accounts of plant diseases in Canada: by the time that Ursuline nun referred to the yellowed crops, wheat had been grown in the geographical area that is now Canada for more than one hundred years. However, it is within the realm of possibility that the grain first grown in this "new" country was remarkably free of disease. There is some support for this theory because nearly another century passed before a reference to diseased grain again appeared in the records of the early settlers. It is also possible that grain growing, or farming in general, was such an insignificant aspect of the lives of the first settlers that they did not consider disease problems worthy of mention in their diaries or correspondence. The first settlements in this country were not established for the purpose of developing agriculture. France and the British Isles were more than self-sufficient in agricultural produce and preferred to exchange wheat and flour for furs and fish. Colonial agriculture was tolerated but not encouraged for a long time after the first seeds of European grain were sown in Canada.

In 1741, the wheat crop in a district on what is now Prince Edward Island was badly damaged by rust,[4] and in 1751 the crop was reported to have been "totally scalded."[5] The author of that report, a military appointee making a census-inventory of the Island, did not describe or otherwise explain what he meant by "scalded."

Six years later, James Monk, writing from Halifax, Nova Scotia, and comparing the climate and crops of that region with those of New England, was a bit more descriptive when he commented on the wheat and other grains near Halifax being "mildewed or blasted."[6] Here again there is much uncertainty as to the true nature of the disease. In some early English writings on the subject, mildew was synonymous with smut,[7] whereas other authors, including Shakespeare in *King Lear*, which was widely read at that time, used mildew in a much broader sense to include almost any unknown factor that lowered the quality of the grain. By 1790, many New England farmers were using mildew as a synonym for rust;[8] because Monk had been in New England shortly before writing about the mildewed grain in Nova Scotia, he may therefore have been referring to a rusted condition.

In spite of the initial lack of encouragement, and because of wars in Europe and the consequent demand for more food, the farming communities of the early English and French colonies expanded and more and more grain was grown. In 1817 the province of New Brunswick encouraged grain growing by offering a bounty for every bushel produced within two years of clearing the land on which it grew.[9] For a relatively brief period around 1802, Quebec was exporting a million bushels of wheat each year.[10] The period was brief because the combined effects of rust, smut, and the "Hessian fly" began to devastate the wheat crops, and by 1835 the situation had become so bad that nearly five hundred thousand bushels of wheat had to be imported into Quebec.[11]

DISEASES OF WHEAT

Smut Diseases

In 1835 William Evans, who published one of the first Canadian books to deal exclusively with agriculture, stated that smut was the disease most injurious to the wheat crops in Lower Canada.[12] That noteworthy but relatively unknown agriculturist was, in several ways, far ahead of his time with respect to plant diseases. In one chapter he described four kinds of blight in wheat, one of which, he wrote, was due to "the propagation of a sort of fungus." Evans was of the opinion that grain smut was a fungus and that it should be called *"uredo foetida"* (his italics). That was a remarkable conclusion for a man in Montreal to arrive at in 1835: few biologists in Europe or elsewhere were convinced of the fungal nature of smut until sometime after 1847, when Louis-René Tulasne (1825–85) and his brother Charles (1816–84) published their studies on those fungi in France.

Beginning late in August or early September, when diseases were manifesting themselves on maturing grain, the newspapers of the day would publish comments about them, together with letters to the editor suggesting how rust, smut, or "the fly" might be controlled. For example, the *St. Francis Courier and Sherbrooke Gazette*, on 3 January 1832, had this comment: "we are much troubled with smut ... if seed free from it cannot be procured, it should be steeped 24 hours in lime water, which will correct the evil."

Steeping grain in salt water, lime water, and various other substances for the prevention of smut was not new to such knowledgeable agriculturists as William Evans. While he was editor of the *Agricultural Journal and Transactions of the Lower Canada Agricultural Society* he often referred to statements made by European authors concerning diseases of grain. He may have known that Richard Remnant, in Britain, had experimented with various "steeps" for the prevention of smut and written about them as early as 1637.[13] Thus it is not surprising that Canadian farmers, many of whom were avid readers of European publications, albeit a few months late, would also do some experimenting along similar lines. To encourage such experiments, the *Gleaner* of Miramichi, New Brunswick, published, on 5 October 1844, the formulae for several steeps, including a weak solution of copper sulphate (blue vitriol or bluestone) that was recommended by Sir John Sinclair, president of the Highland Society of Agriculture, Scotland.

Occasionally smut caused great destruction in the wheat fields of eastern Canada. In 1822 John Young (1773–1837), who was widely known for his authoritative letters on agriculture, written under the pseudonym "Agricola," was warning his Nova Scotia readers to be cautious about using "saltpetre and sulphur dissolved in water" as a steep for seed wheat, even though it was reputed as "an infallible prevention of smut." He suggested they continue to use brine or urine, followed by dusting with quicklime.[14] In 1850 Henry Y. Hind (1823–1908) stated that farmers in Canada West (Ontario) had a remedy for the disease, "which consists of steeping the seed in some liquid which will destroy the vegetative powers of the fungal seeds." He suggested that "five pounds of sulphate of copper dissolved in ten gallons of water" was the best solution in which to steep the grain.[15] The significance of his statement is that, by the time wheat growing became an important aspect of prairie agriculture, a reasonably satisfactory control of smut was well known and widely practised in eastern Canada.

After the railway from the east reached Manitoba in 1879, there was a spectacular increase in the production of wheat. One authority estimated that nearly a million bushels were being grown there in 1880 and that production increased to about fifteen million bushels by 1890.[16] It was during that decade that the government of Canada established experimental farms at Brandon, Manitoba, and at Indian Head, in what was then the Northwest Territories. The superintendents of those farms, S.A. Bedford at Brandon and Angus Mackay at Indian Head, told of trials for the control of smut in their first reports of experimental work. Presumably both superintendents knew of the Central Experimental Farm bulletin no. 3, in which James Fletcher (1852–1908), Dominion botanist-entomologist, recommended the use of copper sulphate, because the first experiments for smut control at both farms included that compound.

Apparently Mackay was so confident in the use of "blue vitriol dissolved in water" that he did not mention any other treatment for smut control in the report of his first year as superintendent. Bedford, in his first report of the Brandon Farm, referred to three different seed-treatment experiments carried out in 1890: bluestone, Jensen's hot water treatment, and a salt-brine treatment, all three of which were moderately successful.

Maybe it was because prairie farmers were primarily concerned with grasshoppers, or because smut was not a serious problem every year, that many of them neglected to treat their seed, and some of those who did were still using salt water, urine, and various other old but often ineffective steeps. Apparently neither Fletcher's bulletin nor the experimental farm reports were widely read by prairie farmers: newspapers such as the *Nor'-West Farmer and Miller* were bemoaning the "immense damage caused by smut, and the subsequent trouble and difficulty experienced in handling smutted grain."[17]

Incidentally, that periodical was one of the first Canadian farm papers to provide a good outline of Jensen's hot water treatment for the control of smut, which it did in 1891.[18] The editor may have learned of it from J. Hoyes Panton, professor of natural history at the Ontario Agricultural College, who had described the hot water treatment in OAC bulletin no. 56, a year earlier.

The great damage due to smut in 1891 stimulated the Manitoba Department of Agriculture and Immigration to issue bulletin no. 32, *Smut in Wheat*, in February 1892, recommending a sulphate of copper solution as the best treatment of grain for the control of smut. That treatment was used by many agriculturists until at least the second decade of the twentieth century.

The fact that there was more smutty grain in Manitoba and Saskatchewan in 1905 than there had been for several years aroused a suspicion in the minds of many regarding the quality of the bluestone used in treating wheat. To ascertain what foundation there might be for that suspicion, Frank T. Shutt (1859–1940), Dominion chemist from 1887 until he retired in 1933, tested samples of bluestone collected from farmers and dealers in those areas, and did not find any adulteration.[19] One of the reasons for suspicion regarding the quality of the sulphate of copper goes back to about 1890 when there was offered for sale, on the prairies, a so-called "Agricultural Bluestone," which on analysis proved to contain a very large proportion of sulphate of iron. Shutt, who had been conducting experiments on the prevention of smut since 1890, had demonstrated that it was much less effective in smut prevention than bluestone.[20]

The chief inspector of weeds for Saskatchewan, T.N. Willing, also collected samples of bluestone from all three prairie provinces and had them tested by the provincial bacteriologist and analyst, Dr G.A. Charlton. Their report, in Saskatchewan Department of Agriculture bulletin no. 2 (1906), also showed that the bluestone was generally unadulterated. That report made one of the earliest Canadian claims that the smut in oats and barley are distinct species that do not affect wheat, and it recommended formalin as the best treatment for smut of oats. Formalin had been under test on the Dominion Experimental Farms, and *Tests of Formalin and Massel Powder as a Prevention of Smut in Barley and Oats* was the title of report no. 394 of the Central Experimental Farm in 1899.

Angus Mackay, who seemed to have confidence in blue vitriol in 1889, experimented on various methods of smut control at the Indian Head Experimental Farm over the next several years without confining himself to research on seed treatments. In the station report for 1908 he told of sowing two bushels of smut spores before grain was sown, in each of five plots of soil eight feet square, and reported, "The results obtained go to show that smut dust, when in any considerable quantity in the soil, produces smut in the grain, no matter how treated."

Around the middle of the nineteenth century, grain smut was of sufficient economic importance in eastern Canada to prompt the Caledonia Foundry and Machine Works, on the Lachine Canal near Montreal, to manufacture smut mills. An advertisement in the *Sherbrooke Gazette and Eastern Townships Advertiser*, 15 March 1851, showed that smut mills could be purchased for $50.00 each. Farmers in eastern Canada had long accepted the presence of some smut in their fields of grain. Most of the time it was of minor concern, but over the years there was always someone who could recall at least one "bad smut year." In 1891, Ontario made it illegal for anyone to sow smut-infested grain without first treating the seed.[21] That law was one of the factors that prompted Frank Shutt to do so much experimenting on smut prevention in the 1890s. Because farmers in Quebec often neglected to treat their seed for smut control, Leonard Klinck (1877–1969), professor of cereal husbandry at Macdonald College, began selecting grain that would be less susceptible to smut in 1910.[22]

Twenty-five years later, Elzéar Campagna (1898–1987), a member of the teaching staff of the Faculty of Agriculture, Laval University, at Ste-Anne-de-la-Pocatière, Quebec, reported finding what he called a new white smut of wheat in that area.[23] A few years later he and René-O Lachance (1909–92) provided a list of the Ustilaginales (smuts) found in Québec.[24]

The Dominion botanist, H.T. Güssow, in an effort to be of service to all farmers in Canada, prepared Central Experimental Farm bulletin no. 73, in which he included most of the then-known smut-control treatments. That was a unique bulletin in several respects. In addition to its relatively modern treatment of the subject, it was well illustrated, and, in an appendix, directions were given for observing the germination of smut spores in artificial culture. Güssow's annual report as Dominion botanist for 1927 included instructions for a hot water treatment to control wheat smut that was being recommended by Richard R. Hurst (1895–1961) BSA, officer in charge, Dominion Laboratory of Plant Pathology, Charlottetown, as a result of his tests there.

In western Canada the smut problem appeared to have been largely solved, because of the success of some more or less standard seed treatments. Evidence for this is found in the 1910 bulletin no. 2 of the Manitoba Agricultural College, which reported that "Thanks to the march of progressive agriculture, the farmer who does not treat his grain for smut is almost unknown, and treatment if effective nine times out of ten." The onset of the First World War changed that optimistic outlook, and J.H. Evans, deputy minister of agriculture for Manitoba, in his annual report for 1921, bemoaned the "abandonment of the practice of pickling [steeping] seed grain, due to shortage of labor and scarcity of formaldehyde and bluestone during the war period." He expressed the view that these factors were "directly responsible for the marked increase in losses from smut."

One of the earliest scientific studies of the smut diseases in western Canada was that of William P. Fraser (1867–1943), AB, MA, who began a study of the biology and control of smut on western rye grass in 1918.[25] After an interval of two or three years, he, in cooperation with Prye M. Simmonds (1897–1973), BSA, who later earned MS and PH D degrees, conducted experiments at Saskatoon on the control of cereal smuts with relatively new seed treatment substances.[26] That work was continued by Ibra Lockwood Conners (1894–1989), BA, MA, at the Field Laboratory of Plant Pathology, Brandon, in 1923 and 1924, and then, until 1929, at the Rust Research Laboratory, Winnipeg. Conners's research included tests on the use of mercury compounds as seed treatments for smut control.[27] Such innovative research was deemed necessary at that time because farmers were becoming increasingly sceptical about the effectiveness of some heretofore recommended seed treatments, such as the use of bluestone, for smut control.

Guy Richards Bisby (1889–1958), B SC, AM, PH D, professor of plant pathology, University of Manitoba, in writing Manitoba Department of Agriculture and Immigration circular no. 85 in 1927, advocated the use of formalin for the control of oat smut and covered smut of barley, and either formalin or copper carbonate for bunt of wheat. According to him, bluestone should never be used on oats or barley.

To stimulate renewed efforts on smut research, an associate committee of field crop diseases (western section) was appointed by the National Research Council of Canada, in 1928, to coordinate and promote investigations into diseases other than rusts.[28] The recommendations of that committee led to the appointment of William F. Hanna (1892–1972) to the staff of the rust laboratory in Winnipeg, where he was given charge of the smut investigations. A Nova Scotian by birth, Hanna had earned a BA degree from Dalhousie University before the First World War, of which he was a veteran, and B SC and M SC degrees from the University of Alberta. He did some studying at the University of Manitoba, the Imperial College in England, and the University of Minnesota, before returning to the University of Manitoba to learn more about mycology with Professor A.H.R. Buller. It was there, in 1928, that he received the first earned PH D degree to be awarded by a university in western Canada.

At Winnipeg, Hanna was assisted in his research on the distribution of the smuts on wheat, oats, and barley, by William Popp (b.1897), an M SC graduate of Macdonald College. Because it is sometimes difficult to confidently distinguish "loose smut" from "covered smut" of oats in the field, they attempted to determine the reason for this difficulty. In the course of those studies they learned that the oat smuts were hetero-thallic, and they obtained hybrids between two species of *Ustilago*.[29] They also studied physiological forms of loose smut of wheat and methods for its control.[30] Hanna was the first to discover the concurrent invasion of the ovaries of wheat by the bunt type of smut and the ergot fungus,[31] and he studied the effect of vernalization on the incidence of smut.[32] Hanna's work with the smuts came to an abrupt halt when he volunteered for military service at the beginning of the Second World War, but his colleague William Popp continued those studies through the war years and after.

Olaf S. Aamodt (b.1892), BS, MS, PH D, associate professor of genetics and plant breeding at the University of Alberta from 1928 until 1935, bred spring wheats and compared varieties for resistance to smuts. He also tested the resistance of some varieties to freezing temperatures.[33]

Environmental factors had a prominent place in the research on grain at the University of Alberta around that time. The report of the board of governors and president, 1932–33, notes that a "chinook" machine had been developed in the Department of Field Crops to test plants for resistance to hot winds and drought. The effect of freezing on the subsequent growth of wheat seedlings was also studied by Arnold W. Platt (b.1909), B SC AGR., M SC, who was an assistant in cereal and forage crop breeding at the Swift Current Research Station in Saskatchewan. The results of that work, done as part of his research for the master's degree at the University of Alberta, and of his studies on the effect of soil moisture and endosperm condition and variety on the frost reaction of wheat, oats, and barley seedlings were published in 1937.[34] Four years later Platt and J.C. Darroch published the results of their research on the seedling resistance of wheat varieties to artificial drought in relation to grain yield.[35] It was around that time, in Alberta, that William C. Broadfoot (b.1899) B SC, MS, PH D, began his unique study of the value of powdered pitchblende as a means of smut control.[36]

In British Columbia, W.R. Foster (b.1905) revised the provincial Field Crop circular no. 10, "Cereal smuts," in 1937, and stated that the treatments most commonly used for smut control in that province were New Improved Ceresan dust, dry copper carbonate dust, formalin, and hot water.

Rust Diseases

In the early days, very few farmers understood or believed in the fungal nature of plant diseases; consequently there were almost as many published theories about the causes of mildew, rust, and smut as there were remedies. For example, a writer in the *Gleaner* of Miramichi, New Brunswick quoting the *New York Morning Post*, ascribed rust and mildew to "a loss of sap through the splitting of the straw under the hot sun immediately after a sudden drenching shower or heavy fog."[37] The directors of the Saint John agricultural society published the theory that rust of wheat resulted from "the tap root coming in contact with the cold clay or a sour subsoil."[38] More cautious or conservative agriculturists did not pretend to know the cause or the cure for any of the grain diseases. Perhaps typical of this group was the Hon. Adam Fergusson, who, in his address to members of the Provincial Agricultural Association of Upper Canada (Ontario), 22 October 1846, referred to rust as "that mysterious scourge which has so often prostrated the faith and well-grounded hopes of the farmer, and which still remains without any satisfactory remedy or preventive."[39]

If the members of that society had been reading the *Transactions* of their sister society in Lower Canada, they would have become much more knowledgeable about grain rust through the writings of William Evans, who when commenting on "The Devastating Growth of Puccinia" stated, "the obvious method of guarding against rust of grain is to endeavor to produce the earliest varieties."[40] But this good advice about growing early varieties was largely ignored for the next decade or more. It was not mentioned by Judge James H. Peters (1811–91) in his book, *Hints to the Farmers of Prince Edward Island*, in 1851, one of whose "hints" was to use charcoal to "check rust in wheat and mildew in other crops, and in all cases mitigate their ravages when it does not altogether prevent them."[41] Peters, who had married the eldest daughter of shipping magnate Samuel Cunard and managed the latter's large estates on the Island for many years, seems to have tried to be the Island counterpart of the famed Nova Scotian agriculturist Agricola (John Young). In fairness to Peters, it must be recorded that in many of his "hints to farmers" he was quoting someone else but giving the proposal his qualified support.

Farmers who wanted to sow their grain early, whether purposefully to avoid the worst of the rust or just to have an earlier harvest, were always worried lest the crop be injured by frost – a physiological disease that could be as devastating as rust. Some of this concern is seen in the *Series of Conversations on the Natural History of Lower Canada*, as recorded by P.H. Gosse in his 1840 book. After a preliminary comment on frost damage, including fall frosts that may kill the grain "when yet in the milk," he quoted a friend who was of the opinion that "grain may be gradually inured to a severity of cold which would kill it if it were exposed to its violence without any preparation. For example, if frost came light at first, but every night gradually increasing in intensity, a heavy frost may be then sustained without any injury."[42] That rather astute commentary was not put to a good scientific test for several decades.

Following the economic depression of 1847, there was a general revival of agricultural prosperity, stimulated by the European demand for wheat created by the Crimean War, and by the prospect of increased trade with the United States, with which Canada had a reciprocity treaty. During the 1850s the geographical area that is now Ontario became the major wheat-growing area of Canada. Nevertheless, the wheat crop was so uncertain throughout Ontario, Quebec, and the Atlantic provinces that the governments and agricultural societies of the day were offering prizes of money for the best essays on ways and means to ensure a better and more reliable harvest.

James Johnston, who had been engaged by the government of New Brunswick to report on the agricultural capabilities of that province, wrote that rust was "a worse foe to the farmer than even the midge, because while the insect destroys only the grain, the fungus injures or destroys both straw and grain together."[43] Incidentally, he advised early sowing, good drainage, and the introduction of more hardy varieties of wheat, "or such as from some peculiarity are less subject to be rusted," as the best means of ensuring a good crop.[44] The theme of early sowing on well-prepared land and the use of hardy varieties also appeared in one of the first prize-winning essays on the topic, that of A.F. Scott of Brampton, Ontario. Scott's essay, written in 1853, was not published until 1856, when it appeared in the first volume of the *Transactions of the Board of Agriculture of Upper Canada*.[45]

Henry Y. Hind, professor of chemistry and geology at Trinity College, Toronto, also proposed "early sowing, with well prepared seed, to escape the time when the climatical conditions occur favourable to rust," and listed several varieties of wheat that had been recommended to be at least partly "rust proof." Hind won first prize for his essay, only one of whose seven chapters dealt with plant diseases. However, that one chapter gave an account of rust, smut, pepper brand, and ergot, and had line drawings of several fungi complete with references, one of which was to the 1853 edition of the Encyclopedia Britannica.[46] Hind's knowledge of mycology and plant anatomy appears to have been about equally limited. His illustration of "smut found on rotten potatoes" is obviously a Mucor, and his comment about the cuticle of plants being composed of "a single row of thin-sided cells," external to which is "an exceedingly delicate transparent membrane called the epidermis," provides evidence of his ignorance of both subjects, or of lax editing.[47] In his comments on smut, Hind said it is common practice in Canada for growers to soak their grain "in brine and chamber ley." Presumably, "chamber ley" was Hind's delicate way of referring to human urine, which he considered to be "a quickener of germination when the moistened seed is dried by sulphate of lime, or gypsum, or charcoal."[48]

Hind's essay was one of several that had been solicited by the Board of Agriculture for Upper and Lower Canada, which offered prizes of forty, twenty-five, and ten pounds sterling each for the three best essays "furnished to the Bureau by the 15th day of January 1857." The second prize that year was awarded to the Rev. George Hill, rector of Markham, Ontario, and the third went to Emilien Dupont, Esq., of St Joachim, Quebec, who was in fact Abbé Léon Provancher (1820–76), using the name of his beadle as a pseudonym when he submitted his essay in the contest.[49]

John William Dawson (1820–99), who later became a renowned principal of McGill University and its first professor of agriculture, included a few comments on grain diseases in his 1853 pamphlet titled *Scientific Contributions Towards the Improvement of Agriculture in Nova Scotia.* While he was superintendent of education in Nova Scotia, Dawson enlarged upon some of his early writings and combined them into a book, *Contributions Toward the Improvement of Agriculture in Nova Scotia,* which was not published until 1856 when he had become principal of McGill. In that book, which, Dawson suggested, "should be largely introduced into the schools and school libraries," there is a chapter on grain crops "and the blights and diseases to which they are liable." He commented that wheat, though the most important of the grain crops, "has acquired the character of being a precarious crop," and he considered it necessary to make a detailed inquiry into its diseases.[50]

In spite of his good intentions, Dawson's "inquiry" into the means of controlling grain diseases was a superficial rehash of what others had been writing. It was in the realm of the etiology of "rust or mildew" that he introduced some new ideas, one or two of which were erroneous. For example, he wrote, "it is probable that when the grain of rusty wheat is sown ... the crop may be more easily affected by the disease, because the seeds of the rust fungus many be attached to the seed."[51] It would appear, from such a statement, that Dawson assumed the rust fungus had a disease cycle similar to that of smut, about which he had more accurate information. While principal of McGill University, Dawson published another book, in 1864, titled *First Lessons in Scientific Agriculture for Schools and Private Institutions.* Although it was updated in several respects, there was relatively little change in the section dealing with grain diseases.

An association between barberry bushes or shrubs and the rust disease of wheat had been suspected by farmers for no one knows how long, but certainly since before 1785, when William Marshall, a farmer in Britain, published a confirmation of these suspicions in the report of his experiment. He transplanted a barberry bush to the centre of a large field of wheat and showed that the rusted area in his field "resembled the tail of a comet, the bush representing the comet itself ... On one side the effect did not reach more than three of four feet, on the opposite side it was obvious to the distance of ten to twelve yards."[52] Yet in spite of this and other published evidence in European literature, the idea that the barberry bush could have a relationship to wheat rust was not generally accepted by Canadian agriculturists for another hundred years.

In 1886, J. Hoyes Panton (1847–98), professor of natural history at the Ontario Agricultural College (OAC), Guelph, Ontario, commented that the barberry family "has a bad reputation for being a source of the rust we find on wheat."[53] Two years later, in the first bulletin devoted to a plant disease produced by the OAC, Panton asserted that the "aecidium" from barberry leaves germinates on wheat to produce rust,[54] thus becoming one of the earliest of the authoritative Canadian agriculturists to be convinced of the barberry wheat-rust relationship. But James Fletcher, Dominion botanist-entomologist, showed that he was not convinced of the importance of barberry in relation to rust when he commented, "I believe if you destroyed every barberry tree in the country you would still have the rust."[55] Perhaps Fletcher was somehow responsible for the view expressed by his long-time associate and superior, William Saunders (1836–1914), who, in a letter to Fred Foyston, Minesing, Ontario, dated 10 March 1899, stated, "with regard to the question of Barberry causing the growth and spread of rust in grain: I think more is made of this than need be. There is no doubt that in the early stages barberry is one of its host plants ... but it is also certain that many other plants, beside the barberry are host plants ... I hesitate to advise you to destroy your barberry bushes under the circumstances."[56] Fletcher, like so many of his time, displayed a much more accurate knowledge of the smut diseases of grain than of the rusts, as is evidenced in his bulletin no. 3, *Smuts Affecting Wheat*, though Fletcher acknowledged that much of that 1888 bulletin came from a book by Worthington G. Smith, published four years earlier in England.

The first wheat to be exported from Canada was grown near Montreal, but the most productive wheat-growing area gradually moved westward, to Ontario, where that crop was the engine of economic growth for many years. By 1880 Ontario was producing 84 per cent of all Canadian wheat,[57] but farmers were complaining that it could not be grown there as well as it formerly could.[58] The stage was already set for another westward movement of the major wheat-growing area.

Some wheat was being grown on the Canadian prairies as early as 1674 or 1675 when the Hudson's Bay Company ordered that a bushel each of wheat, rye, oats, and barley seeds be provided to some of its posts in Manitoba, and encouraged their production in the hope that the trading posts would become self sufficient in food, especially bread.[59] However, for various reasons, the first attempts to grow wheat on the prairies were often unsuccessful. Miles Macdonald, in a letter to Lord Selkirk, 17 July 1813, commented that both the wheat sown in September 1812 and that sown in the spring of 1813 were total failures.[60] In contrast to that, "the Scotch," who sowed wheat in the spring of 1813, had a crop that "turned out exceedingly well" in the Red River settlement.[61]

Frost, grasshoppers, and smut were serious obstacles encountered by the early wheat growers in western Canada, but rust was virtually unknown. Henry Y. Hind, in his book, *Narrative of the Canadian Red River Exploring Expedition of 1857*, stated that he had "heard no complaint of rust, nor did I see a single instance of its presence."[62] By the time effective methods of smut control were coming into vogue on the prairies, however, rust of wheat was becoming a problem. In 1893, the editor of the *Nor'West Farmer* could comment that, "In this country rust has of late attracted no great attention because its ravages have been comparatively small."[63] Only three years later S.A. Bedford was commenting, in the Experimental Farms Report for 1896, that rust of wheat caused considerable damage in Manitoba.

Beginning around the turn of the century, with a few exceptions, the emphasis shifted from smut to rust as the major disease problem of prairie wheat. The first official record of grain rust dates from 1891, when Bedford reported that some wheat varieties suffered from rust at Brandon. A somewhat similar report was made by Angus Mackay the following year for both wheat and oats in the Indian Head region. In 1904, the majority of agriculturists who wrote about prairie crops remarked on the considerable losses due to rust. These included the report of Frank Shutt, Dominion chemist, and those of all superintendents of the experimental farms in Manitoba and the Northwest Territories.

In spite of frost, grasshoppers, smut, and rust, the volume of grain produced on the prairies steadily increased, and by 1910 Saskatchewan had become the premier wheat-growing province of Canada with a production of 224,312,000 bushels, or 57 per cent of the total crop. Manitoba was second, Alberta third, and Ontario had slipped to fourth place.[64]

Nova Scotians were early Canadian leaders in rust research, and Alexander H. Mackay (1848–1929), BA, of that province was the first to make a collection of rusts, including *Puccinia graminis,* and one of the first Canadians to use their Latin names, as he did in his 1902 publication, *Nova Scotia fungi: A provisional list.*[65]

Collecting fungi had become a fashionable diversion for learned Nova Scotians ever since Thomas C. Haliburton, the "Sam Slick" of literary fame, included a list of fungi collected by the vice-president of King's College, Windsor, in his book, *An Historical and Statistical Account of Nova Scotia.*[66] Thus it was not surprising that William P. Fraser, who was then science master at Pictou Academy, should also make a collection of fungi. After an initial period of general collecting, with early emphasis on the mildews, Fraser was unique in that his collection was composed largely of rust fungi and their host plants. In his *Rusts of Nova Scotia,* published in 1910, Fraser described ninety-two species and two forms, several of which had not been previously reported from North America. He also noted that a parasitic fungus, *Darluca filum,* is often present in the uredinia or tilia of the rusts.[67] That article, embodying the results of field, cultural, and microscopic studies in Nova Scotia beginning in 1908, may well be considered as the beginning of "scientific" research on the rust fungi in Canada. Fraser, who was destined to become internationally recognized as Canada's leading authority on the rust fungi, accepted the position of lecturer in biology at Macdonald College in 1912, where he taught mycology and plant pathology until he resigned in 1919 to study grain diseases in western Canada as an employee of the federal government.

Incidentally, one of Fraser's students was Margaret Newton (1887–1971), the first woman to complete all requirements of the degree course in agriculture at Macdonald College, which she did by graduating with a BSA degree in 1918.[68] Her research on the rust fungi, under Fraser's guidance, indicated that they were working with at least two pathologically distinct strains of stem rust of wheat. Theirs was the first such discovery in Canada.

The Dominion Department of Agriculture was well aware of the great loss of wheat due to rust on the prairies in 1916, and made a small effort that year to do something about it. In November the Dominion botanist, Hans T. Güssow, sent a questionnaire about the rust to several hundred farmers in the three prairie provinces. It was designed so that the answers might reveal where and when the rust first appeared, how far it had spread, and how much damage had been done in the various regions. He received 560 replies and hired W.P. Fraser to analyse them. Fraser was appointed in February 1917, on a part-time basis, as officer in charge of grain disease investigations.

While still an employee of Macdonald College, Fraser studied grain diseases in western Canada during the summers of 1917 and 1918, from temporary laboratories set up at Brandon and Indian Head. Those studies continued on a more or less full-time basis after 1919, when he was appointed officer in charge of a new Dominion laboratory of plant pathology in Saskatoon, and until he left that position to become a full-time professor of biology, University of Saskatchewan, in 1925. On 12 December 1919, the Canadian Division of the American Phytopathological Society (on 19 December 1929 it became the Canadian Phytopathological Society) named Fraser and W.H. Rankin a committee of two to consider the taking of a plant disease survey of the Dominion of Canada and petitioned the Dominion botanist to pay the cost of publishing an annual survey. The first report of that committee appeared in the annual report of the Division of Botany, Experimental Farms Branch, for 1920.

In spite of his teaching and other responsibilities at the university, Fraser never ceased to collect plants infected with rusts and to study grain diseases, until forced to do so by ill health after he retired from the university in 1937.[69]

Several representations had been made to the federal government urging positive action with regard to wheat rust, the most noteworthy of which was by A.H. Reginald Buller (1874–1944), professor of botany at the University of Manitoba, in the form of a *Memorandum on the rust diseases of wheat*. That memorandum, dated 1 June 1917, commented on the great losses due to rust in 1904 and again in 1916 and recommended a cooperative plan of action between the federal Department of Agriculture and the Agricultural College at Winnipeg, with E.C. Stakman in charge, if he could be persuaded. Buller sent copies of the memorandum to J.H. Grisdale, director of the Central Experimental Farm, Ottawa, and to the presidents of the University of Saskatchewan and his own University of Manitoba. Doubtless Buller's initiative was one of the major stimuli responsible for what later became known as the First Cereal Rust Conference, held in Winnipeg, 16–18 August of that year.[70] Buller's bold plan and the recommendations that came from the rust conference were not implemented to any significant extent for several years, however, because Canada was at war and there was a shortage of trained people and finances.

While Fraser was surveying the rust situation in the summers of 1917 and 1918 he also began an investigation of the smut fungi in western Canada and made some initial studies on the epidemiology of rust diseases. In 1919, as officer in charge of the Dominion Laboratory of Plant Pathology, in Saskatoon, Fraser "laid the foundation for plant disease work generally, and in particular for research work on the cereal rusts" in western Canada.[71] Early evidence of this "foundation" work was his establishment of what became known as the uniform rust nurseries in 1920. These were plots in which the resistance of grain varieties to rust could be evaluated. They were also convenient places from which to collect rust fungi for physiologic-race studies. In 1921, Fraser began a study of the "wintering over" of rust spores and initiated studies on the root rots of cereal grains. He noted that root rot of wheat was common in some districts of Saskatchewan.[72]

At the University of Saskatchewan, Walter P. Thompson (b.1889), professor of biology, had been breeding rust-resistant varieties of wheat for several years and studying the cytology of crosses between different species. In the rust year of 1916, he crossed a variety of wheat that was "extremely resistant to rust under field conditions" with several standard varieties and grew the first hybrid generation the following year.[73] One of his selections proved to be rust-resistant in Saskatoon but not in Rosthern, an area less than eighty kilometres away. In discussing this unexpected event with Fraser, he learned of the evidence that Fraser and his former student, Margaret Newton, had found to indicate that the rust fungus was composed of more than one strain. That knowledge prompted Thompson to induce Newton to do the research for her doctor's degree at Saskatoon, where she could work with him and continue her rust research in cooperation with Fraser, who had recently moved there, and do her course work during the winters in Minnesota with Professor E.C. Stakman. This she did, and obtained her PH D degree from the University of Minnesota, with a thesis on rust research, in 1922, the year in which she joined the staff of the Department of Biology, University of Saskatchewan.[74] By that time she and Fraser had identified more than a dozen physiologic races of stem rust on cereal crops, and Fraser was experimenting on the culture of several heteroecious species.[75] Although Thompson, who eventually became president of the university, became so involved with lecturing and administrative responsibilities that he had little time to continue his basic studies in genetics and cytology, he did accumulate a rich legacy of genetical and cytological information that became valuable to others.[76]

J.H. Ellis, experimentalist in the Field Husbandry Department, Manitoba Agricultural College, had several hundred plots under observation for evidences of grain rust during the summer of 1916. Seeing rust in some plots and not in others led him to believe that some conditions of soil management tended to make a crop more resistant to rust, even though he saw that some varieties of wheat had less rust than others. He noted the effect of the rate of seeding and the effect of manuring on the amount of rust, and he noticed that differences in the amount of rust damage were related to the date of seeding. The information gleaned from those field plot experiments was published in Manitoba Department of Agriculture and Immigration Extension bulletin no. 41, in 1919.

The outbreaks of grain rust on the prairies in 1919, 1921, and again in 1923 stimulated a renewed interest in the problem. On 18 January 1924, a committee on rust research was appointed by the National Research Council of Canada, and five months later the Select Standing Committee of Agriculture and Colonization of the House of Commons was recommending, at the behest of the Dominion botanist, Hans Güssow, that the federal government should appropriate funds for the erection of research facilities for a thorough study of the rust problem. Eventually $25,000 was voted as a special grant toward that purpose.[77]

Understandably, there was controversy over where the proposed laboratory should be located. E.S. Archibald, director of the Experimental Farms Service, in a note dated 7 March 1924 to J.H. Grisdale, deputy minister of agriculture, wrote, "Mr. Güssow apparently is rather favoring the University of Manitoba ... I still maintain that Morden or Brandon or Indian Head should be the site of the laboratory."[78] The two latter sites were proposed because in the spring of 1917 field laboratories for the study of grain diseases had been established at those locations. However, the site originally proposed in Buller's memorandum of 1917, and the one favoured by Güssow, received the approval of the Hon. W.R. Motherwell, minister of agriculture. In the summer of 1925 two greenhouses were erected on land provided by the Manitoba Agricultural College, and a building referred to as the Dominion Rust Research Laboratory, or simply as "the rust lab," was constructed adjacent to the greenhouses in the winter of 1925–26.[79] That new laboratory was placed under the general direction of Dixon L. Bailey (1896–1984), who had been in charge of a field laboratory of plant pathology located at the Agricultural College since April 1923.[80]

Bailey, a BA graduate of Queen's University who had done some advanced study with Professor H.H. Whetzel at Cornell University, was, after some additional work, awarded a PH D degree by the University of Minnesota in 1924.[81] That was the year in which Frank J. Greaney (1897–1976), a native of Birmingham, England, who had earned a BSA degree from the University of Toronto in 1922, became his assistant in the field laboratory.

In the spring of 1925, Cyril H. Goulden (1897–1981), BSA, MSA, University of Saskatchewan, was appointed to take charge of the cereal-breeding program. He was assisted in that work by John N. Welsh (b.1894), BSA, M SC, who soon became involved in a successful oat-breeding program. Both of them were on the staff of the Dominion cerealist but stationed in the rust laboratory for close cooperation with the plant pathologists. Goulden was able to incorporate some of his breeding research into a thesis for the PH D degree that he earned from the University of Minnesota in 1925. Destined to became an internationally recognized authority on statistics in agriculture, Goulden was applying mathematics to his study of disease resistance in 1929,[82] at a time when very few, if any, young plant pathologists were competent in the mathematics of a statistical analysis when comparing results of different treatments for disease control. In 1937 Goulden came to their rescue by publishing fully worked out examples of statistical methods that were very useful to research workers.[83] These were but a prelude to the many books and papers he was to produce in the 1940s and more recently.

The selection of staff for the rust laboratory was much easier in the mid-1920s than it would have been in 1917 when the first rust conference was held. That was because, after the First World War, many young Canadians were attracted to the study of plant pathology and plant breeding, and there were a number of scholarships or other inducements for them to do advanced studies in those fields.

The research team at the Rust Research Laboratory was joined by John H. Craigie (1887–1989), AB, MS, and Margaret Newton, in the summer of 1925. Three additional appointments were made later that year: Ibra L. Conners, to continue the work he had been doing on smuts at Brandon; Thorvaldur Johnson (1897–1979), B SC, BSA, M SC, who earned a PH D degree five years later, to collaborate with Margaret Newton in work on physiological specialization in the rust fungi; and William L. Gordon, to do research on stem rust of oats.[84] Years later, Johnson wrote *Rust research in Canada and related plant-disease investigations*, published as Canada Department of Agriculture publication 1098 in 1961. That publication provides the most complete coverage of grain research in western Canada up to that date.

The tests by Greaney, with assistance from Bailey, in controlling rust by dusting the grain plants with sulphur were among the highlights of the work of the Rust Research Laboratory between 1925 and 1932.[85] Although C.V. Kighlinger had demonstrated the effectiveness of sulphur in rust control at Cornell University in 1924, the work at Winnipeg was the first large-scale effort to assess the practicability of controlling the disease by chemical means. The element of Greaney's research that caught the attention of agriculturists throughout the country was the application of the sulphur by aeroplane. The results of the trials involving an aircraft in 1927, 1928, and 1930 showed that the cost of the method was too high for profitable returns, except perhaps in years of severe rust. They also provided some of the earliest estimates of the amount of injury actually caused by leaf and stem rusts under field conditions. Those results were so conclusive that there was little need for further study on the effectiveness of sulphur as a protective cereal-rust fungicide. Nevertheless, that method of rust control was never widely adopted in western Canada. This was partly due to the cost, but also because rust-resistant varieties of wheat, and those that ripened early enough to escape much of the rust, were available or became available soon after the sulphur trials were completed.

The first Canadian breakthrough in the development of a wheat variety of general suitability, and which was not severely rusted, resulted from the pioneering work at Ottawa of William Saunders and his sons Percy and Charles that led to the development of Marquis wheat. From 1910, Marquis was the yardstick by which other varieties of bread wheats were measured in terms of yield, baking quality, and length of time to maturity. Although Marquis was not a rust-resistant variety, it had the potential to escape rust injury because of early ripening.

Canada's rise in prominence as one of the world's great producers of spring wheat has been due in no small measure to the success of the cooperative efforts of plant breeders and plant pathologists. It is doubtful if any country in the world has benefited more directly than Canada from the work of these scientists in their running battle against adverse environmental factors and diseases. The rise in importance of the wheat-growing industry in western Canada was made possible by the selection or introduction of early maturing varieties into which the plant breeders incorporated genetic material that made them resistant to diseases, especially rust, smut, and root rots. Early in the 1920s, in the prairie provinces alone, grain rust research, or the manitenance of plots in which rust could be studied or compared, was being carried on at the Dominion Experimental Farms at Brandon, Morden, Indian Head, and Rosthern, and at the universities of Saskatchewan, Alberta, and Manitoba, in addition to the Rust Research Laboratory in Winnipeg. Some rust research was also being carried on in other provinces.

To ensure and promote coordinated as well as cooperative action in the development of new varieties of grain, committees referred to as Associate Committees were set up under the National Research Council of Canada and the Dominion Department of Agriculture. They were composed of representatives from the universities, the provincial and federal Departments of Agriculture, and invited authorities on the problems to be studied. After meeting for the third time, in 1926, the Associate Committee on Grain Rust met at least once a year, until the beginning of the Second World War, to review the results obtained by the several research groups and to make recommendations for further cooperative efforts. Other committees, such as the Associate Committee on Grain Research and one on Field Crop Diseases, each had a role to play in the development of new crop varieties.

The meetings of those committees proved to be very valuable because they made all interested institutions, and their relevant individuals, parties to whatever decisions were reached. All new varieties of grain, after 1928, had to be approved by the Associate Committee on Field Crop Diseases (which concerned itself with smuts, root rots, and, in general, all diseases other than rusts) and the Committee on Grain Research (the function of which was to coordinate the quality aspect of the product resulting from research) before being recommended for a licence. That requirement led to the plan of conducting cooperative field trials of the most promising new varieties at the various universities and experimental stations, and on a number of selected private farms.[86] The triple objectives of high yield, disease resistance, and excellence in baking quality required the assistance of chemists, food specialists, and others, including commercial milling companies. However, their valuable contributions to the development of Canadian grain varieties of the highest quality are beyond the scope of this essay.

Some of the varieties of wheat developed for rust resistance had unsightly discolourations that took the form of black chaff or a sort of melanism of their glumes, necks, and internodes which disfigured otherwise desirable hybrid lines. A study of this condition was undertaken by Walter A.F. Hagborg (1908–82), a gold medallist BSA graduate of the University of Manitoba. Hagborg was first appointed as a temporary plant disease investigator in the summer of 1928, and, except for two seasons when he was ill, he worked as a summer employee until 1933, when he was taken on full-time at the Winnipeg laboratory. The grain discolouration study, carried on largely at Winnipeg, became the major component of his thesis research for the PH D that he was awarded, by the University of Toronto, in 1935. Hagborg continued those studies at Winnipeg through the war years, and in doing so became an authority on bacterial pathogens of cereals.[87]

The most noteworthy and far-reaching research on grain diseases at the University of Manitoba, and the Rust Research Laboratory on its campus, was that of John Hubert Craigie, who confirmed the long-suspected function of the pycnia (spermagonia) of a rust fungus.[88] Craigie, a Nova Scotian and a nephew of W.P. Fraser, spent the early

Staff of the Dominion Rust Research Laboratory with Professor
A.H.R. Buller on the occasion of his retirement. Back row
(left to right): Frank J. Greaney (1897–1976), John E. Machacek
(1902–70), Walter A.F. Hagborg (1908–82). Middle row:
John N. Welsh (b.1894), R. Peturson, D.B. Waddell, Thorvaldur
Johnson (1897–1979), Archibald M. Brown, William Popp
(b.1897). Front row: Bjorn Peturson (1894–1984), William
F. Hanna (1892–1972), Margaret Newton (1887–1971),
Arthur H.R. Buller (1874–1944), John H. Craigie (1887–1989),
Cyril H. Goulden (1897–1981).
Courtesy Agriculture Canada Research Station, Winnipeg

years of the First World War with the British Army in India. After that
war he earned an AB degree from Harvard University in 1925, and, following a period of employment with the United States Department of
Agriculture, an M SC from Minnesota in 1926. At Winnipeg, Craigie
took charge of the investigations on the epidemiology of the rust disease that had been initiated by Fraser in 1918. He reviewed several
years of that work when he left Winnipeg in 1944.[89] The research that
led to Craigie's historic discovery of the function of the pycnia was part

of his program for the PH D that he earned, while under A.H.R. Buller's nominal supervision, from the University of Manitoba in 1930. Buller, in his role as research supervisor, claimed to have provided the idea that inspired Craigie's discovery.[90] When the question was raised why Buller had not made the discovery himself, if this was so, I.L. Conners remarked, in a communication with the present author, that the greenhouse Buller had to use was so exceedingly dry that he could never get satisfactory infection of the barberry, whereas Craigie worked in a greenhouse that was much more suited to that kind of research.

Craigie's discovery, plus accumulated evidence that a greater variety of rust races occurred in collections of the fungal spores derived from aecia on barberry than in those taken from the rust pustules on wheat, led to a reassessment of the importance of barberry in the control of the rust disease. This, in turn, led to a revival of campaigns for the eradication of that shrub, campaigns in which Buller often played a leading role.

The legal basis for these and earlier crusades against barberry had already been established. Ontario, with its Act Respecting the Barberry Shrub of 30 April 1900, was the provincial leader in barberry control legislation. That act provided a penalty "not exceeding $10.00, besides the cost of conviction" for planting barberry, and compensation for its removal if it had been planted prior to the date of the act. The federal government, by a 1916 revision of the Destructive Insect and Pest Act of 1910, prohibited the distribution of barberry plants from nurseries and gave the necessary authority to remove, without compensation, those that had already been planted. In the prairie provinces, the destruction of barberry was authorized by amendments to existing provincial laws relating to weeds. (See the essay on legislation, below.)

Dixon L. Bailey resigned from the Winnipeg Rust Research Laboratory in 1928, to become a professor at the University of Toronto, and Craigie became officer in charge. In 1926 Kenneth W. Neatby (1900–58) joined Goulden and became largely responsible for the wheat-breeding program there. Neatby had BSA and MSA degrees from the University of Saskatchewan, where he was the first graduate student of James B. Harrington (b.1894), BSA, M SC, PH D, professor of field husbandry. Before leaving the university, Neatby had the misfortune of losing his thesis, and much of the data on which it was based, in a fire that burned the building in which the field husbandry department was housed.[91] Neatby went to Minnesota for the PH D that he received in 1931.

In 1937 the rust-resistant variety Renown was released from the Winnipeg laboratory. It was largely the creation of Neatby and Rudolph F. Peterson (b.1900), BSA, MS, PH D, who succeeded Neatby when he resigned in 1935 to head the Department of Field Crops, University of Alberta. Peterson had been transferred from the experimental farm at Brandon where he had been assistant superintendent since 1933, the year in which he obtained his PH D degree. Peterson's story, "Twenty-five years' progress in breeding new varieties of wheat for Canada," was published in *Empire Journal of Experimental Agriculture*, volume 26, in 1958. Except for Regent, released in 1939, no additional rust-resistant varieties of spring wheat, developed largely at the Winnipeg laboratory, became available to farmers until Redman was licensed in 1946.

At the University of Alberta, Neatby studied the genetic relationship between the reactions of wheat to various fungal diseases. The initial research for one of his papers on that general topic was done in Winnipeg, continued while he was at the Plant Breeding Institute, Cambridge, England, on a National Research Council (USA) fellowship in 1933–34, and written in Alberta.[92] Neatby left the university in 1935 to become director, Line Elevators Farm Service, North-West Line Elevators Association, Winnipeg.

At the Manitoba Agricultural College, cereal breeder W.T.G. Wiener, BSA, developed, through selections of material from Minnesota, a pure amber strain of a durum wheat variety named Minden. It was rust resistant and of such high macaroni-making quality that it became a standard of excellence for decades.[93]

In the meantime, at the University of Saskatchewan, James B. Harrington was engrossed in a breeding program that had started about the same time and with somewhat similar objectives as those of Goulden and his colleagues in Winnipeg. Harrington, who has been credited with having coined the term "agrologist" in the 1920s, studied the relationship between various factors and the expression of rust in the second generation of varieties and species that he crossed. It was he who produced the rust-resistant variety Apex in 1937. Harrington also studied the effect of low temperature, including spring frosts, on yields of grain.[94]

The selection and breeding of rust resistant cereal grains became so popular that the grain growers themselves were getting into the act. Perhaps the most noteworthy of the many farmer-breeder-selectors was Seager Wheeler (1868–1961), a grain grower near Rosthern, Saskatchewan, who won five world grain championships. His selection of a few heads with red kernels from a field of Australian White Bobs became the famous Red Bobs variety, which, in the early decades of this century was widely used by farmers of Alberta and northern Saskatchewan.[95] But the selection of outstanding wheat plants by Canadian farmers goes back to the early 1840s, when David Fife, who farmed near Peterborough, Ontario, received a small sample of wheat from a friend in Scotland, who had obtained it from the hold of a ship recently arrived in Glasgow from Danzig. When Fife planted part of that sample only a single plant matured to produce seeds. The few seeds that he obtained were planted the following spring, and they produced plants that were free of rust and otherwise of high quality. The progeny of those seeds became the well known Red Fife variety of hard spring wheat that was popular in eastern Canada for many years. It was somewhat less popular in the west because it matured too late to ensure a good crop every year.[96]

The December 1981 issue of the *Country Guide* noted that a certain Mr Dawson of Waterloo County, Ontario, had discovered a new variety of wheat in 1881, when a violent storm flattened his whole stand, except for one plant which he saved and multiplied. The progeny of that single plant eventually became known as Dawson's Golden Chaff wheat, which in 1950 accounted for half of the winter wheat sown in Ontario. That lone plant was remarkably rust-resistant, although Dawson had not selected it for that reason.

Another farmer who rendered inestimable service to the production of rust-resistant varieties of wheat was Malcolm McMurachy of McConnell (Strathclair), Manitoba. In 1930, he selected an outstanding plant from a field of heavily rusted wheat and sent it to the Department of Agriculture, Ottawa. In the hands of wheat breeders at Ottawa and Winnipeg, it was multiplied and became an exceptionally useful breeding parent for the development of rust-resistant varieties.[97]

When D.L. Bailey transferred to the University of Toronto he did not have to give up his interest in grain rust research. Dorothy F. Forward (b.1903), one of the graduate students in the Department of Botany there, had just completed a master's thesis on certain aspects of the parasitism of the rust fungi and was keen to continue rust research with the added support of Bailey. Forward's research for the PH D that she was awarded in 1931 developed a hypothesis concerning the relation of modifications of the rust infection type to the dark metabolism of the grain plant. The results of much of her thesis research, performed under the joint supervision of Bailey and plant physiologist G.H. Duff, was published in *Phytopathology* the following year. Forward, a charter member (student) of the Canadian Phytopathological Society, became a member of the academic staff of the Department of Botany, University of Toronto in 1935 and remained there until some time after the Second World War.

Root Diseases

The realization that a number of cereal disease symptoms observed in the field could be traced to some disorder originating below the surface of the soil directed the attention of plant pathologists, soil scientists, plant breeders, and others to nutritional factors and micro-organisms in the soil that might be responsible for deficient root development or to root diseases.

Some of the earliest studies of grain root diseases in western Canada were begun by W.P. Fraser at the Dominion Laboratory of Plant Pathology, Saskatoon, in 1920.[98] In the course of his surveys each summer, Fraser found the browning root-rot disease of wheat, and, in 1923, the take-all disease in several locations in that province.[99] This latter discovery so focused the attention of his summer-student assistant, Ralph C. Russell (1896–1964), on the problem that he made a study of the disease and its inciting fungus the major topic of his research for several years. It also became the subject of his thesis for the PH D degree, which he was awarded by the University of Toronto in 1934. Russell, who became a charter member of the Canadian Phytopathological Society, had earned BSA and MS degrees from the University of Saskatchewan in 1924 and 1926 respectively.[100] In 1927 he showed that the damage from take-all could be reduced by a simple crop rotation.[101] Fraser, together with fellow investigators of the take-all disease, gave a brief progress report of the work at Saskatoon up until 1926.[102]

In 1927, Frank Greaney and D.L. Bailey wrote an account of root rots and foot rots of wheat in Manitoba that was published as Dominion Department of Agriculture bulletin no. 85. Shortly after that, Greaney teamed up with John E. Machacek (1902–70) to begin a study of root rots of both wheat and oats. Machacek, a native of Czechoslovakia who was brought to Canada as a child, earned bachelor's and master's degrees from the University of Saskatchewan in 1924 and 1925 respectively. He then did some teaching at Macdonald College, where he studied associations of plant pathogens with each other and with saprophytic organisms in the rots of fruits and vegetables. In 1928 he became the first student to earn a PH D degree in plant pathology from McGill University.

When the Canadian Phytopathological Society sponsored a symposium on seed-borne diseases, it was Greaney who provided a background survey of the topic as it applied to cereal grains and acted as leader of the discussions in which Prye M. Simmonds and Howard W. Mead (1899–1973), B SC, MS, PH D, were the principal speakers. Notes from the talks of all three were published in the *Proceedings* of that society in 1944. Simmonds was such an internationally recognized authority on root diseases of grasses that he had been invited to write an article titled "Rootrots of cereals" for *Botanical Review* in 1941.

In studying the pathogens involved in root rots, Greaney and Machacek experimented with ultra-violet radiation to produce "saltants" of fungi isolated from diseased roots.[103] In 1946, Greaney, who was acting head of the laboratory for a while in 1944–45, resigned to succeed Kenneth W. Neatby as director of the Line Elevators Farm Service while Neatby became director, Science Service, Dominion Department of Agriculture, Ottawa. But before leaving the rust laboratory, Greaney and Machacek had found such great variations within species of fungi that it was decided specialization would be necessary. Machacek concentrated on studies of Helminthosporium, and his colleague W.L. Gordon took on the formidable task of identifying species of Fusarium.

Henry A.H. Wallace (1907–82), who had a B SC(Agr.) degree from Alberta and an MS degree from Minnesota, worked with both Greaney and Machacek. From 1935 until 1942 he and Greaney tested varietal susceptibility to kernel smudge in wheat at several stations in Manitoba, Saskatchewan, and Alberta, and between 1937 and 1942 he and Machacek tested more than three thousand samples of cereal seeds, from all parts of Manitoba, for seed-borne diseases.[104]

Root-rot studies were also being carried out by Nova Scotia-born Guthrie B. Sanford (1890–1977), BSA, M SC, at the Dominion Laboratory of Plant Pathology, Saskatoon, where he had succeeded W.P. Fraser. When Sanford was transferred to Edmonton in 1928, to head a new laboratory there, the work he had initiated at Saskatoon was continued by his successor, Prye M. Simmonds, and his colleagues, especially Bryce J. Sallans (1901–74), BSA Manitoba, BA McMaster, M SC Saskatchewan, where he had been one of Fraser's students. Both Simmonds and Sallans eventually earned PH D degrees from Wisconsin, following several years of root-rot research. Simmonds, who proposed the term "common root rot" to distinguish root rots caused by *Helminthosporium sativum* and *Fusarium* species from take-all and browning root rot, published a review, in 1939, of the research on cereal root rots in western Canada to that date.[105] He continued to study root rots for the next ten years or more. Simmonds, Sallans, and R.J. Ledingham had two papers dealing with their research on *Helminthosporium sativum* in the *Proceedings of the Canadian Phytopathological Society* for 1949.

Others who made significant contributions to the research that made Saskatoon the major centre for root disease studies included Ralph C. Russell and Howard W. Mead, mentioned earlier, and Robert J. Ledingham (1912–83), who had B SC and MS degrees from the University of Saskatchewan. While studying root diseases, Russell found a new species of nematode in the roots of wheat in eight widely separated fields in the Humboldt district,[106] and, with Simmonds and Sallans, he compared different types of root rot.[107] Mead and Gordon

A. Scott (1895–1946), BSA, M SC, who worked at Saskatoon from 1925 until he resigned in 1929, studied seed-borne infections by common root rot fungi, and various seed treatments for their control. That work with fungicides led to Mead's development of a biological method for detecting the presence of fungicides in seeds.[108] Ledingham also worked on root diseases of wheat, but in the early 1940s he digressed from that and studied the effect of seed treatment and dates of seeding on the emergence and yield of peas.[109]

At the University of Saskatchewan, Thomas C. Vanterpool (1898–1984), BSA, M SC McGill, followed the lead of W.P. Fraser, the man he eventually succeeded, by getting his students interested in cereal-root disease problems, particularly the browning root rot of wheat. They learned that a deficiency or imbalance of nutrient elements in the soil may predispose a crop to attack by pathogenic organisms, and Vanterpool discovered the fungus *Asterocystis radicis* in the roots of cereal grasses in 1930.[110] Three years later, he reported on a toxin being formed by species of Pythium parasitic on wheat.[111] Much of Vanterpool's work on the browning root rot disease is recorded in a series of papers, the sixth of which was published in 1940.[112]

At Edmonton, Sanford continued his pioneering studies on root diseases,[113] and, with his colleagues, particularly W.C. Broadfoot, put emphasis on the biotic rather than the abiotic components of the soil. In doing this they demonstrated the importance of microbiological antagonism in suppressing the activities of disease-inducing fungi.[114] Thus they were pioneers in work leading to biological control of root diseases.

Sanford, in his review of some soil microbiological aspects of plant pathology, suggested that the reason disease is usually reduced by summer-fallow may be an adjustment of the general soil flora and fauna that is unfavourable to root parasites.[115] Broadfoot, using some of the organisms being studied by his colleagues, found that many of those that exercised a marked degree of *Ophiobolus graminis* control, on take-all disease in wheat roots in soil, were not antagonistic to that fungus in culture media.[116] He thus showed that the growth reactions of various organisms in culture media are not reliable indications of how they will react in natural soil. In his wide-ranging research on root diseases, Broadfoot studied the effects of crop rotations on foot rot and found that there was less rot when wheat followed oats, and more when wheat followed wheat.[117]

Others who did some research on root diseases of grain at Edmonton prior to the end of the Second World War include H.T. Robertson, for a period between 1930 and 1932,[118] and his replacement, Lawrence E. Tyner (b.1902), B SC and MS Alberta, PH D Minnesota,[119] who had two papers pertaining to root rots of cereals in the

1945 *Proceedings of the Canadian Phytopathological Society* and one on the effect of formalin on root development of wheat seedlings in the *Proceedings* of 1947. Melville W. Cormack (b.1908), BSA Manitoba, MS Alberta, PH D Minnesota, studied root diseases of wheat seedlings at Edmonton with Sanford,[120] but he became known mostly for the thoroughness of his studies on root diseases of forage legumes. (See below.)

Within a few metres of Sanford and his colleagues, but working independently, New Brunswick–born Arthur W. Henry (1896–1989), in the Department of Field Crops, University of Alberta, was also studying the natural microflora of the soil in relation to root diseases and their control. He showed, in 1931–32, that a root rot of wheat seedlings was suppressed by a number of fungi and bacteria in the soil and that changes in temperature produced changes in the degree of microbial antagonism against the disease incitants.[121] Prior to this, he and William R. Foster (b.1905), the first student to register for the M SC degree in plant pathology at the University of Alberta, provided the first Canadian report of Leptosphaeria root rot of wheat.[122] Henry and his graduate students found abundant evidence that cereal root-rot pathogens are poor competitors with the natural microflora of the soil and that they are so suppressed by the microflora as to decline in natural soil. A student who provided much of this evidence was Ralph A. Ludwig (1915–77) whose 1939 thesis for the M SC degree, under Henry's supervision, was titled "Studies on the microbiology of sterilized soil in relation to its infestation with plant pathogens."

Henry was a long-time promoter of seed treatments for the control of soil-borne pathogens of cereals. This is shown in the thesis research of several of his students, notably that of Jack G. Grimble, whose 1941 thesis for the M SC was titled "Studies on the chemical seed treatment of grasses." Nevertheless, Henry was generally opposed to the use of formaldehyde.[123] He had earned his BSA degree from the University of Saskatchewan, and then went on to become the first student to receive an MSA degree from that university, in 1920. His thesis dealt with a species of Helminthosporium that had not been previously reported in Canada. After that, he studied plant pathology under E.C. Stakman at the University of Minnesota, where his research for the PH D that he received in 1923 was based on root rots of wheat.

Canadian students, and others interested in advancing their knowledge of grain diseases and methods of grain disease research, were, for several years, attracted to the University of Minnesota, where Stakman's work on wheat rust was receiving international acclaim, as was H.K. Hayes's program of breeding rust-resistant wheat. Henry, who stayed on as an employee for three additional years of research, was at Minnesota the same time as Margaret Newton, Dixon Bailey, Guthrie

Sanford, Ibra Conners, and probably others. After that he spent a year at various laboratories in Europe, on a Rockefeller Foundation Fellowship. While in Europe, he accepted the position of assistant professor of plant pathology at the University of Alberta, in 1927.[124]

Robert Newton (1889–1915), the man who hired Henry, had earned M SC and PH D degrees from the University of Minnesota in 1922 and 1923 respectively, after obtaining a BSA from McGill University in 1912, and undertaking a period of military service. He and his graduate students at the University of Alberta did some outstanding innovative work on frost and drought resistance in wheat.[125]

At the University of Manitoba, Russian born Michael I. Timonin (1900–91), who had a Diploma in Agriculture from Czechoslovakia and a BSA from Manitoba, was studying soil fungi while working for a master's degree under the supervision of plant pathologist G.R. Bisby and bacteriologist N. James. It was during this research that Timonin became the first person to show that *Trichoderma lignorum* could protect wheat from such root-rotting fungi as *Fusarium culmorum* and *Helminthosporium sativum*.[126]

Timonin worked for a while as an assistant of the Dominion bacteriologist, Allan Grant Lochhead (1890–1980), in Ottawa. Lochhead, a graduate of McGill University, from which he received his M SC in 1912, was studying in Germany for his doctorate when the First World War broke out. After being interned for four years, he returned to Canada, and in 1919 McGill awarded him the PH D degree, largely on the basis of his work at the University of Leipzig in Germany. In 1923 he became head of a newly created Division of Bacteriology in the Experimental Farms Service of the Department of Agriculture. He and his associates, beginning around the time he was made a Fellow of the Royal Society of Canada in 1940, published a monumental series of papers on the associative and antagonistic phenomena of microorganisms in the soil and in the rhizosphere of plants. They showed that the rhizosphere bacteria differed markedly from others in the soil and that they had an influence on the resistance of the associated plants to root pathogens. That work, and the work of others who had been dealing with rhizosphere microorganisms in relation to root disease fungi, was reviewed by Lochhead and published in 1959.[127] Much of what Lochhead and his associates discovered about rhizosphere microorganisms had a direct bearing on the work of those who were studying the root rots of grain.

DISEASES OF OATS

Until about the beginning of the First World War, the production of oats in Canada was greater than that of wheat, the oats being used mainly as food for horses. Living in an age in which we have become almost completely liberated from a dependency on mammalian muscle power, it is well to recall that oats were prime sources of energy in Canada for nearly a century.

Throughout eastern Canada, from about 1850, the oat was considered to be a more reliable crop than wheat and, except for smut, less prone to destructive diseases. Rust of oats was occasionally prevalent but not considered to be a serious threat to production, especially in Prince Edward Island, where the disease was seldom seen until 1840.[128] James Fletcher, Dominion botanist-entomologist, was concerned about a red leaf or blight of oats in 1890,[129] but it did not prove to be a serious problem. He and his staff were conducting tests on the use of formalin and massel powder for smut control in oats and barley at the Central Experimental Farm in 1899, the results of which were published as report no. 394. Charles E. Saunders (1867–1937), BA, PH D, who had been appointed experimentalist at the Central Experimental Farm, Ottawa, in 1903, worked on the accumulated material resulting from crosses made by his brother Percy, and those made earlier by his father, and commenced a program of breeding oats that led to the development in 1920 of the variety Legacy.[130]

In the 1920s the agronomists and plant pathologists at Macdonald College were cooperating in experiments for the control of oat smut,[131] and graduate student W.L. Gordon assessed the injury to oats following smut disinfection.[132] Plant pathologist John G. Coulson (1893–1974), BA, MA, and agronomist Emile A. Lods inquired into the relationship between oat smut infection and the size of seed,[133] and Bertram T. Dickson (1886–1982), chairman, Department of Plant Pathology, discovered, described, and named a fungus that was parasitizing the smut fungus *Ustilago levis* on oats.[134] Dickson's graduate student William Popp studied crown rust of oats in eastern Canada,[135] and later, in Winnipeg, he and W.F. Hanna studied the physiology of oat smuts.[136]

In the summer of 1939, when I.L. Conners was making a survey of cereal diseases in Quebec, he was assisted by David Leblond (1907–81) BA, B SC, a teacher in the School of Agriculture at La Pocatière. Leblond was interested in the various oat diseases they encountered, particularly a speckled leaf blotch, the study of which became the major subject of his research for the M SC degree that he earned at Macdonald College in 1942.

In Nova Scotia, Arthur Kelsall (b.1892), BSA, entomologist in charge of insecticide investigations at the Dominion Entomology Laboratory, Annapolis Royal, was experimenting on the use of various chemical dust formulations for smut control in general, but especially oat smut, in the 1920s.[137]

The Ontario Agricultural College produced a bulletin on smut and rust of grain in 1915, and John E. Howitt (1881–1966), MS, professor of botany, experimented with the use of formalin for the prevention of oat smut.[138] He and Roland Elisha Stone (1881–1939) B SC, M SC, PH D, associate professor of botany, cooperated with the Crop Protection Institute of the United States Department of Agriculture in trying various methods to control oat smut for at least five years.[139] Stone did not confine his smut control research to oats. In 1928 he reported in *Phytopathology* that formalin applied to foxtail millets for the control of smut tended to prevent seed germination.

During the Second World War, John D. MacLachlan (1906–87), a BA graduate of Queen's with AM and PH D degrees from Harvard, who joined the staff of the botany department at the OAC in 1939, studied the manganese deficiency disease of oats and boron deficiency in turnips.[140] In 1948 he became head of the Department of Botany, following the retirement of Howitt.

In western Canada, the first reference to severe losses from smut in oats appeared in Superintendent Bedford's 1894 report of work at the Brandon Experimental Farm. In the following year he commented, "At no time in the history of the province has there been so much smut among oats as prevailed this year."[141] Bedford's report for 1898 and those for several subsequent years told of successes in the control of smut when the seed oats were treated with formalin.

With smut more or less under control, it was not long before oat stem rust become an important grain disease, second only to wheat stem rust. At Winnipeg, D.L. Bailey recognized this, because he had done some research on the problem as early as 1921, and made physiological specialization in oat stem rust the topic of his research for the PH D degree in 1924.[142] Consequently, projects on oat stem rust and varietal resistance of oats to crown rust were begun at the Dominion Field Laboratory of Plant Pathology, Winnipeg, before the breeding program for the development of rust-resistant wheat got started.[143]

The study of physiologic races of oat stem rust at Winnipeg was largely taken over by W.L. Gordon, with the assistance of cerealist John N. Welsh (b.1894),[144] BSA, M SC, who also studied the effect of smut on the development of rust in oats from 1925 until 1932.[145] With the help of plant pathologists, especially W.F. Hanna, Welsh introduced the first Canadian-produced, rust-resistant oat variety, Vanguard, in 1937. After 1932 the work on physiologic races of oat rust became the responsibility of Margaret Newton and T. Johnson, while studies on smut diseases of oats were carried on by W.F. Hanna and W. Popp.

In 1927, the National Research Council reported that no adequate study of crown rust of oats had yet been made.[146] As a consequence of that report, Bjorn Peturson (1894–1984) was appointed to the staff of the Winnipeg laboratory as an assistant plant pathologist, 18 June 1929, to study crown rust of oats and to assist the oat breeders in bringing resistance to that rust into their varieties. Peturson had just obtained an M SC degree in plant pathology and botany from the University of Minnesota, with a thesis on physiologic specialization in *Puccinia coronata avenae*, following a BSA degree from the University of Manitoba in 1928. He later obtained a PH D from the University of Minnesota. Between 1929 and 1935 he showed that more than a dozen races of crown rust occurred in Canada.[147]

W.P. Fraser, who studied a wide range of grain diseases, made some experiments in 1922 to determine the host range of the crown rust fungus in Saskatchewan, which he continued for several years. In 1925 that work became largely the responsibility of Mabel Ruttle, who also made a cytological study of the disease.[148] In 1926–27 G.A. Ledingham took over and, with Fraser, published a summary of the research in 1933.[149] When Fraser retired, his successor, T.C. Vanterpool, continued studies of fungi in the roots of grains and other grasses and discovered a fungus that he thought could be a contributing factor in the root-rot complex of oats.[150] Root diseases of oats were also studied by P.M. Simmonds at Saskatoon, who published an account of his investigations on common root rot of oats in 1928,[151] and G.B. Sanford at Edmonton, who discovered a previously unreported root rot of oats.[152]

A somewhat mysterious disease of oats known as blindness or blast was troublesome at times and in various places across Canada. Leonard C. Huskins became concerned about the problem in 1924, while he was with the Department of Field Crops, University of Alberta. He worked on the disease for several years and, when he became an associate professor of genetics at McGill University, concluded that there was genetic resistance to the disease and that it should be possible to breed blast-resistant varieties.[153] That study was taken up by Russell A. Derick (b.1894), M SC, senior cerealist in charge of oat breeding at the Central Experimental Farm, Ottawa, in 1932. Derick, who had been assistant agrostologist at the Dominion Experimental Farm, Brandon, Manitoba, was assisted in his continuing studies of oat blast by John L. Forsyth (b.1908), BSA. They concentrated on the causes of blast and in doing so studied the effects of such variables as seeding date, soil moisture, and light.[154] When Forsyth left to do postgraduate studies in Minnesota, Derick was helped by graduate assistant Donald G. Hamilton (b.1917) in further studies on oat blast.

Derick and R.M. Love, who had made cytological studies of oats his PH D research topic at McGill in 1935, used X-rays and hybridization methods to produce gene changes in oats in their efforts to find a genetic basis for either dwarf oats or blast.[155] The general topic of blast studies was reviewed by T. Johnson and Archibald M. Brown (b.1890) of the Winnipeg laboratory in 1940.[156] Brown had joined the staff in 1926 to assist Johnson and Margaret Newton in their research on wheat stem rust. He quickly became a valuable technician and photographer, and was promoted to assistant plant pathologist in 1927. In 1937 he had two papers in the *Proceedings of the Canadian Phytopathological Society*, and three years later he published on the sexual behaviour of several plant rusts in the *Canadian Journal of Research*.

During the summer of 1932, the Ontario Research Foundation was asked by farmers in South Simcoe County to investigate the cause of the repeated failure of their spring grain crops. In response to that request, research fellows Donald F. Putnam (b.1903), BSA, and Lyman J. Chapman (b.1908), BSA, were given the task of determining the cause of the failures. Because the region had been settled for a long time and been more or less specialized to the raising of cereal grains, it was suspected that the exhaustion of soil nutrients might be responsible. Extensive soil tests did not support this suspicion, and attention was directed toward the investigation of root-inhabiting parasites. In July 1933, while examining the roots of oat plants growing in "oat-sick soil" in a greenhouse, they discovered the presence of nematodes that were later identified as the oat strain of *Heterodera schachtii*, the first record of its occurrence in North America. A survey of fifteen fields in which the oats were very poor or patchy showed that these nematodes were consistently found on the roots of diseased plants. In publishing the results of their survey, Putnam and Chapman included a description of the symptoms and effects of nematode damage and suggested methods for its control.[157]

In western Canada, the oat breeding program that led to the production of Vanguard in 1937 continued, and the varieties Ajax and Exeter were released in 1941. However, new races of the stem rust fungus appeared on the prairies in the 1940s and several oat varieties that were resistant to the stem rust fungus were found to be susceptible to crown rust. The breeding program designed to combine both stem rust resistance and crown rust resistance in one variety of oats produced the variety Gary, which was released to farmers in 1947. But Gary had a serious defect, one that could not have been anticipated when the breeding program began. It was susceptible to a root rot incited by the fungus *Helminthosporium victoriae*. That fungus was unknown to western researchers before 1944, when it was found to cause a root rot on oat varieties derived from crosses with the crown-rust-resistant variety Victoria. The disease became known as Victoria blight. This stimulated renewed efforts on the part of oat breeders and plant pathologists that culminated in the development of a new Victoria-blight-resistant Gary, early in the 1950s.[158]

D I S E A S E S   O F   B A R L E Y

Barley diseases have been important to some farmers from the beginning of agriculture and the use of malt in brewing. The Book of Exodus (9:31) in the Bible mentions barley being "smitten" by disease before wheat was tall enough to be damaged.

In Canada, barley has been grown since 1605, when Champlain introduced it as a crop to meet the needs of his brewers. Because it was never considered as important as wheat or oats in the general economy of Canadian agriculture, the study of barley diseases was relatively low on the list of research priorities. Nevertheless, James Fletcher, at the Central Experimental Farm, reported on blight of barley and oats in 1894.[159] It was there that Charles Saunders made a relatively disease-free selection of barley, which he called Manchurian. However, to Saunders, at that time, freedom from disease was probably secondary to the development of a high yielding variety that would appeal to the malting industry. That may also have been true for Charles A. Zavitz (b.1863), BSA, D SC, who selected the long-time favourite OAC 21 variety at the OAC in 1910, and for John A. Clark (b.1879), BSA, who selected Charlottetown 80 at the Experimental Station in Charlottetown in 1912.

Research on diseases of barley at Macdonald College was intermittent for many years, although it was constantly on the minds of the agronomists as they worked to develop new cultivars. Interest in barley diseases was revived when plant pathologist Ralph Anthony Ludwig (1915–77), B SC, M SC, joined the staff in 1944, and it was not long before he and his graduate student Thomas Simard were presenting papers on the subject to members of the Quebec Society for the Protection of Plants, two of which are in the thirtieth annual report of that society. Another of his graduate students, Dean B. Robinson (1922–61), studied barley diseases for the M SC degree that he was awarded in 1948.

On Prince Edward Island, the artificial hybridization of standard varieties of barley and wheat to produce high-yielding disease-free cultivars began during the winter of 1927–28.[160] Also around that time, J.E. Hewitt and R.E. Stone, at the OAC, were experimenting on methods for the control of the barley stripe disease.[161] Shortly after that, their colleague E.C. Beck, professor of bacteriology, studied the application of serological methods in the differentiation of closely related smut fungi, studies which became part of his thesis for the PH D from the University of Toronto, and which were published in 1938.[162]

Because of almost chronic wheat surpluses in the 1920s, agriculturists in western Canada were giving thought to the economics of other cereals, and a good malting barley was an attractive alternative for many of them. However, new varieties suitable to local conditions had first to be developed by the plant breeders. Whereas disease resistance was largely incidental to barley breeding programs in eastern Canada, the perspective was quite different in the west where rust and smut were prominent in the minds of all grain growers. Consequently, there was much more emphasis on the breeding of disease-resistant varieties.

S.J. Sigfusson had begun an important wheat and barley breeding program in 1924 at the Brandon Experimental Farm, but his tragic death in 1933 put a stop to that work until Walter H. Johnston (b.1908), BA, B SC, M SC, was appointed cerealist in 1936. Johnston's work became the cornerstone of barley research on the Canadian prairies after the new variety Plush was licensed in 1939.[163]

At the University of Alberta, plant breeder Olaf S. Aamodt (b.1892), BS, MS, PH D, with graduate assistant W.H. Johnston, published accounts of their research into the reaction of barley varieties to covered smut infections, and of their efforts to breed disease-resistant varieties.[164]

A.M. Brown and Margaret Newton, whose primary responsibilities at the Rust Research Laboratory in Winnipeg were those associated with diseases of wheat, became apprehensive about the discovery of a dwarf leaf rust on barley that was found to be quite abundant in southwestern Manitoba and at Indian Head, Saskatchewan.[165] Newton, together with W.J. Cherewick, was also concerned with the little-studied mildew disease of grain. In 1947, they published a comprehensive review of the fungus *Erysiphe graminis* in Canada, and expressed their belief that powdery mildew was the most destructive barley disease in Ontario and Quebec.[166]

A type of barley smut investigation that roused considerable interest at the time was embryo examination as a means of predicting the amount of loose smut in the field. In 1946, P.M. Simmonds showed that the smut fungus could be seen in detached embryos of barley and wheat by microscopic examination.[167] It was postulated that if the percentage of infected embryos in a given lot of seed was closely related to the percentage of smutted plants produced by that seed, the embryo examination would provide a useful substitute for the determination of smuttiness by field examination. In 1950, R.C. Russell showed that there was close agreement between the embryo tests and the field examinations for smuttiness of barley, and that it was not such a reliable test for wheat.[168]

DISEASES OF FLAX

The cultivation of fibre flax in Canada began during the colonial period when it was used to provide linen for hand spinning in the home. Those early farmers learned that they could almost always get a good first crop on newly broken land, but they commonly got a somewhat lower yield on older cultivated land. That was one of the reasons why flax seed production moved westward with the settlement of new lands. Those early farmers had no way of knowing that a disease known as flax wilt was the primary cause of the decreased yields on previously cultivated land. Flax wilt and flax canker were noted in the annual reports of the Saskatchewan Department of Agriculture in 1911 and 1912, but those reports have no references to research being done to control either disease. Virtually all of the early research for their control was done outside of Canada.

Linseed flax became an economically important crop during the First World War because of a need for linseed oil, but there was little, if any, flax disease research or flax breeding for resistant varieties, anywhere in Canada, until the 1920s. Then, Canadian plant pathologist W.P. Fraser, working in Saskatchewan, found the browning and stem break disease of flax in 1923. His assistant, R.R. Hurst, initiated some research on the disease in Saskatchewan and continued it in Prince Edward Island when he was transferred there.[169] A.W. Henry, of Alberta, published on the reaction of flax species of various chromosome numbers to both rust and powdery mildew in 1928.[170] In a very real sense, it was Henry who laid the foundation for a program of breeding for rust resistance in flax while he was working in the United States between 1923 and 1930.[171]

Cerealist William G. McGregor (b.1900), BSA, MS, PH D, began work in Ottawa on a rust disease resistance program in 1936,[172] but the paucity of rust on flax in the Ottawa area was an obstacle to the success of his work there. In some years it was virtually impossible for him to find any rust. For that reason a flax nursery was also established in Winnipeg, where rust was likely to be found almost every year. It was from Winnipeg that Bjorn Peturson and Margaret Newton made a survey, in 1940, of the races of flax rust in Canada. This was the beginning of an annual survey that was intended to chart the distribution of the various races that should be tested at both Ottawa and Winnipeg.[173] The first flax variety to be developed in Canada with moderate resistance to rust was Royal, selected in 1939 by J.B. Harrington, Department of Field Husbandry, University of Saskatchewan.

Although the majority of Canadian growers of flax were using seed of wilt-resistant varieties developed elsewhere, the wilt disease was still considered a potential threat to flax production. It appeared to be even more so after W.C. Broadfoot discovered that the wilt-inducing fungus, *Fusarium lini,* was composed of pathogenetically distinct strains.[174] Nevertheless, there was relatively little flax disease research done in Canada prior to the Second World War, and what had been done in Saskatchewan was reviewed by T.C. Vanterpool.[175]

Cooperative tests of various treatments applied to sound and fractured flax seeds were conducted from 1944 to 1946 inclusive, at thirteen research stations in Canada and eight in the northern part of the United States. A summary of the results of those experiments was provided by J.E. Machacek and F.J. Greaney in 1948.[176]

Pasmo, a fungus-induced flax disease, began to pose a threat to flax production late in the 1930s and early 1940s. McGregor added that disease to his flax-breeding program at Ottawa, in 1944. Waldemar E. Sackston (b.1918), BSA, M SC, agricultural assistant at the rust laboratory, Winnipeg, established a pasmo nursery there in 1945.[177] Reference is made to nine of his papers on flax diseases, and one on sunflower diseases, in the *Proceedings of the Canadian Phytopathological Society* for 1949. Sackston, who earned a PH D degree from Minnesota that year, became the most widely acknowledged authority on flax and sunflower diseases, especially the latter, after the period covered by this essay. For several years Sackston worked closely with Eric D. Putt (b.1915), who, later, became known as the father of sunflower breeding in North America. Incidentally, Guy R. Bisby, then professor of botany, Manitoba Agricultural College, made tests of the rate and manner of spread of sunflower diseases in the 1920s.[178]

Small amounts of fibre flax have been grown in the province of Québec from almost the beginning of agriculture there. The amounts increased dramatically during the American Civil War, 1861–65, in an effort to compensate for the sharply reduced supply of cotton from the south. The government of the then Province of Canada imported flax-scutching machines and assigned them to groups of farmers who had been growing small amounts of fibre flax, and some local governing bodies gave grants or made loans for the purchase of high quality seed or otherwise induced farmers to grow more flax. However, by the time that ambitious program was fully operational the Civil War had ended, as had the demand for fibre flax.

There was another stimulus to flax production, mostly in eastern Canada, during the First World War, and again during the Second. By 1943 nearly three quarters of the fibre flax being grown in Canada was in Quebec.[179] Although there were no major diseases associated with the fibre flax crop, anthracnose posed a threat to the desired increase in production during wartime. For that reason, it was deemed desirable to know much more about that disease. In working toward that objective, the most noteworthy studies during the war years were made jointly, in 1943 and 1944, by René-O Lachance (1909–92), BA, BSA, M SC, PH D, who was at that time a member of the Laval University Faculty of Agriculture, at Ste-Anne-de-la-Pocatière, and Albert Payette (1911–90), BA, BSA, M SC, plant pathologist at the science service laboratory in the same community. They had found anthracnose on flax seed from crops grown in at least two counties in Quebec. In preparation for measures to control the disease, if it became an economic problem, they tested various seed treatments, including heat and fungicides.[180] Both Lachance and Payette continued their research on diseases of flax into the 1950s.

During the Second World War the National Research Council of Canada sponsored some research, by H.R. Sallans, G.D. Sinclair and R.K. Larmour, on the spontaneous heating of flax seed and sunflower seed stored under adiabatic conditions. Their results were published in NRC publication no. 1233, in 1944, and in the *Canadian Journal of Research*, series "F," of that year.

DISEASES OF INDIAN CORN

When North America was first being settled by Europeans, the word "corn" was synonymous with the word "grain" to many of the British, the Germans, and several other nationalities. Consequently, when the first European settlers saw a distinctive new grain being cultivated by the native Indians, they called it Indian corn, meaning Indian grain. Following the American propensity for shortening words, Indian corn is now commonly referred to simply as corn.

Indian corn, botanically known as maize (*Zea mays*), has been an important crop in eastern Canada for more than 150 years. In 1822, John Young (1773–1837) of Halifax, the famed agriculturalist who wrote under the pen name "Agricola," was telling his readers how to select cold-hardy seeds so that corn imported from the United States would become "completely naturalized" in Nova Scotia.[181] In 1857 Robert Russell, writing about the excellent crops of Indian corn that he saw in the Dumfries district of Ontario, did not mention seeing any diseases.[182]

The well-known author and nature-lover Catharine Parr Traill (1802–99) was one of the earliest writers in Canada to mention a corn disease, in the summer of 1883, when she wrote about "a remarkable instance of smut in our corn," and commented on the effect the disease was having on the developing cobs.[183] During the following half century, several authors made more or less casual comments about corn smut in their publications, but there are very few references to research on corn diseases in Canada prior to W.F. Hanna's studies, begun in Minnesota in 1928 and 1929 but continued in Canada.[184] There are equally few references to research on corn diseases between 1928 and the end of the Second World War. One of those few was by Gordon P. McRositie (b.1887), BSA, MSA, PH D, professor of field husbandry at the OAC, who published an account of his studies on the thermal death point of corn from low temperature in 1939.[185]

DISEASES OF FORAGE CROPS

Very early in the development of Canadian agriculture, the value of forage legumes was recognized, largely because their importance in crop rotations had been amply demonstrated in Europe. Nevertheless, there was a tendency for farmers to neglect these crops.

Leading agriculturists were advocating the use of high-quality seed for better production of red clover and other forage legumes as early as 1836. It was during that year that the Central Agricultural Society of Prince Edward Island, "being desirous of inducing Farmers to turn their attention to the cultivation of Red Clover," advertised in the *Royal Gazette* that they had imported, "direct from Boston, 15 casks Red Clover seed." They offered this seed for sale at cost price by way of encouraging local farmers to grow high-quality clover.[186]

The noted Canadian educator-agriculturist John W. Dawson was one of the first authorities in this country to comment on "clover-sick" soil and to suggest ways to prevent the so-called winter-killing of clover. He advocated cutting the grain straw so high that the "scythe or sickle shall not touch the clover leaves." This alone, he said, would help to shelter the clover in winter. Furthermore, he suggested, the ground should be rolled in the spring to press the clover roots into the soil, and clover should not be sown several years in succession on the same land.[187] But comments such as those were merely discussions of the problem. There was little or no research on forage crop diseases until the early decades of the twentieth century.

In western Canada, where agricultural authorities were encouraging farmers to grow alfalfa, there were early reports of problems with dodder parasitizing the plants. In 1909, a grower complained that the alfalfa in one of his cooperative field experiment plots was almost completely overrun with that parasitic plant. Nevertheless, he remained optimistic, and commented, "Given clean seed ... I believe alfalfa will do well in this district."[188]

Periodically, over the years, dodder has posed a threat to crop production. It was such a problem in British Columbia that the first circular to deal with a plant disease, published by the government of that province, was devoted to, and titled, *Clover dodder.*[189] During the Second World War considerable concern was felt when dodder was found on flax in thirty-nine farms in Ontario.[190]

For many years agriculturists debated whether the winter-killing of forage legumes was due to frost damage or fungal diseases. R. Newton and W.R. Brown, working in Alberta, were among the earliest Canadian researchers to indicate that the winter-killing was due to a species of Sclerotinia, rather than frost.[191] In Manitoba, G.R. Bisby, who had written about forage crop diseases in 1925, was assisted by I.L. Conners in a more thorough study of forage legumes. They considered *Sclerotinia sclerotiorum* to be a cause of injury to both clover and alfalfa, and they were the first to notice both powdery mildew and rust of clover in that province.[192]

Downy mildew on alfalfa, *Peronospora trifoliorum,* was included among the most important diseases in the dry belt of British Columbia by provincial plant pathologist J.W. Eastham when he wrote bulletin no. 68 for that province's Department of Agriculture in 1916.

On Prince Edward Island, R.R. Hurst was concerned about mildew on forage legumes and commented in 1930 on the resistance that he observed in varieties and strains of clover to that fungus.[193] Almost twenty years later, William J. Cherewick (1904–83), BSA, MS, who was an agricultural assistant in the Dominion Laboratory of Plant Pathology, Winnipeg in 1936, became the first to report Rhizoctonia rot of sweet clover.[194] Cherewick, who earned a PH D degree at the University of Minnesota in 1943, is best known for his more recent work on diseases of wheat and barley.

The most comprehensive early research on diseases of forage grasses and legumes in Canada was that done by G.B. Sanford and his colleagues, W.C. Broadfoot and M.W. Cormack, working in Alberta. In 1926 Sanford found an undescribed fungus in diseased roots of sweet clover and he later showed that it was capable of producing a root rot in that legume and in related crops.[195] Broadfoot's interest in the diseases of grasses led to a study of the so-called "snow-moulds" that had been causing considerable damage to turf on golf courses, bowling greens, and lawns in Alberta,[196] and he, with Cormack, found one of them to be a low-temperature basidiomycete that caused early spring killing of grasses and legumes.[197] Incidentally, B.T. Dickson had described, in the 1927 annual report of the Quebec Society for the Protection of Plants, a peculiar winter injury to turf which he thought might have been due to snow moulds. One of his references dealt with that topic.

A chapter could be written about the work of Cormack, because he was the one who did most of the early research on forage crops in Alberta. He began that more or less specialized endeavour around 1930 and soon became an internationally recognized authority on forage crop diseases. Cormack published results of his research in the leading scientific journals from that date until long after the period covered by this essay.[198] In 1945 he was named associate plant pathologist at the Edmonton laboratory, and three years later he moved to Lethbridge as officer in charge.

During the decades when so many plant pathologists and others were researching diseases suspected of being due to mineral deficiencies, the forage plants were included in a number of their studies. Chemist C.G. Woodbridge cooperated with plant pathologists H.R. McLarty and J.C. Wilcox in showing that yellowing of alfalfa was due to a deficiency of boron.[199] That was a digression from McLarty's major work in relation to suspected mineral deficiencies in fruit trees in the Okanagan Valley.[200] That early work on boron deficiency of alfalfa in British Columbia was followed up by investigators at the OAC who began some preliminary studies on possible boron deficiency in alfalfa around the end of the Second World War.[201]

During the height of an early interest in plant virus diseases at Macdonald College, R.A. Boothroyd, in his research on mosaic diseases for a master's degree, included experiments on clover even though most of his research was on tobacco.[202] Also at Macdonald College, Joseph N. Bird (b.1899), BA, MA, MSC, assistant professor in agronomy, studied the improvement of red clover for several years in the latter part of the 1940s.[203]

There has never been as much interest in the diseases of forage grasses as in those of forage legumes. However, some early research was done on diseases of grass by investigators in various parts of the country. Probably the first of them was W.P. Fraser, who studied the rusts of both wild and domesticated grasses in Nova Scotia in 1906.[204] Fourteen years later he was doing research on a smut disease of a wild grass in Saskatchewan,[205] where his colleague T.C. Vanterpool studied root rots of various grasses.[206] Vanterpool's graduate student at the University of Saskatchewan, John T. Slykhuis, investigated diseases of brome and crested wheat grass for his master's thesis in 1943. Slykhuis published, in 1947, his studies of a blight of crested wheat and brome grass seedlings.[207]

On Prince Edward Island, R.R. Hurst was attempting to select strains of timothy grass that would be resistant to rust, late in the 1920s,[208] and in 1934 J.N. Bird was assessing the influence of rust injury on the vigour and yield of timothy grass at Macdonald College.[209] On the other side of the country, Walter Jones (1890–1957), B SC, M SC, plant pathologist at the Dominion Laboratory of Plant Pathology, Saanichton, British Columbia, who had established himself as the authority on mildew and other diseases of hops in the 1930s, made a relatively small study of grass diseases late in the 1940s and reported a previously unknown downy mildew on orchard grass in 1949.[210]

CONCLUSION

One of the truly great accomplishments in Canadian agriculture has been the conquest of smut and rust. For many years they were the major diseases of grain crops. Although the conquest was largely based upon a knowledge of the life cycles of the causal organisms, most of which had been worked out by European investigators, Canadians, in conjunction with their American counterparts, played important roles in determining the genetic makeup of the causal fungi. More important, they made good use of the knowledge they had acquired in the development of disease-resistant grain crops, and to a lesser extent, forage crops.

By the middle of the twentieth century, the causes or the inciting factors of the major diseases of grain and forage crops had been determined, as had most of the basic principles for their control. In the continuing search for means of disease control, studies were being made of the nutritional and optimal temperature requirements of disease-inducing fungi. For example, J.D. Gilpatrick of the Ontario Agricultural College and A.W. Henry of the University of Alberta cooperated in a study of the effect of nutritional factors on the development of *Ophiobolus graminis*,[211] while L.E. Lopatecki, in a branch laboratory of the Dominion Department of Agriculture's Division of Chemistry, at Saanichton, British Columbia, was studying the nutrition of species of Phytophthora.[212] In other words, a firm foundation had been laid for further developments in grain and forage crop disease control that could be built upon in the second half of the century.

5 Early Forest Pathology in Canada

Prior to the first settlements by Europeans, the land mass that is now Canada was sparsely inhabited by nomadic and semi-nomadic people who, lacking metal tools, wheeled vehicles, and beasts of burden, made very little impact on the vast forests.

To the pioneer European settlers the forest was a mixed blessing. Although it provided materials for their shelter and warmth it also harboured livestock predators and was a great impediment to the development of agriculture and to communication by land. The forest was often looked upon as something to fight against and to burn, rather than as a valuable asset to be protected. The treeless and barren aspect of large areas of the south shore of Nova Scotia is directly attributable to the burning of trees and soil by people searching for gold.[1]

The conviction that the forest was an obstacle to the development of the economy began to change late in the eighteenth century and the early decades of the nineteenth, when a growing shortage of timber in Europe stimulated a few Europeans to take an interest in the great forests of North America. The French took an early lead in the importation of Canadian timber. They were soon followed by the British, who were in dire need of timber when their traditional supply from the Baltic region was cut off when the French, after the defeat of their navy by Nelson at Trafalgar, attempted to bring down the British economy by excluding British shipping from European ports.[2] By the third quarter of the nineteenth century the forests of eastern Canada had become so depleted of their prime timber trees that people began to express concern about the lack of plans for the future of this great resource. An example of this concern in the minds of some leading agriculturists may be seen in the 1879 report of the Ontario Fruit Growers' Association, in which it was declared that members should "put forth their best efforts to husband our provincial and Dominion resources in their timber limits."

All members of the Fruit Growers' Association were not owners of orchards. Many citizens whose main interest was the general advancement of agriculture, in its broadest sense, were enthusiastic members. Entomologists, mycologists, botanists, and others from Canada's fledgling scientific community tended to join progressive associations, such as that of the Fruit Growers, so their voices would be more clearly heard. At that time the Ontario Department of Agriculture published the proceedings of the fruit growers' Association with its own annual reports. Such free publicity gave speakers at association meetings a very wide audience.

When the great explorer-naturalist John Macoun addressed members of the Royal Society of Canada about Canada's forest resources in 1894, he remarked that "it seems the aim of governments and individuals to annihilate [them] as quickly as possible." He went on to warn that when the public "awakens to the truth it will be appalled at the enormous waste and loss that has been going on for more than a generation."[3] Such comments were manifestations of a remarkable wave of public enthusiasm for forest conservation that expressed itself in the form of fire prevention legislation, tree planting on "arbour days," and the establishment of tree nurseries and plantations. Additional evidence of the rising interest in forest conservation is seen in the prizes awarded for the best essays on forestry by the Agricultural and Arts Society of Ontario in 1881, and in the appointment, by the government of Ontario, of three delegates to attend the American Forestry Congress held in Montreal, Quebec, 21–3 August 1882.[4]

One of those delegates was William Saunders (1836–1914), who later became director of the Central Experimental Farm, Ottawa, and whose concern for the preservation of Canada's forest heritage became well known among influential people when he published a paper the following year on that general theme in the first volume of the *Proceedings and Transactions of the Royal Society of Canada.* Sixty-one papers were listed on the program of the congress, and at least eighteen of them were by Canadians. Saunders read a paper titled "On the growth of poplar trees for manufacture of paper and charcoal." There was only one presentation, by an American, that dealt with tree diseases, and it was largely mycological.[5] Obviously, the pathology of forest trees was of little interest to that group.

It may be useful to recall here that forest pathology did not develop as a distinctive science until after Robert Hartig of Munich, Germany, began publishing on the topic late in the nineteenth century. Hartig has been referred to as "the father of forest pathology" because of his pioneering studies on the causes of tree diseases and how to control them. Most of his work in this field was done between 1870 and 1900.[6] Thus, one cannot expect to find many references to tree diseases in Canada, other than mere observations, before the beginning of the twentieth century.

One of the earliest published comments about tree diseases in Canada was by James Monk, who, in writing from Halifax, Nova Scotia, in September 1757, remarked that, on some islands near Halifax, "all the forest trees (except the evergreens) appeared scorched."[7] Another early one was John McLean's 1836 report that the woods in New Caledonia "are decaying rapidly, particularly several varieties of fir, which are being destroyed by an insect."[8]

Prior to about the second decade of the twentieth century, most of the tree research in Canada was mycological in nature rather than pathological. Many species of fungi found on living or dead trees, or on wood, were described and named, but they were seldom studied as causal agents of disease or decay. Nevertheless, those early studies were important preliminary steps to the emerging field of forest pathology, the science which, when considered in its broadest sense, has as its objective the production and maintenance of a healthy population of trees. It includes not only the prevention and control of diseases of forest, shade, and ornamental trees, but also the microbial deterioration of wood. For various reasons, it does not normally include the diseases of orchard trees, except when one is a host or an alternate host of a fungus that attacks forest trees.

In Canada, prior to 1950, forest pathology did not included tree surgery, or much research on the treatment of wounds, even though they have long been known to be portals of entry for various saprophytic and parasitic fungi. This lack of interest in the treatment of wounds is strange, because as early as 1910 the Rev. Robert Campbell, an influential amateur mycologist in Montreal, was urging "wood rangers to have their eyes open for wounded or diseased trees and take steps to protect any wounded tree by painting the affected part." He further suggested that maple sugar makers "should see that the notch or auger hole by which they tap their maples [is] afterwards so treated as to ward off the attack by fungi."[9]

When the experimental farms were established in 1887, their organization largely followed the recommendations of William Saunders, one of which was: "At the central station there would be required, in addition to the director, a ... superintendent of forestry, who would direct all forestry experiments, and inquire into all questions relating to tree culture and tree protection in the Dominion."[10] One year later

William T. Macoun (1869–1933), son of the botanist John Macoun, was appointed as an assistant to Saunders and foreman of forestry. Young Macoun was an advocate of forest conservation, especially farm forests or woodlots. His enthusiasm for this aspect of conservation is seen in the illustrated address that he delivered to members of the Farmers' Institute of Ontario, the text of which was published as an appendix to the annual report of the Ontario Department of Agriculture for 1897. Although Macoun had no university training, he did much to develop an arboretum at Ottawa; however, he did little, if any, research on forest tree diseases. After becoming Dominion horticulturist in 1898, he made some valuable contributions to plant pathology, but not in the realm of forest tree diseases. It is noteworthy here that the first Bulletin from the Central Experimental Farm was one dealing with the treatment of forest tree seeds to prevent them from becoming infected by fungi, and the second one also dealt with forestry. They were both written by Saunders and published in 1887.

Before the prairie provinces were founded – Manitoba in 1870, Saskatchewan and Alberta in 1905 – and until 1930, the federal government was responsible for the protection and administration of the forests in those areas. One of the early steps in doing this was the organization of the Canadian Forestry Service, which began with the appointment of Elihu Stewart as chief inspector of timber and forests in 1899.[11] Within a year of his appointment, Stewart called together a group of interested men to consider the formation of an association for the promotion of forestry in Canada. As a result of their recommendations the Canadian Forestry Association was formed, and its first meeting, chaired by Sir Henri Joly de Lotbinière, was held in Ottawa, 8 March 1900. A major objective of the association was to inform the public of the value of forests, and the importance of protecting them from fire, insects, and diseases.

Up to 1900, no technically trained foresters were employed anywhere in Canada. However, it was around this time that Norman M. Ross (b.1876), a BSA graduate of the Ontario Agricultural College (OAC), went to the United States to inquire about technical forestry in the New England forest area. In 1901 he was appointed chief of the Division of Tree Planting, in the Dominion Forestry Branch, Department of the Interior, Ottawa, where he supervised the production and distribution of trees for the western provinces.

In 1903 the first sods were turned for the establishment of a Dominion Forest Nursery Station two miles southwest of the town of Indian Head, Saskatchewan. The first buildings were erected there during the following year. Eight years later, another forest nursery station was established at Sutherland, Saskatchewan, as a branch of the one at

Indian Head. In 1931, after the federal government transferred juris-diction of the natural resources of the prairies to the provinces, those stations became part of the Dominion Experimental Farms System.[12]

It was Ross who provided a history of tree planting on the prairies, in Forestry Branch bulletin no. 1, in 1910, and later, in his 1923 book, *The tree-planting division: Its history.* Ross retired in 1941, after nearly forty years of service as superintendent of forest nursery stations in western Canada, with headquarters at Indian Head. He was succeeded by John Walker, who had been professor of horticulture, University of Manitoba.

During 1907 and 1908 a large provincial nursery was established in Quebec, at Berthierville, a town on the north bank of the St Lawrence River about midway between Montreal and Trois Rivières. Another tree nursery of that decade that is worthy of mention here was owned and operated by the Canadian Pacific Railway, at Wolsely, Saskatchewan. That nursery produced the 1,356,200 trees that were planted along the main railway line between Winnipeg and Calgary before the end of November 1912.[13]

Those, and the many other tree nurseries that were in operation in Canada prior to 1945, must have experienced disease problems, but, except for R.W. Lyons's comment on the use of formalin and hydro-chloric acid in the sterilization of soil to prevent the damping-off dis-ease of tree seedlings,[14] René Pomerleau's discussion of the cause of damping-off and other seedling diseases,[15] and L.P.V. Johnson and G.M. Linton's account of their experiments on chemical control of the damping-off disease of pine seedlings,[16] nursery diseases were seldom mentioned in the early Canadian literature of forest pathology.

Elihu Stewart was a strong advocate of schools for the training of for-esters. As a result of his appeal, and those of many like-minded people, two forestry schools were established, one in Toronto, Ontario, and one in Fredericton, New Brunswick, in the first decade of the twentieth century. The one in Toronto got approval for its establishment from the Board of Governors, University of Toronto, on 14 February 1907, and in the following month, Bernhard E. Fernow (1851–1923), a former colonel in the Prussian Army who had studied forestry in Germany, was appointed as its dean. From the standpoint of a history of forest pathology, it is noteworthy that the original curriculum for the 1907 course in forestry included, among other topics, diseases of trees.

Forestry instruction began at the University of New Brunswick in October 1908, with Yale graduate Robert P. Miller (1875–1965) in charge of the forestry department. Graduates of those schools, and ones that soon followed at Laval University and the University of British

Columbia,[17] added their voices to those of a growing number of individuals expressing concern over the rapid depletion of the nation's forests through exploitation, fire, and the ravages of insects and disease. Their calls for action were usually in that order of descending concern. The control of forest tree diseases was at the bottom of their list of priorities, even though all of the new departments, schools or faculties of forestry included forest pathology in their course of studies.

In the spring of 1898, Manning W. Doherty (1875–1938), BSA, MA, associate professor of natural science at the OAC, described a fungus from diseased fir trees around Guelph.[18] Doherty did not do much research on that disease, however, because he became interested in politics and resigned his professorship in 1902.

One of the most comprehensive of the early accounts of tree diseases in Canada was that written by William P. Fraser (1867–1943) when he was teaching mycology and plant pathology at Macdonald College, Quebec. In 1912, Fraser, already an authority on the rust fungi, described eighteen tree diseases induced by rusts, and an additional eight or nine due to other fungi. He also named several fungi that attack dead or dying trees and suggested methods for the control of certain parasitic species.[19]

Fraser's work was useful in drawing the attention of mycologists and plant pathologists to a number of relatively unknown tree diseases, but the man who provided much of the early stimulus for Canadian students to study forest tree diseases was Joseph Horace Faull (1870–1961), a citizen of the United States who got most of his early education in Ontario, including a degree in arts from the University of Toronto in 1898. Faull later studied mycology at Harvard, under Professor R. Thaxter, for the PH D that he obtained in 1904. In the meantime he had accepted a lectureship in botany at the University of Toronto in 1903, where he continued his teaching until he resigned to accept a professorship at Harvard in 1928.

While at Toronto, Faull became so interested in tree diseases that he went on leave in 1909–10 and studied forest pathology for several months with Robert Hartig in Germany. He also attended summer courses in forestry at the College of Forestry, Syracuse University. Those studies were soon followed by a series of publications on tree diseases and their causal fungi, the research for several of which was done in the Lake Timagami Forest Reserve of Ontario. It was on that reserve, in cooperation with the provincial and federal governments, that Faull began a survey of forest tree diseases and wood-rotting fungi, and it was there that he established a field laboratory in 1918. In the early phase of those studies, Faull was surprised to discover that many of the common tree diseases on the reserve had not been researched at all, and that very little work had been done on some of the most destructive ones, particularly the butt rots.[20]

In the relatively short interval between the time Faull studied forest pathology with Robert Hartig in Germany and the date he left the University of Toronto, he had more graduate students whose theses dealt with forest pathology, including wood decay, than did all the other schools of forestry in Canada for the first two decades of their existence. Faull saw great economic potential in the large stands of spruce and fir in eastern Canada, and his interest in this regard was reflected in the thesis research of his graduate students, three of whom, namely Lillian M. Hunter, Hugh P. Bell, and Ezra Henry Moss, became widely known for their mycological work pertaining to the rusts of fir trees.

Other workers with an interest in the diseases of fir trees at about this time include Ronald Elisha Stone (1881–1939) and William H. Rankin (b.1888). Stone, an associate professor of botany at the OAC, studied witches' broom of Canada balsam while investigating the incubation period of *Cronartium ribicola* on white pine.[21] Rankin studied butt rots of the balsam fir in Quebec during his brief term as officer in charge of the Laboratory of Plant Pathology, St Catharines, Ontario, and while writing his book, *Manual of tree diseases*, published in 1923. He asserted that the so-called butt-rot disease of fir trees is a complex of diseases caused by many different fungi.[22] It was also in 1923 that N.L. Cutler published a list of twenty-four species of fungi that he considered to be important tree destroyers in the Vancouver forestry district, and another two dozen species associated with timber rots found there.[23]

In 1927 and 1928, Frank Dickson (1891–1969), assistant professor of botany, University of British Columbia, conducted a study of Alpine fir in the upper Fraser River region on behalf of the provincial forestry service. His report, which included a preliminary survey of Douglas fir diseases on Vancouver Island, constituted the first detailed pathological analysis of over-mature timber in British Columbia.[24] Dickson included forest pathology in his teaching of plant pathology until that subject became a separate course in 1947. He then taught the new course 317, Forest Pathology, for two or three years prior to the appointment of Donald C. Buckland as professor of forest pathology in 1949.In 1932, Richard Mayers, one of Dickson's graduate students, earned his MSA degree largely through research on a fungus causing conifer wood rot. His thesis was titled "Studies on the biology of *Echinodontium tinctorium* E. and E."

Shortly before this, Herbert W. Eades (1893–1969), a Canadian graduate of the University of Washington who had served in the air arm of the Royal Navy during the First World War, began writing about fungus-induced rots and other natural defects of coniferous trees and wood in British Columbia. His valuable 126-page monograph on those topics, written as much as possible in non-technical language, was published in 1932.[25] Eades had been appointed to the Forest Products Laboratory, Vancouver, in 1923, the first forest products pathologist in western Canada. He continued to write about the cause and prevention of wood decay and stains in wood through and beyond the Second World War. Eades was virtually alone until 1940 when P.J. Salisbury became his assistant for nearly four years. He was then alone again until J.W. Roff became his assistant in 1948.

One of Faull's early graduate students was Alan Wilfrid McCallum (1893–1967), who, after receiving his MA with a thesis on the parasitism of trees by the fungus *Armillaria mellea* in 1918, joined the Division of Botany, Dominion Department of Agriculture, in 1920, as its first forest pathologist. During that year he presented a paper to members of the Canadian Division of the American Phytopathological Society on a new disease of poplars in Ontario and Quebec.[26] As a member of the staff of the Division of Botany, McCallum wrote the sections on forest pathology and white pine blister rust that were published with the annual report of the Dominion botanist in 1921. In August of that year he made a field study of the root rots of trees in northern Ontario,[27] and during the summer of 1923 a similar study of decay in balsam fir in a large area of Quebec.[28] McCallum's continued interest in the diseases of fir trees, acquired from Faull, is shown in the first bulletin of a series dealing with forest pathology produced by the federal Department of Agriculture in 1928,[29] and in his 1940 paper, with Carl Heimburger, on balsam fir rot.[30]

McCallum and G. Aleck Ledingham (1903–62), B SC, M SC Saskatchewan, PH D Toronto, investigated the alleged damage to trees and other vegetation which farmers in the state of Washington claimed was due to smoke and fumes from the smelters at Trail, British Columbia.[31] Their studies roused the interest of Nora L. Hughes, a student at the University of British Columbia who researched the effects of sulphur dioxide on coniferous trees for her MA degree in 1934.

Faull did not do much research on diseases of spruce trees, which, he said, "grow like weeds and tend to overrun areas cleared for agriculture,"[32] and neither did anyone else in Canada prior to the 1940s. Robert Bell, in his book *The Forests of Canada*, published in 1886, had written about a disease of the spruce trees in New Brunswick that caused great havoc in the forests of that province. However, except for his comment that the disease was "supposed to be due to a fungus that attacks the roots," he gave no description of the disease. The research of R.B. Thomson and H.B. Sifton, in their study of the resin canals of spruce trees, was not directly related to disease, although they made an oblique reference to it in 1925 when they showed that some of the canals were of traumatic origin.[33]

In 1938 the British Columbia Forest Service requested the Division of Botany and Plant Pathology, Ottawa, to assign a forest pathologist to that province. Consequently, John E. Bier (1909–67) of that division was transferred from Ottawa to the west coast for periods of study in 1938 and 1939. When it was decided that his position there should become permanent, Bier became the first, and for a while the only, forest pathologist in the new forest pathology laboratory that was established in Victoria in 1940. That event is noteworthy because it provided an opportunity to give continuing on-site attention to forest diseases and to plan an orderly program of investigation aimed at their solution. Although Bier was expected to assume responsibility for the investigation of forest tree diseases in all of British Columbia, the size of the province and its mountain ranges, plus the lack of technical assistance, restricted his activities to Vancouver Island and relatively small areas of the lower mainland.

During the early part of the Second World War it became recognized that British Columbia contained the major supply of spruce wood suitable for aircraft production in the British Empire. Therefore, efforts were made to guarantee a maximum supply of the highest quality. Those efforts stimulated more interest in tree diseases and, indirectly, permitted Bier to have technical assistance.

Donald C. Buckland (1916–56), a recent PH D graduate of Yale University, became Bier's research assistant in 1942, and Raymond E. Foster (b.1919), BA, who had logging experience with the British Columbia Forestry Service, became Bier's assistant in 1943, shortly after graduating from the University of British Columbia with a BSF degree. He became interested in the relation of research in forest pathology to the utilization of over-mature timber, and began publishing on that subject in the *B.C. Lumberman* in 1946, the year in which he had three articles in that journal, as did Bier in joint authorship with others. Foster's studies of the decay relationships of western hemlock (*Tsuga heterophylla*) on the Queen Charlotte Islands became the research topic for the PH D degree that he was granted by the University of Toronto in 1949.

Philip J. Salisbury, a 1939 graduate of the University of British Columbia who had been working in the forest products laboratory with Herbert Eades for three or four years, also joined the team as Bier's assistant, sometime in 1943–44. Within the next two years they were joined by R.A. Waldie and G.P. Thomas, who were to study decays in forest trees.

The British Ministry of Supply was particularly concerned about a decay of Sitka spruce on the Queen Charlotte Islands known locally as "pocket rot," about which there were no references in the literature of tree diseases. Bier and his first two assistants worked on that problem and published an account of it in Canada Department of Agriculture technical bulletin no. 783, "Studies in forest pathology. iv Decay in Sitka spruce on the Queen Charlotte Islands," in 1946. Bier and Mildred Nobles published a paper on brown pocket rot of Sitka spruce that same year.[34] One year later Bier left British Columbia to head a new laboratory of forest pathology in Toronto.

After Faull became professor of forest pathology at Harvard, he returned to Canada many times to continue the surveys and research that he had initiated while in Toronto. In the latter half of July 1929, at the request of the Nova Scotia Department of Lands and Forests, he made a "pathological reconnaissance" of some forested areas of that province, and became the first to report white pine blister rust in Nova Scotia and the first to find Vaccinium rust of hemlock in North America. He also found more than a dozen diseases of trees that had never been reported in Nova Scotia.[35] He made a further important contribution to Canadian forest pathology when he published the results of his research on the disease of conifer needles known as snow blight, and the effective control measure that he proposed.[36] His studies on conifer needle rusts culminated in a monograph on the genus Uredinopsis in 1938.[37]

PINE TREE DISEASES

Probably the greatest stimulus to the slowly rising interest in forest pathology in Canada was the discovery that the dreaded white pine blister rust fungus, *Cronartium ribicola*, had somehow got into this country. The life cycle of that fungus had been largely worked out by European biologists before it was found here. They had learned that, like many other rust fungi, the white pine blister rust fungus passes part of its life cycle on another host plant. In this instance the primary host is a species of the genus Ribes, especially cultivated black currants, but also other kinds of currants, and wild and cultivated species of gooseberries.

The fungus was first discovered in Canada on some black currants in a new currant plantation at the OAC, near Guelph, Ontario, in September 1914. David Richmond Sands (1883–1966), a student at the college, found some infected leaves of currants while making the collection of plant disease specimens required for a course in plant pathology.[38] Although Sands became a member of the botany staff at the OAC, it is unlikely that he, as a young student, would have associated the infected currant leaves with the white pine blister rust. Consequently,

John E. Howitt (1881–1966), BSA, the botanist who taught the course, identified the disease and took credit for being the first one to find it in Canada.[39]

The Ontario Department of Agriculture responded to this discovery by calling together, in 1915, various people who were presumed to have some knowledge of white pine blister rust, or currant rust, as it was also known. Those attending the conference, in addition to Howitt, were Professors L. Caesar, entomologist at the OAC, W.A. McCubbin, assistant in charge of the Dominion Laboratory of Plant Pathology, St Catharines, and E.J. Javitz, provincial forester for Ontario. As a result of the recommendations that came out of that conference, the province of Ontario included the white pine blister rust fungus in the terms of its Fruit Pest Act, sent a circular letter to all nurserymen in the province prohibiting the sale of young pine trees, and initiated a survey to ascertain to what extent the disease had spread.[40] The federal government, after some prodding by the Dominion botanist, Hans T. Güssow (1879–1961), had wisely added the blister rust fungus to the list of organisms and diseases to be excluded from Canada by the Destructive Insect and Pest Act, through an Order in Council on 27 February 1911, nearly three years before it was found in Canada.

One of the reasons for the quick legislative response to the presence of blister rust lies in the fact that many influential people had been making the populace aware of the threat that this disease posed to Canadian forests. One of the most learned and persuasive of these was Hans Güssow. As early as 1910 he had communicated with officials at the OAC, warning them of the danger associated with their importation of more than 250,000 white pine seedlings from Germany. The authorities there assured him that all possible precautions were being taken and that all species of Ribes within a considerable distance were eradicated.[41] Güssow, who had studied in Germany, knew that white pine blister rust had caused severe damage to forests in that country, and he was afraid it would become established at or near the OAC, which it did. His annual report for 1911 included a useful descriptive commentary on white pine blister rust.

Several million white pine seedlings had been imported from Europe into the United States and Canada before it was declared illegal to do so. Most of those importations took place between 1906 and 1908, although many seedlings had been brought in earlier. Europeans had imported white pine, *Pinus strobus*, from North America and, because of cheap labour, were able to export pine seedlings to North America at less cost than they could be grown for in Canada or the United States.[42] Professor W. Lochhead of Macdonald College warned of the threat of an invasion of blister rust from the United

States, shortly after it was found there.[43] Canadian authorities were aware of the devastating effect that another fungus from Europe, *Endothia parasitica*, was having on valuable stands of chestnut trees in the United States. The publicity associated with that disastrous disease, and the fact that it had been found in British Columbia in 1913,[44] were additional motivating factors for prompt legislative action on white pine blister rust.

Before about 1900, relatively few tree species other than red and white pine were being cut for lumber in the forests of Ontario, and the rafting of pine timbers on the Ottawa and St Lawrence rivers was part of such a colourful era in the history of Canadian forestry that it became a long-lasting component of the folklore associated with the lumbering industry. Therefore, any menace to pine trees roused emotions in the general public that were beyond mere economic considerations.

As with most plant diseases, the early work on white pine blister rust in Canada consisted largely of field surveys to learn how far the disease had spread. Shortly after the infected currants were found in Guelph in 1914, a hurried survey was undertaken by Walter A. McCubbin (b.1880), MA, assistant plant pathologist in charge of the Dominion Field Laboratory of Plant Pathology in St Catharines, Ontario. He found the disease to be generally present throughout a large part of the Niagara peninsula.[45] The search for diseased currants and infected pine trees was extended in the following year, during which the rust was found on native white pine in several widespread areas. In 1916 thousands of infected currants and pine trees were eradicated, because at that time it was believed the disease could be eliminated by this procedure. The eradication work was continued through 1917, and, in addition to this general work, a strip one mile (1.609 kilometres) wide along the Niagara River was cleared of all species of Ribes. The American authorities cooperated and created a similar Ribes-free zone on their side of the river.[46] This was done in an attempt to prevent the rust from spreading from Ontario into New York State, even though it had been found farther south in that state several years before it was discovered in Ontario.

During August and September of 1917, inspectors, with the help of children from the public schools, found the disease in thirty-eight of the forty-three counties surveyed in Ontario.[47] When that became known, it was realized that white pine blister rust could not be eradicated from southern Ontario. After that, the work was largely confined to a number of control areas five to ten acres (approximately two to four hectares) in size. In those, and for a distance of five hundred yards (457.2 metres) around each one, all species of Ribes were removed, and it was intended that the pine trees in the control areas would be kept under close observation for at least ten years. The objective of that exercise was to see if young pines isolated from Ribes could be grown free of the rust.[48]

In 1916, the blister rust fungus was discovered in the province of Quebec by Professor W.P. Fraser, who found it on gooseberry bushes at Ste-Anne-de-Bellevue and at Oka.[49] This new discovery prompted a limited survey of forested regions in Quebec that year, with emphasis on areas near nurseries that had obtained pine seedlings from Guelph. This was followed by a more extensive search for the disease in 1918. As had been done in Ontario, the federal government, with a special appropriation under the Destructive Insect and Pest Act, defrayed the travel expenses of the surveyors, and members of the respective provincial forest services undertook the actual scouting. The disease was not found on pine trees in 1917, but Henri Roy (1889–1941) of the Quebec Forest Service reported that "at least 275 cases of distinct infection" were found on species of Ribes.[50]

Finding the disease so widespread prompted the convening of an international forest conference in Washington, DC, early in 1917 to consider particularly the danger to the white pine in the United States and Canada. Recommendations from that meeting, and another later in the year, resulted in the convening of a committee representing the various interests, both governmental and private, to study the whole situation in Canada.[51] In 1922 the experiments in forest pathology being carried out by the Division of Botany of the Dominion Department of Agriculture included ones on needle blight of white pine and on white pine blister rust.[52]

For several years blister rust appeared to be confined to the pine-growing areas of eastern North America, but westerners prepared themselves for its invasion there by convening an International White Pine Blister Rust Conference for Western North America, in Portland, Oregon, 23–4 April 1919. At that meeting resolutions were passed regarding quarantines and other actions to be taken if the disease appeared west of the Rocky Mountains. The delegates also set up a "watch committee" to be on the outlook for the disease.[53]

In September 1921, some leaves and twigs of black currants that appeared abnormal to a grower in North Vancouver, British Columbia, were brought to the office of John William Eastham (1881–1968), the provincial plant pathologist, with a request for information. On those leaves, Eastham found what he believed to be the telial stage of *Cronartium ribicola*, the incitant of white pine blister rust. To be more certain of his diagnosis, he sent a sample to Güssow, in Ottawa, for confirmation. In the meantime he wondered what should be done in view of his superiors' advice to go slow in publishing or otherwise announcing the presence of the dreaded fungus. He did not have to wonder very long; within a few days a professor at the University of British Columbia discovered infected currants in his garden and informed his brother, in Oregon, who was a member of the watch committee set up at the 1919 meeting.[54]

The knowledge that the blister rust fungus had been found in British Columbia provoked a series of events, including the convening of another international White Pine Blister Rust Conference in Portland, Oregon, 18–20 December 1921. The Canadian delegation to that meeting included A.W. McCallum, the Dominion government's only forest pathologist, R. Cameron, Dominion forest officer, W.H. Lyne, officer in charge of port of entry inspections for British Columbia, and J.W. Eastham. Conference members soon formulated plans for the eradication of Ribes and infected pine trees. Naturally, British Columbia, the only west coast culprit at that time, was to be the scene of the proposed action. But, to the bewilderment of their American colleagues, the Committee on Canadian Plans brought in a minority report which stated that because white pine in British Columbia was of relatively little economic importance, "we cannot agree to recommend extreme action in the eradication of cultivated black currants at the present time."[55]

From Ottawa, Güssow, whose responsibility included all plant diseases in Canada, sent Alec T. Davidson, of the Division of Botany, to British Columbia in 1922 to cooperate with Eastham and other authorities there in supervising a survey and an assessment of the problem. The surveyors also made an effort to determine the source of the disease. In so doing they traced the records of all importations of pine seedlings between 1910, the earliest date for which records were available, and 1914, when importation was forbidden by federal quarantine regulations. Only one plantation of imported trees was found to have diseased pines. It belonged to Thomas Newman, who had imported one thousand white pine seedlings from France in 1910.[56]

Apparently Davidson's report to Güssow supported the contention of the Canadian committee regarding the relative importance of white pine and black currant; in his report to the minister of agriculture that year, Güssow stated that, because white pine was of relatively minor importance in British Columbia, it would not be a sound economic policy to spend money in controlling the disease there. He probably took into consideration the fact that lumbermen in British Columbia were much more interested in Douglas fir, western hemlock and certain other tree species than they were in the relatively few known pockets of white pine, and that black currants were highly regarded and extensively cultivated in that province. J.W. Eastham told of an irate woman who stopped a federal employee in the act of removing a currant bush from her garden in Revelstoke, and of the acrimonious correspondence between her and federal officials after she billed the minister of agriculture fifty dollars for damages to her garden. Those currant bushes and her claim were eventually debated in the House of Commons.[57]

When Davidson was killed in a crash between his automobile and a Vancouver street-car, 26 February 1927, all white pine blister rust surveying in British Columbia was abandoned, as far as Canada was concerned. But, as a partial appeasement, the American authorities were given a tract of woodland in British Columbia, containing a stand of easily accessible white pine that was beginning to show evidence of blister rust infection, in which to conduct experiments on the development and control of the disease. They were also given special permits to bring into their test area various species of Ribes and pines for susceptibility and resistance testing. From their Portland, Oregon, headquarters, more than a dozen American scientists were sent to British Columbia to study the blister rust problem. Some of those men were there for only short-term learning experiences, but several worked more than one full summer, between the late 1920s and about 1936. They were not restricted in their activities to the specific plot areas that had been assigned to them. For example, at one time, J.L. Mielke was assigned virtually all of the interior of British Columbia for a study of the spread of the disease, while C.N. Partington conducted a similar study in the coastal area and Vancouver Island. T.S. Buchanan was in charge of plots at Hunters Siding, W.F. Cummings had plots near Nelson, and A.A. McCready was at one time in charge of a large plot near Revelstoke. J.W. Kinney and C.J. Nusbaum were primarily responsible for studying the disease on Ribes, although they had other responsibilities.[58]

The major positive action taken by the British Columbia government was to prohibit the movement of either pine or currants from infected to non-infected areas within the province, by legislative action that became effective 1 March 1922.[59] After the disease was known to be well established in some of the west coast states, the American effort in British Columbia was gradually phased out. However, the friendly, cooperative working relationships established at that time between the Canadian and American forestry people did not diminish.

By 1922, white pine blister rust had been found in the three Maritime provinces, Quebec, Ontario, and British Columbia.[60] For several months following the initial discoveries of the disease in Ontario, there was hope that it could be eradicated, but W.A. McCubbin, who was in charge of the early studies on blister rust there, did not sound optimistic about this when he addressed members of the Quebec Society for the Protection of Plants at Macdonald College, 14 March 1916.[61] In his talk he told of efforts to control the disease by searching for and destroying infected pines. He was then working on the theory that the disease-inciting fungus did not over-winter on species of Ribes and that therefore it was wiser and more economical to destroy the relatively few infected pines than to try to find and destroy all wild and cultivated species of Ribes. He also knew how difficult it would be to persuade gardeners to get rid of the highly susceptible but well-liked black currants in their gardens. For this reason, he and his colleagues experimented with various spray mixtures to control the disease on currant bushes while searching for and eradicating infected pine trees.

McCubbin soon realized that although spraying might protect individual plantations, it was unlikely to be of much use in combating the disease as a widespread pest. He studied how far the spores of *Cronartium ribicola* were likely to be dispersed from infected Ribes, and calculated, after measuring their fall in still air, that they could be carried more than two miles (3.2 kilometres) in five minutes if the wind was blowing thirty miles (48.28 kilometres) an hour. He also conducted experiments in an effort to determine how long the spores could remain viable in daylight and in darkness, after being discharged from infected Ribes.[62]

Except for that early work by McCubbin, very little attention was given to the control of rust on currant bushes in Canada until it was revived by J. Frederick Hockey (1895–1980), in Nova Scotia, some ten or eleven years later. Hockey, a 1921 BSA graduate of Macdonald College, who became pathologist in charge of the Dominion Field Laboratory of Plant Pathology, Kentville, Nova Scotia in September, 1924, showed that several fungicides could control the disease on currants.[63] Incidentally, it was Hockey, in 1949, who made the first plant disease survey in Newfoundland.

A search for varieties to use in breeding black currants for rust resistance was begun in Ottawa in 1935, and that breeding and selecting program continued until at least 1943, when A.W.S. Hunter and M.B. Davis published a paper that outlined the work.[64] It was not mere coincidence that their research followed shortly after the discovery of the high vitamin C content of black currants and the consequent market demand for more berries, plus increased reluctance on the part of gardeners to remove them.

Except for the federal program of forest tree breeding that started in 1938, which included a project to develop strains of white pine resistant to blister rust, there were few efforts to develop a rust-resistant white pine prior to the 1940s. In the first decade of the twentieth century, Sir Henri Joly de Lotbinière planted walnut and white pine, from

Frank Dickson (1891–1969)

J.H. Faull (1870–1961)

various sources, on his seigneury near Pointe Platon, Quebec, and some of those pine trees – ones that were free of blister rust in 1940 – were used as parents in breeding rust-resistant trees.[65]

When McCubbin resigned on 31 May 1919, to accept a position with the Pennsylvania Department of Agriculture, Paul Murphy (1887–1938) became officer in charge of the St Catharines Laboratory for a few months until William H. Rankin (b.1888) was appointed to that position in October of the same year. Both men continued McCubbin's white pine blister rust work, and employee George H. Duff (1893–1958) began to play a more prominent role in it. While doing some basic research on the blister rust fungus, initially under McCubbin's direction, Duff exposed some of its spores to light from an ultraviolet arc lamp and concluded that any "toxic effect" sunlight might have on the viability of the spores was due to its ultraviolet component.[66] Duff, who had BA and MA degrees from the University of Toronto, became professor of plant physiology in the Department of Botany of that university in 1919, and earned a PH D there in 1922. At Toronto, he continued the blister rust investigations as a cooperative project with his former supervisors at St Catharines. Duff supervised the work of a number of young men who were employed as surveyors, both for diseased pine trees and for species of Ribes to be eradicated, and one of those men was Ibra L. Conners (1894–1989), who thus began what was to become a long and noteworthy career in plant pathology.[67]

In the meantime, A.W. McCallum, under the authority of the Dominion botanist, had been given nominal supervision of all white pine blister rust work sponsored by the federal government. His brief history of blister rust in Canada, up to 1921, was published in the annual report of the Dominion Department of Agriculture that year.[68] One of McCallum's early priorities was the establishment and maintenance of a good working relationship with Edmund J. Zavitz (1875–1968), provincial forester for Ontario, who was in charge of the blister rust work sponsored by that province for a number of years. Almost from the beginning of his blister rust work, McCallum was not convinced that all the money being spent to control the disease was necessary. This scepticism is revealed in his review of the work that had been done up to 1921, when he wrote, "it is difficult to appreciate the viewpoint of those who regard this fungus (*Cronartium ribicola*) as such a deadly enemy of white pine forests."[69] McCallum knew, as did many others by that time, that early predictions of the dire consequences of the blister rust had not materialized. He also knew that most of the great stands of white pine in the Atlantic provinces had been harvested before the rust fungus appeared in Canada. Doubtless the memory of the enormous amount of earlier white pine logging, such as that of the great John R. Booth,[70] lingered on in the minds of the general public and any threat to the logging was perceived as a potential disaster.

Ontario, which had long been termed "The White Pine Province," was, understandably, more concerned about the potential ravages of blister rust than the other provinces. The Dominion Bureau of Statistics for 1933 showed that the value of the white pine lumber sawn in Ontario that year was greater than that of all the other tree species combined. It also showed that Ontario produced eight times as much white pine lumber as British Columbia, nine times as much as Quebec, and ten times as much as Nova Scotia and New Brunswick put together. But, although blister rust was disastrous to a few stands of susceptible pines, and to a number of tree nurseries, it did not decimate the pine trees in large areas as *Endothia parasitica* had done to the chestnut trees in the USA. White pine blister rust is not a spectacular disease; it works slowly, both in becoming established in a given locality and in killing infected trees. With the realization that the disease was unlikely to ruin the waning white pine lumber business in eastern Canada came a diminution in the amount of money and effort put into a study of the disease and means for its control.

McCallum's interest in pine tree diseases was not confined to blister rust. In 1929 he reported the presence of the gall-forming woodgate rust in Canada, and estimated that it had been in the Ottawa area for about forty years.[71] Apparently he was not sufficiently interested in the disease to do any publishable research on it.

A conference on forestry research, called under the auspices of the National Research Council of Canada (NRCC) to take stock of the general forestry situation in the country, was held in Ottawa in 1935. Among the papers presented was one by Dr Güssow, the Dominion botanist, titled "Forest pathology." One of the results of that conference was the initiation, in 1937, by the Dominion Forest Service in cooperation with the Dominion Department of Agriculture and the NRCC of a white pine breeding program, the objectives of which were resistance to blister rust and satisfactory growth form and growth rate.[72] Another less direct result was the construction of a laboratory of forest pathology at Petawawa, Ontario, in 1939.

Except for E. Silver Dowding's work on a mistletoe of jack pine and its parasite,[73] J.H. Faull's studies of the needle blights of white pine,[74] and a note on the occurrence of needle blight and late fall browning in red pine by W.R. Haddow and M.A. Adamson,[75] there was virtually no other research on the diseases of pine trees by Canadians until after the Second World War.

In 1943, when politicians and others were anticipating an unemployment problem after the war industries closed and thousands of demobilized soldiers would be seeking employment, J.R. Dickson, Dominion Forest Service, Ottawa, tried to revive a campaign to eliminate the threat of the blister rust disease. He advocated this as a source of postwar employment and provided an outline of some of the unique features of his proposal, one of which was to zone the province into white pine districts and Ribes districts, according to which host plant was of controlling local importance. The next step in his plan was to employ trained scouts to locate, and workers to destroy, all of the Ribes and infected pine trees within their respective districts.[76] At that time there were only six forest pathologists in Canada, of whom four were employees of the Dominion Department of Agriculture and two were provincial employees: one in Quebec and one in Ontario.

Just prior to the end of the war, Donald Buckland made an inquiry into the relationship of basal scar injury to the management of western white pine at Revelstoke, British Columbia. An account of that work is in a mimeographed report of the Dominion Laboratory of Forest Pathology, Victoria, for 1945. Buckland's contribution, with Foster and Vidar J. Nordin (b.1924), to the series of papers titled "Studies in forest pathology" was published in 1949, in the *Canadian Journal of Research* "c". Buckland resigned in 1948 to become professor of Forest pathology at the University of British Columbia, where, in 1949–50, he taught four courses: General Forest Pathology, Applied Forest Pathology, Problems in Forest Pathology, and Advanced Forest Pathology.

It is appropriate to mention here that several American workers, in addition to those mentioned above, did some valuable research on pine diseases in British Columbia and thus contributed to the development of forest pathology in Canada. For example, two American forest pathologists discovered *Cronartium comptoniae* on several species of pines there in 1924,[77] and American pathologist H.G. Lachmund studied the damage to *Pinus monticola* by *Cronartium ribicola* in that province over a period of several years in plots that were laid out in 1922.[78]

BEECH TREE DISEASES

At the beginning of the twentieth century, beech trees, species of Fagus, were a dominant feature of the forests in many areas of the Maritime provinces, as several place names, such as Beechwood and Beech Brook, indicate. The gathering of beech-nuts in the autumn or early winter was traditional with many Maritime families until the beech trees died, first in Nova Scotia around 1920 or earlier, and eventually throughout the Maritimes.

Canadian forest pathologists were not involved in the initial stages of research on the devastating disease that became known as the beech bark disease. Entomologists associated an insect with the disease shortly after it was discovered in the Halifax area,[79] and John Ehrlich, an American authority on fungi of the Nectriaceae, provided a very comprehensive account of the disease and its history, supplemented by twenty-nine tables.[80] He declared that 90 per cent of the beech trees in some stands that he examined in New Brunswick and Nova Scotia were infected by Nectria, the cause of the diseased condition. Ehrlich gave credit to J.H. Faull for having been the first person to refer to the "bark disease of beech" and of its importance in Nova Scotia. By 1926 beech trees throughout the central and southern parts of Nova Scotia were reported to be in very serious condition, with many of them already dead,[81] and ten years later living beech trees had "almost completely disappeared" from that province.[82]

The sixteenth annual report of the National Research Council of Canada, 1932–33, tells of the council's involvement in the beech tree disease from 1930 until 1933. In this instance, the cooperating organizations were the Nova Scotia Department of Lands and Forests and the Arnold Arboretum of Harvard University, represented respectively by Mr O. Schierbeck and Dr J.H. Faull. The 1936 report of the Fredericton, New Brunswick, Dominion Entomological Laboratory, of the Division of Forest Insects, refers to the destruction that had taken place in Nova Scotia and the fact that beech trees had been killed throughout Prince Edward Island and in southern New Brunswick, but listed a study of an associated insect as only a minor project.

No forest pathologist was assigned to a study of the beech bark disease until after the Second World War, when most of the beech trees had died throughout the Maritimes. In 1946, plant pathologist Arthur J. Skolko (1912–89) was assigned the nearly impossible task of being responsible for all forest pathology in the three Maritime provinces. For most of the next two years he was given some space in the Department of Biology, University of New Brunswick, Fredericton, in exchange for which he taught forest pathology in that department. In June 1950, a new unit of forest pathology, with Vidar J. Nordin as officer in charge, was established in Fredericton to serve the Maritimes. Thus began a new era in forest pathology for that region.

BIRCH TREE DISEASES

For more than a century after the great white pine trees had virtually all been shipped out of the Maritime provinces, the yellow birch was the most majestic tree in the Acadian forest region. But except for the brief account by David Penhallow (1884–1910), professor of botany and vegetable physiology at McGill University, of "a remarkable tumor growing upon a white birch,"[83] and the occasional reference by mycologists to the bracket fungi or conks and other evidences of fungal growths, there was little, if any, study of birch tree diseases in Canada prior to the 1930s. Probably the major reason for such a paucity of pathological studies is that foresters were, in general, more interested in the coniferous trees than in the deciduous ones; also, there were very few forest pathologists in Canada at that time.

In the early 1930s, foresters and entomologists in the area began commenting on the number of birch stands in central and southern New Brunswick that had a distinctly unhealthy appearance, with some dead trees among them. At first there was no clear picture of the disease because widespread injury to foliage by insects tended to obscure an overall view of the problem. The first official comments on this situation may be seen in the annual reports of forest insect investigators, from the Dominion Entomological Laboratory, Fredericton, beginning from 1932. In the report for that year, L.J. Simpson remarked that for the previous four seasons birch stands throughout the Maritime provinces, Quebec, and eastern Maine had been attacked by insects to such an extent that the leaves turned brown and fell off in early August. By 1935, the growing economic importance of the yellow birch, and the clear evidence of its declining vigour, was cause for concern among foresters. The general public, however, showed little interest in the problem: it was traditional for Maritimers to think of their forest heritage in terms of coniferous trees – pine, spruce, and fir. They knew the hardwoods were useful for winter fuel but did not worry about the loss of a few birch trees.

Because insects appeared always to be associated with the problem, virtually all of the early reports of the disease were written by entomologists. The 1940 report from the Dominion Entomological Laboratory, Fredericton, includes comments on the considerable increase in the deterioration and death of large numbers of birch trees and states that all species of birch were affected. In that report, R.E. Balch used, for the first time, the term "dieback" with reference to the deteriorating condition of the birch trees. Balch and J.S. Prebble were concerned with the role of the bronze birch borer in dying trees, but they saw many trees with advanced stages of dieback that had not been attacked by the borer. This prompted them to look for fungi and bacteria that could be responsible for the disease, but they found only species of saprophytic organisms on the dead parts of the tree crowns, and none that could have been responsible for the dying. They could not assign a supportable reason for the dieback disease.[84]

In 1943, R.F. Morris, agricultural assistant, and a temporary member of the entomological laboratory staff, conducted some root injection and fertilization studies on both white and yellow birch trees near Fredericton, in an effort to control the dieback. That work constituted only a part-time project and was discontinued, after a second series of injections in 1944, because dieback had reached such an advanced stage in the test area that suitable trees showing initial disease symptoms were not available.[85]

Forest pathologist René Pomerleau (b.1904) B SC, M SC, PH D, a keen observer of the disease situation in all hardwood trees, had reported on a general decline of certain hardwood species in Quebec, including birch but primarily white ash, as early as 1925, shortly after graduating from Laval University with a bachelor's degree. Perhaps it was because he was a young summer employee of the Dominion Department of Agriculture that his report did not arouse much interest, and the alleged problem was not investigated by plant pathologists or forest pathologists until after the Second World War, when, for several years, Pomerleau was one of those most actively involved.[86]

POPLAR TREE DISEASES

The poplars, which for this essay include the aspens, grow throughout the forested regions of Canada and are subject to a number of diseases. The first to be described by a knowledgeable Canadian was a rust disease found by James Fletcher, the Dominion entomologist-botanist, when he was travelling across the prairies in 1900.[87] Several authoritative Canadians have written about the diseases of poplars, or the fungi that incite them, since that time. However, there was no research on them until J.H. Faull got his students interested in poplar diseases nearly twenty years later. The first of them was A.W. McCallum, who published an account of a new poplar disease in 1920.[88] The next was E.H. Moss, who was asked by Faull to examine some Lombardy poplars that had been planted in 1920 by the Harbour Commission of Toronto. He studied those, and other poplars in the general area of Toronto, and found that two canker-producing wound parasites on the Lombardy poplars were "not of uncommon occurrence in southern Ontario."[89] By the time Moss's paper was published, in 1922, he had become a member of the staff of the Department of Botany, University of Alberta.

Soon after Herbert S. Jackson (1883–1951) became professor of mycology at the University of Toronto in 1929, he interested himself in studies of the fungi that cause tree diseases and those that destroy wood. That interest reflected itself in the research of his students, one of whom was George E. Thompson, who studied poplar diseases for his master's degree[90] and retained an interest in those diseases until at least 1939.[91] A second student, John E. Bier, studied poplar diseases for his PH D thesis research[92] and did some research on Septoria canker of poplars and Hypoxylon canker of maple[93] while he was a graduate assistant working at the Central Laboratory, Division of Botany and Plant Pathology, Ottawa. His study of Hypoxylon canker of poplars was part of a series of studies in forest pathology and was published as Canada Department of Agriculture publication no. 691, technical bulletin 27, in 1940.

Bier, together with Clifford G. Riley (1899–1989), conducted a general investigation in 1936 into heartwood decay of poplars on the reserve of the Petawawa Forest Experiment Station, Ontario, where it had been noticed that lumbermen taking out poplar for match splint stock had a tendency to reject whole trees even if they displayed only a single sporophore (fruiting structure) of the fungus *Fomes igniarius*. Bier and Riley showed that the amount and position of heart rot in standing trees could be estimated to a useful degree of accuracy, that the mere presence of one sporophore did not necessarily indicate a cull tree, and that a good log could often be taken from an infected tree.[94] Riley's interest in poplar diseases was rekindled, following his period of military service during the Second World War, when he made that topic the subject of his thesis research for the PH D that he was awarded by the University of Toronto in 1946. His thesis title was "Studies on decay in poplar caused by *Fomes igniarius*."

In the prairie provinces, where poplars have always been very plentiful, W.A. Munro, BA, BSA, noted that Russian poplars started from cuttings at Rosthern, Saskatchewan, in 1912 were "continuing to die of canker" in 1929.[95] Canker was considered to be such a serious menace to the numerous poplar shelterbelts on the prairies that John L. Van Camp, B SC F, MA, the only forest pathologist in the prairie provinces for several years, studied the disease on Russian poplars at Indian Head,[96] and G.B. Sanford studied it at Edmonton in 1929.[97]

The loss of poplar trees from disease had been of concern to foresters from at least 1909, when J.R. Dickson, BSA, MSF, assistant inspector of forest reserves, reported that the loss of poplars in the Riding Mountain Forest Reserve was serious.[98] Thus it is understandable that one of the early projects of forest tree breeder Carl C. Heimburger, B SC F, PH D, was the creation of poplar hybrids that were resistant to disease, a project in which he was a Canadian pioneer.

The Heimburger family grew up in St Petersburg, Russia, under the tsars. At the time of the Russian revolution they moved to Denmark, where Carl attended the Royal Veterinary and Agricultural College and obtained a degree in forestry. In 1925 he migrated to Canada and got a second degree in forestry at the University of Toronto, and, later, a PH D from Cornell. Heimburger's initial tree-breeding program in Canada was done, unofficially, at the Petawawa Forest Experiment Station, where his job was to describe the forest types. In May 1937 he and geneticist Frank H. Peto (b.1905), who made significant contributions to the poplar-breeding program, prepared a statement about the possibilities of using auxins in tree-breeding. By 1939 an official tree-breeding program had begun, under the auspices of the National Research Council of Canada and the Dominion Forest Service. Although the Second World War interrupted that program, it was not stopped until the end of the war, when the research council terminated its work on tree breeding. After that, Heimburger initiated a tree-breeding program at the Southern Research Station, Maple, Ontario, under the Ontario Department of Lands and Forests.[99]

In Quebec, at a time when the poplars were considered to be of relatively little economic importance, forest pathologist René Pomerleau, who became director of the Laboratory of Forest Pathology of the Quebec Department of Lands and Forests, reported on the presence of the ink-spot disease on different species of poplar in that province in 1935, and published an account of his studies on the disease in 1940.[100] There was no additional work, in Canada, on diseases of poplar trees until several years after the Second World War.

WOOD DECAY

After food crops, wood has long been Canada's most valuable renewable resource. Wood is one of the most useful materials for basic construction, for ornamentation by carving, and for paper making. It has the triple properties of strength, workability, and durability, but a major weakness is its susceptibility to rot in a moist environment. In other words, it is naturally biodegradable under certain environmental conditions. The prevention of post-harvest wood rot has a long history, and one century of it has been documented by Sir Harold Boulton, the first president of the British Wood Preserving Association.[101]

In Canada, F.W. Cumberland, writing about timber preservation, commented on the fact that William Tompkins got a patent from the governor-general – for an apparatus to get sap out of logs and replace it with a wood preservative – on 16 May 1856;[102] that same year, on 20 December, Joseph Robinson read a paper on the preservation of timber from decay to members of the Canadian Institute in Toronto. Robinson referred to a French invention which was alleged to expel the sap from a log and replace it with "a preserving liquor." He was aware of the value of copper sulphate in the control of several plant diseases and stated, in that paper, that "the most efficacious solution [for wood preservation] is composed of sulfate of copper and water mixed in a proportion of 1 to 100."[103]

Liquid creosote, obtained from tar by distillation, was well known and commonly used as a wood preservative by the time Nova Scotian medical doctor A.P. Reid was writing about its use, and the problem of getting it and other preservatives to penetrate wood, in 1874.[104] A few years later, W.S. Finch expressed the opinion that the loss to Canada from wood decay would pay the interest on the national debt, and went on to introduce a wood preservative that "could be applied by unskilled labor, with the use of a broom or brush, or by saturating in a tank," for which he was awarded a diploma at the Dominion Exhibition, Toronto, in 1887. He commented that the true cause of wood decay was "the result of the growth of a fungus of the mushroom type, ... not propagated by seeds but by mycroscopic sphores [sic] which are blown by the wind."[105]

One of the very few records of instruction having been given by a university professor on wood decay and its prevention is that which appeared in the published examination questions for botany students at Queen's University in 1896. In Professor E.A. Atwell's plant physiology examination that year, he asked, "What wood preserving processes are employed to prevent decay?" – thus indicating that he dealt with that topic in lectures to his students.

The cause of wood rot was well known by the time David P. Penhallow wrote about the dry rot fungus in 1905,[106] or Professor A.H.R. Buller of the University of Manitoba wrote the third of his papers on wood-destroying fungi in 1909.[107] Nevertheless, theirs may have been the earliest reports by scientists in Canada on wood rot.

In the early days of paper making, the roofs of the pulp and paper mills were made of wood that had to be renewed every eight to ten years because of rot. In 1920, Roy J. Blair, a graduate of McGill University who had read a paper on "Common Wood-Destroying Fungi" to members of the Natural History Society of Montreal on 25 February 1918, suggested ways to inhibit the rotting and thus prolong the "life" of the roofs.[108] At that time, Blair was the only plant pathologist on the staff of the Canadian Forest Products Laboratory that had been established in Montreal by the federal Department of the Interior, in cooperation with McGill University, and officially opened in December 1915.

A Division of Wood Preservation, within the Department of the Interior, had been organized in October 1914, with W.G. Mitchell as chief, and W. Kynock as assistant chief. The scope of that division included the study of wood preservatives and methods of treating wood to prolong the life of railroad ties, paving blocks, and structural timbers in general. Before the division was established, Mitchell had written Forestry Branch bulletin no. 49, "Treated wood-block paving," which did not get published until 1916.

In 1918, a sub-laboratory was set up in Vancouver, in cooperation with the University of British Columbia. In 1923 Herbert W. Eades was appointed to the Vancouver laboratory, where attention was directed to the natural discolourations, fungal stains, and decay of wood. His treatise on those topics provided a wealth of information in that field.[109]

J.H. Faull, at the University of Toronto, made a study of a timber-destroying fungus in 1915,[110] and he supervised the studies of various wood-rotting fungi by his students, one of whom was James Herbert White (1875–1957), the first graduate in forestry and an early appointee to the teaching staff, whose PH D thesis on that topic was published in 1919.[111] White had obtained his bachelor's degree in honour science at Toronto in 1904, his MA in 1907, and his B SC F in 1909.

Two more of Faull's doctoral students, Clara W. Fritz (1889–1974) and Irene Mounce (1894–1987), became known internationally as authorities on wood-destroying fungi. Clara Fritz, a Canadian woman pioneer in studies of fungal deterioration and staining of wood, published her PH D thesis, which included a useful key for the identification of wood-destroying fungi.[112] She joined the Montreal staff of the Forest Products Laboratory in 1925, after spending a post-doctoral year at the University of Toronto where she was an instructor in the Department of Botany. Her work in eastern Canada was somewhat parallel to that of Herbert Eades in British Columbia. When the Forest Products Laboratory moved from Montreal to Ottawa in September 1926, she was joined there by Ernest A. Atwell (1900–73), M SC, who isolated a Cadophora fungus from stained sapwood of several species of coniferous trees[113] and assisted her with long-term studies of the decay, under service conditions, of red-stained jack pine railway ties. They showed that the red stain, caused by *Fomes pini*, had practically no effect upon the strength or serviceability of the wood.[114]

Early in the great economic depression of the 1930s, when it became necessary to hold over large supplies of wood in storage piles because of decreased activity of the mills, the question of how rapidly the stored wood might deteriorate became a pressing one. In the summer of 1931 the pulpwood committee of the Pulp and Paper Association asked the cooperation of the Forest Products Laboratories to determine the rate of deterioration due to decay. Shortly thereafter, Fritz became involved in that difficult research, and in 1936 she outlined some of the problems associated with the investigation and gave reasons for the general lack of definite answers.[115] She also researched the causes and the prevention of decay in creosoted and untreated railway ties, and other aspects of wood deterioration, including the influence of mine depth on the rate of decay in mine timbers, through and beyond the Second World War. She retired in 1954.[116]

Irene Mounce began her studies of wood deterioration at the University of British Columbia, where her thesis, for an MA degree in 1920, dealt with factors affecting the commercial value of spruce wood. Her aroused interest in fungi prompted her to go to the University of Manitoba where she earned an MS degree under A.H.R. Buller's supervision. From there she went to the University of Toronto and earned a PH D in 1929, the thesis for which was largely mycological. It was a study of the biology of *Fomes pinicola*, a fungus that produces brown heart rot in several species of coniferous trees. While working toward her doctorate, she and her supervisor, J.H. Faull, determined the cause of red heart rot of balsam.[117]

After she joined the staff of the Central Laboratory, Dominion Department of Agriculture, Ottawa, in 1924, accounts of Mounce's cultural studies of wood-destroying fungi appeared in the annual reports of the Dominion botanist for several years, beginning with that of 1925. By 1932 she had become so well known, and her work on the production of sporophores (spore-producing structures) in culture as an aid to the identification of wood-destroying fungi so highly esteemed, that one of her papers on that topic was presented to members of the Royal Society of Canada, by H.T. Güssow, and published in the *Transactions* of that society.[118] Mounce was transferred from the Ottawa laboratory to the one in Saanichton, British Columbia, in 1938. From that time onward her studies of wood deterioration soon declined as she became involved in the identification of fungi in vegetable seeds. She got married and resigned her position in 1945.[119]

Incidentally, Mildred K. Nobles, who became a world leader in cultural identification of wood-destroying fungi, worked in the laboratory with Mounce during the summer of 1929. Under the supervision of H.S. Jackson, she earned an MA degree in 1931, and a PH D in 1935, the theses of which were largely mycological in content. Nobles, who joined the staff of the Ottawa laboratory in 1935, made her most outstanding contributions to forest pathology, and there were many, after the period covered by this essay.

Commercially treated wood for fence posts was not readily available to farmers in most parts of the country. Therefore, J.A. Roy outlined some economical methods for prolonging the durability of wooden posts when he addressed a meeting of the Quebec Society for the Protection of Plants, at Macdonald College, 1 March 1921.[120] In many areas of western Canada, there were few trees from which to get fence posts or poles; consequently, when a few were obtained it was important to keep them free of rot for as long as possible. Being well aware of this, the superintendent of the Dominion Forest Nursery Station, Indian Head, Saskatchewan, conducted a series of experiments involving creosote, zinc chloride, creozol, and copper sulphate as preservatives of fence posts in 1926. An account of that early research was written by John Walker and forest engineer Charles Edwards.[121]

Clifford G. Riley, a veteran of both world wars, who was a research assistant with the British Columbia Forest Service from June 1928 to September 1930, discovered a root rot of Douglas fir on Vancouver Island in 1929, a year after having graduated with a master of forestry degree from Yale University. That discovery was noted by Irene Mounce, J.E. Bier, and Mildred K. Nobles, in 1940.[122] Riley, who was destined to become one of Canada's most highly respected forest pathologists, gained a wealth of experiences in his studies of tree diseases and the decay of wood before becoming involved in the Second World War. For example, he conducted an investigation into the rate of deterioration of pulpwood in log jams, at the request of the Woodlands Section of the Canadian Pulp and Paper Association in 1933,[123] investigated the rate of deterioration of insect-killed spruce on the Gaspé Peninsula in 1935,[124] and studied the deterioration of piled pulpwood in 1940.[125]

After serving as a captain in the Canadian Army during the Second World War, Riley attended the University of Toronto for the PH D that he obtained in 1946, following his research on a decay in poplar caused by *Fomes igniarius*. In June 1948, Riley established a forest pathology unit at Saskatoon, as part of the Division of Plant Pathology, Dominion Department of Agriculture, to serve the prairie provinces. A forest pathology unit had been established in Toronto in 1947, to serve the forest interests of Ontario.

Major events in forest products research in Canada, up to 1930, were reviewed by T.A. McElhanney.[126] After that, the director of forestry provided a progress report of work on the identification of the destroyers of jack pine railway ties and of the various means of combating the decay of ties, poles, and fence posts, in his annual reports. In addition to those sources of information, a history of the first fifty years of the Pulp and Paper Research Institute of Canada was written by Charles A. Sankey.[127]

EPILOGUE

The development of forest pathology in Canada lagged far behind the pathology of field and garden crops. This was largely because of the vastness of the forest resources and the opinion, long held by many influential people, that those renewable resources were virtually inexhaustible. Serious consideration of forest conservation, including the control of forest tree diseases, had to wait until there was fear of the consequences of their depletion or loss. In the early days of agriculture in eastern Canada, the forests were looked upon as of secondary importance, and they were readily sacrificed in the interest of arable land expansion by the farmers, many of whom regarded forests as their natural enemies.

The export of pine, for ships' masts, was the first export trade in forest products to be of economic importance in the Maritime provinces, and up until about the beginning of the twentieth century little else but red and white pine was being cut for lumber in the forests of Ontario and Quebec. In other words, "to speak of the lumber industry before 1900 is to speak of the pine, chiefly white pine industry."[128] Because of the great importance of pine trees to the economy of the developing nation, forest pathology in Canada received its initial impetus from white pine blister rust, a major diseases of those trees, and from chestnut blight, even though the latter was of relatively little importance to Canada. The vast amount of publicity associated with the devastating effect the chestnut blight disease had on chestnut trees in the United States made Canadians fear a similar disastrous effect on their pine trees when the white pine blister rust disease was discovered in their midst. Their appeal to the government for leadership in efforts to ward off the impending catastrophe produced an almost immediate demand for people with a knowledge of the dreaded disease. This demand was largely responsible for the early development of forest pathology in Canada.

Entomologists were among the first to focus the attention of the scientific community on such forest tree diseases as beech dieback and the loss of yellow birch trees in New Brunswick and Nova Scotia, and they began to play a more cooperative role in forest pathology after the Second World War. An example of this cooperation is the work of L.S. Hawboldt, provincial entomologist for Nova Scotia, and associate forest pathologist A.J. Skolko, in an investigation of yellow birch die-back in 1947.[129] They were assisted by forestry students A.G. Davidson, J.T.B. Kingston, H.G. MacGillivray, and D.R. Redmond. That study was continued the following summer, at the end of which Skolko transferred to Ottawa to study seedborne diseases in the Division of Botany and Plant Pathology of the Department of Agriculture.

Prior to the research on tree diseases by Joseph H. Faull and his graduate students, mostly in the 1920s, virtually all of the forest tree disease research in Canada was really forest mycology. It embodied little more than surveys to determine the presence or absence of specific fungi or diseases in certain areas. Faull and his students initiated a "scientific" phase of forest tree pathology, and also of post-harvest tree decay. He was interested in, and studied, the disease of trees and not just diseased trees or the fungi that were in or on them. Thus Joseph H. Faull may justifiably be regarded as the father of forest pathology in Canada.

Opportunities for the training of forest pathologists in Canada lagged far behind those that were available for plant pathologists. This was due, in part, to the fact that it took a long time for the general public to see the need for forest pathologists. When university courses in forest pathology did become available, the early graduate forest pathologists were considered, by foresters, to be plant pathologists, while plant pathologists thought of them as foresters. Both were right, because most forest pathologists in Canada are professionals in both fields.

A great stimulus to research on tree diseases resulted from the discovery in 1944 of the Dutch elm disease in Quebec by Dr René Pomerleau, who wrote a history of the disease in that province.[130] However, because the elms were noted more for their beauty in urban situations than as forest trees of economic importance, and because there were little more than surveys of their geographical extent before the end of the Second World War, that disease is not dealt with in this history of forest pathology. Nevertheless, the Dutch elm disease focused the attention of thousands of urban dwellers on tree diseases and thus, indirectly, stimulated more studies in forest pathology. As a consequence of this and other factors, the number of forest pathologists employed in Canada, and the amount of research in forest pathology, had their greatest expansion in the decade after the end of the Second World War, when new laboratories were opened in Fredericton in 1946, Toronto in 1947, and Saskatoon in 1948, to name only three of that period. In British Columbia, the number of people engaged in forest pathology increased, in the 1940s, from two to twenty-five over a period of less than five years.[131]

Because studies on tree diseases are generally long-term projects, few of those new employees had time to publish results of their research before the end of the 1940s, which is the end of the time period covered by this essay.

6 Early Nematology in Canada

Through some accident in its evolution, plant pathology in Canada has not, since about the 1950s, generally included the study of plant diseases induced by insects. However, throughout North America, it does include diseases induced by nematodes. There seems to be no logical explanation for the generally accepted view that a plant gall resulting from the action of a nematode comes within the realm of plant pathology, whereas one induced by an insect does not.

As a discipline, nematology might well be expected to embrace studies of all forms of nematodes, but a more restricted definition has developed. The study of nematodes that parasitize humans and animals has become the field of study by parasitologists and veterinarians, almost to the exclusion of the plant parasitic and so-called free-living nematodes. Both of these latter are the ones that are commonly considered to be within the realm of "nematology," as the term came to be used in the second decade of the twentieth century – that is to say, shortly after Nathan A. Cobb (1859–1932), in the United States of America, created the word "nematology" as a contraction of nematodology, which, he stated, was that branch of science that deals with the plant parasitic and free-living nematodes.[1] In general, the education of a plant pathologist does not include an in-depth study of zoology, at least not to the extent that zoology is studied by entomology students. Nevertheless, after Cobb's definition of the term, plant pathologists took an interest in the field of nematology, and in some instances tended to dominate it to such an extent that the expressions "plant nematology" and "phytonematology" began to be applied to their work with nematodes in relation to plants.

In the early days, entomologists were in the forefront of nematology in Canada. Their early interest in nematodes may be seen in the minutes of the 6 March 1896 meeting of the Microscopical Section of the Entomological Society of Ontario. They show that "Mr. Gammage presented specimens of roses the roots of which were infected with a worm-like pest. Samples were put under the microscope and proved to be what they considered to be a species of *Heterodera*."[2] This may well be the first report of parasitic nematodes in association with plants in Canada. In October of that year, John Dearness (1852–1954), president of the Entomological Society of Ontario, reported finding nematodes in diseased roots of peas and considered them to have been one of the causes of the disease.[3] A few years later, in Quebec, Jean C. Chapais (1850–1926), a lawyer-cum-agriculturist of St-Denis-en-bas, conducted a small investigation of a nematode-incited disease of geraniums. The account of his study, with its statement about control measures that included the use of hot water, are recorded in the first annual report of the newly formed Quebec Society for the Protection of Plants, published in 1909.[4]

The first Canadian report of nematodes in association with a cereal crop came about as a result of the keen observations of a farmer in Raymond, Alberta, who sent injured wheat plants to the Dominion entomologist, Gordon Hewitt, in September 1912. Hewitt reported that "A microscopic examination ... disclosed the presence of numerous nematodes, commonly called eelworms."[5] In about midsummer of the following year, John Musgrave of Cowley, Alberta, wrote to Hewitt about a problem in his wheat. In this instance, Hewitt instructed entomologist Edgar H. Strickland (1889–1962), B SC, MS, who had been hired by the Division of Entomology a few months earlier, to investigate the problem. Strickland, whose laboratory was then in a corner room on the second floor of an implement shed at the Lethbridge Experimental Station, found nematodes in stems of the affected wheat.

Those particular nematodes were later found to be of species not considered harmful to plants.[6] Nevertheless, Strickland's interest in nematodes had been so aroused that he made collections from the root zone of various plants, and drawings of several different kinds of those fascinating microscopic worms. Some of his drawings are identifiable as stylet-bearing species of Aphelenchoides, Aphelenchus, and Tylenchus, in addition to non-stylet-bearing species of Chiloplacus, Plectus, etc. That work by Strickland, begun in 1913, constitutes the earliest attempt to record, illustrate, and identify plant-parasitic nematodes in Canada.[7] His continuing work as an entomologist, after serving overseas in the Canadian Army in the First World War, did not involve nematodes.

An early nematode control method is mentioned in the report of the Entomological Society of Ontario for 1917. It is in a note under the heading "Greenhouse Insects," which tells of nematodes causing damage to tomatoes and of how they had been effectively controlled, "in places where they had been bad the previous season," by removing the soil to a depth of about eight inches (20.3 cm).[8] In the report of that society for 1918, there is an account of cyclamen being severely injured by nematodes in a Hamilton greenhouse.[9]

Entomologists and zoologists were not the only ones interested in, or concerned about, nematode damage to plants. Dominion botanist Hans T. Güssow (1879–1961), writing about common tulip diseases in the 1911 report of the Horticultural Societies of Ontario, commented that the myriads of eelworms commonly found in decaying bulbs were not responsible for their decay or failure.[10] When he published an account of wheat diseases, in 1919, Güssow included ear "cockle," a disease that he knew was caused by nematodes.[11] In 1922 someone in the Division of Horticulture, Dominion Department of Agriculture, sent specimens of carrots to Güssow that were "badly infected with eel worms." He quickly recognized the worms as root knot nematodes, which, at that time, were referred to as the species *Heterodera radicicola*. Güssow learned that those carrots had followed a crop of nematode-infected tomatoes and, therefore, that the nematodes producing root knots in tomatoes could also produce them in carrots.[12]

The professor of pomology at the Ontario Agricultural College (OAC), J.W. Crow, was an early advocate of the use of steam sterilization as an effective means of eradicating "eelworms" from greenhouse soil. Crow's account of this effective method of nematode control was included in his annual report for 1911.[13] In 1918, plant pathologist Walter A. McCubbin (b.1880), BA, MA, at the Field Laboratory of Plant Pathology, St Catharines, Ontario, expressed the opinion that eelworms were not uncommon in the roots of greenhouse-grown tomatoes, and that soil sterilization was the only way to eliminate them.[14] Later, at the same laboratory, plant pathologist Alexander A. Hildebrand (1896–1969), BA, MA, PH D, photographed nematodes and their eggs in roots of strawberry plants,[15] and his accompanying comments included one of the earliest suggestions that nematodes may play an important role in the root rot complex of those plants. That work was expanded, in cooperation with L. Ward Koch, BA, MA, PH D, to include tobacco root diseases, in which nematodes were also implicated.[16]

In 1926, assistant plant pathologist Ralph C. Russell (1896–1964), BSA, MS, working out of the Dominion Agricultural Research Station, Saskatoon, discovered nematodes on the roots of wheat plants.[17]

Samples were sent to the Bureau of Plant Industry, Washington, where their identity as a plant-parasitic species of the genus Heterodera was determined by American nematologist Gerald Thorne. In the following spring, Russell sent additional samples of affected wheat to Thorne, who subsequently named the nematodes *Heterodera punctata,* a new species.[18] Also in Saskatchewan, Robert J. Ledingham (1912–83), B SC, M SC, working as a plant disease investigator while stationed at the Dominion Laboratory of Plant Pathology, Saskatoon, found nodules or galls on the seminal roots of wheat seedlings collected near Fish Creek on 4 June 1937. Although he likened them to galls found on legumes, there must have been an element of doubt in his mind, because he concluded that further observations on that point were necessary. Ten years later, plant pathologist Thomas C. Vanterpool (1898–1984), BSA, M SC, Department of Biology, University of Saskatchewan, in reporting on his discovery of a root-gall nematode on a few plants of Thatcher wheat sent to him by a farmer near Radisson, Saskatchewan, speculated that the nodule-like galls referred to by Ledingham could have been caused by nematodes.[19]

In eastern Canada, Richard R. Hurst (1895–1961), BSA, officer in charge of the Laboratory of Plant Pathology in Charlottetown, Prince Edward Island, reported the presence of "many round worms or nematodes" in alfalfa plants from several fields he had examined in 1926.[20] Root knot nematodes were found in the roots of strawberries, at Weymouth, Nova Scotia, in 1926, by entomologist John P. Spittall (b.1888), and in roots of other plants the following year.[21] The root knot nematode was not discovered in Quebec until 1929, when mycologist Ivan H. Crowell (b.1904) reported finding it on greenhouse tomatoes at Macdonald College, Ste-Anne-de-Bellevue.[22] Because root knots are easily seen with the unaided eye, reports of them soon came from various places in Canada. One was by Donald J. MacLeod (1894–1990), plant pathologist at the Dominion Laboratory of Plant Pathology in Fredericton, New Brunswick, who discovered root knot nematodes in potatoes in 1930.[23]

In the summer of 1932, farmers of South Simcoe County, Ontario, appealed to the Ontario Research Foundation for a study of the cause of the repeated failures of their spring grain crops. Research fellows Donald F. Putnam (b.1903) and Lyman J. Chapman (b.1908) conducted a thorough investigation of the problem and, in July of that year, discovered nematodes on the roots of oat plants grown in "oatsick" soil. Some of those nematodes were eventually identified as the oat race of *Heterodera schachtii,* thus making Putnam and Chapman's the first report of that race in North America.[24] Chapman also found nematodes of that species on winter wheat.[25] Several years later, the botany department of the OAC was asked to find the cause of the failure of some oat crops. When the affected area was investigated, high numbers of nematodes were found in fifteen fields. Samples of the suspect nematodes were sent to Putnam, who identified the parasites as *H. schachtii.*[26]

In 1926, Odery J. Robb (b.1888), BSA, assistant in horticultural research, Ontario Horticultural Experiment Station, at Vineland Station, gave an illustrated lecture on nematodes to members of the Vegetable Growers' Association of Ontario, in which he commented that "these insects or animals are very small, being invisible to the naked eye."[27] How to categorize those invisible "insects or animals" posed problems for many early Canadian biologists, and it was not uncommon for the Entomological Society of Canada to include nematodes in its published lists of insects. At that time it may have been difficult for the entomologists to create a special category in which to include them, and it may have been for this reason that the *Canadian Insect Pest Review* included nematodes in its reports for several years – for example, in the note about Mr L. Daviault's discovery of nematodes associated with gladiolus plants at Berthier, Quebec, in 1934.[28]

It was in 1934 that Thomas W. Cameron, B SC, MS, PH D, D SC, director of the newly established Institute of Parasitology at Macdonald College, Quebec, demonstrated his awareness of the rising interest in nematode diseases of plants by lecturing on that topic.[29] It was also the year in which research student E.H.J. Marchant, under the direction of Professor R.A. Wardle, Department of Zoology, University of Manitoba, published results of the first study of nematodes in the soils of that province.[30]

The 1937 annual report of the OAC has a comment on a problem with greenhouse-grown chrysanthemums in London, Ontario, that had been brought to the attention of members of the botany department. In this instance, the difficulty was found to be due to an infestation of foliar nematodes, pests that had never been seen in Ontario greenhouses. That report also showed that it was the Department of Field Husbandry, in cooperation with the Ontario Department of Agriculture and the County of Waterloo, that began the investigation of the reported problem of nematodes inhibiting the production of oats in some sections of Ontario. One year later, that department produced circular no. 47, titled *A study of the oat nematode,* which provided some of the results of their investigations. The oat nematode study, by employees of the field husbandry department, continued until 1941, according to the OAC report for that year.

In August 1946 a farmer from near St-Janvier, Quebec, brought some abnormal carrots to plant pathologist Roger Desmarteau, an em-

Robert J. Hastings (1891–1971)

A.D. Baker (1894–1974)
Courtesy Agriculture Canada
Research Branch

ployee of the provincial Ministry of Agriculture. Desmarteau recognized the nematode problem and made an investigation of it on the owner's farm. His report of that incident, and his findings, are recorded in the annual report of the Quebec Society for the Protection of Plants for 1947.

By 1940, many agronomists, horticulturists, plant pathologists, and other knowledgeable people had become aware of the fact that nematodes were posing a potential threat to the production of high-yielding crops in Canada. Nevertheless, for many years after the events mentioned above, federal authorities did not consider nematodes to be of sufficient importance to warrant an inspection for them by someone with training in nematology, at ports of entry for plant material imported from foreign countries.

In 1931 Hans Güssow again demonstrated his knowledge of nematodes by preparing a pamphlet illustrating *Tylenchus dipsaci* and its damage to narcissi.[31] He was prompted to do that by the results of a survey of the areas devoted to narcissus production in the Lower Fraser Valley and the southern part of Vancouver Island, British Columbia. That survey, late in the 1920s, provided convincing evidence that the decline in production was due to parasitism by *T. dipsaci*, the nematode species illustrated by Güssow. The horticulturists in that area, with the full support of William Newton (1893–1973), BSA, M SC, PH D, officer in charge of the new Dominion Laboratory of Plant Pathology at Saanichton, made an urgent appeal to the Dominion Department of Agriculture for an investigation of the problem, and the department soon authorized Newton to look into the bulb diseases in several areas of British Columbia.

One of the people employed by Newton to work on that challenging problem was Formosa-born Robert J. Hastings (1891–1971), who had no training in nematology beyond what he may have got during his studies for a bachelor's degree at the Ontario Agricultural College. Nevertheless, he soon became the one who contributed most to an understanding of nematode problems associated with narcissi and how to solve them. Hastings joined the staff of the Dominion Laboratory of Plant Pathology, Saanichton, in 1929, and was assigned to the newly authorized study of narcissus and other plant-bulb problems. His research colleague, Jack E. Bosher, whose title was head greenhouse-man until at least 1946, had even less formal training. However, they were fortunate in having the guidance and supervision of William Newton, a well-trained scientist, who had seen the need for nematological research in that area, encouraged its undertaking, and arranged for Hastings and Bosher to do the work.

Hastings, Bosher, and Newton were joint authors of several papers in 1933, the first of which was "The nematode disease of narcissi and crop sequence." In that short paper, they stated that characteristic symptoms of nematode infections appear in barley, oats, and wheat seedlings growing in soil infested with *Tylenchus dipsaci* that had come from narcissus bulbs.[32] Because a transfer of nematodes from narcissus to grain was unknown, or denied, elsewhere in the world, their report stimulated controversy and additional research. As if to reinforce their claim, the title of their second paper that year was "Nematode infestation symptoms on barley as a means of determining the efficiency of chemicals as lethal agents against *Tylenchus dipsaci* Kuhn." In that investigation they screened more than a hundred chemical compounds for nematicidal properties and used barley as a test plant to prove the presence of nematodes that survived contact with the chemical solutions on infested bulbs. Surviving nematodes invaded the barley seedlings and produced typical symptoms – white spots along the midrib of the leaves, which were easily seen by the time the seedlings were four weeks old.[33] The question of how the bulb nematodes reached the bulb-growing areas of British Columbia was never positively answered, but an indication that they may have come via bulbs imported from Holland was provided in the annual report of the British Columbia Department of Agriculture for 1935. It told of intercepting Dutch irises that were infested with the nematodes.

Within ten years, Newton, Hastings, and Bosher, publishing individually or in various combinations, produced more than a dozen papers on their nematode investigations. Their work was the first in Canada to provide continuity in nematological research. They posed questions and exposed new problems, and then persisted with research until answers to the questions, or solutions to most of the problems, were found.

The work of Hastings, who retired in 1951, was considered to have been so valuable to their industry that the Northwest Bulb Growers' Association broke a precedent and dedicated the 1952 volume of their *Proceedings* to him. He was a pioneer in studies of the bulb nematode of the iris, and, with his associates, worked out a satisfactory method for its control. In 1934 he and his colleagues learned that the pre-adult was the most heat-resistant stage in the life cycle of the bulb nematode, which, in a dry state, could survive five hours at 180° F (82.2° C).[34] That discovery explained the many failures by those who had been following the previously recommended hot water treatment of three hours of immersion at 110° F (43.3° C). Most reports of nematode survival, following the application of heat or chemicals, were based on evidence of motility – if the nematode moved it was alive; if not, it was dead. Research at Saanichton, published in the first volume of the *Proceedings of the Helminthological Society of Washington* in 1934, showed that factors other than death could induce quiescence in nematodes, and that the quiescence could be mistaken for death.[35]

Hastings's research played an important role in convincing other nematologists of the distinctiveness of the iris nematode from the narcissus form. He also learned that the bulb nematode, *Ditylenchus dipsaci*, could be transferred to potato plants.[36] Bulb nematodes on potatoes were well known in parts of Europe, but the phenomenon had not, to his knowledge, been previously reported in North America, though actually there were earlier, obscure reports of bulb nematodes on potatoes from Prince Edward Island and New Brunswick. Nematode-infested tubers had been intercepted upon entering the United States from the Island in 1931 and from New Brunswick in 1934; when some of the infested tubers were sent to the Division of Nematology, Washington, the nematodes were identified as *Ditylenchus dipsaci*.[37]

Although the reputations of Hastings and Bosher as Canada's most productive research nematologists were based largely on their work with nematodes that parasitize horticultural plants with bulbous "roots," they also did some pioneering work in studying the combined effects of nematode and fungal parasitism. They discovered that reductions of plant growth resulting from the action of mixed cultures of parasitic nematodes and the fungus *Cylindrocarpon radicicola* were greater than the sum of the two parasites acting individually. In other words, they found a synergistic effect and in doing so were the first in Canada to find such an effect in a plant disease involving nematodes.[38]

Hastings was also a pioneer in another aspect of plant-nematode relationships, the susceptibility of plants to nematode penetration or parasitism in relation to the pathogenicity of the parasitizing nematodes. When he experimentally infected several kinds of plants with *Pratylenchus pratensis*, he found that more nematodes entered the roots of oat seedlings than any of the other test plants, but the growth of the oats was affected the least. Thus he showed that the susceptibility of a plant to nematode penetration is not always directly related to the pathogenicity of those nematodes.[39] Bosher's first work, with Newton, was on the host preference of the root knot nematode.[40] They knew that steam sterilization of the soil was a way to control nematodes in greenhouse soil beds, but because few greenhouses in southern British Columbia were steam heated they worked out an effective chemical method for the control of root knot and other greenhouse pest-nematodes.[41]

In August 1931, Harold D. Brown, head of the Research Department of the Canada and Dominion Sugar Company, Chatham, Ontario, was looking over some experimental plots on a sugar beet field at Glencoe and noticed a number of unhealthy beets. An examination of the roots disclosed the presence of nematodes. He sent some infected beets, through the Dominion Entomological Branch, to the United States Department of Agriculture, where the nematodes were diagnosed as *Heterodera schachtii*, a new sugar beet parasite for Canada.[42] But except for that one field, no further infestations of sugar beets in Ontario were reported until 1939, when a new area of infestation was discovered.[43] This latter discovery prompted an extensive survey of the whole sugar-beet-producing region that was conducted jointly by the Ontario Agricultural College, the Dominion Department of Agriculture, the Ontario Department of Agriculture, and the Dominion Sugar Company. The annual report of the OAC for 1940 provides a brief account of those surveys, and of a series of experiments conducted by the college "covering everything which could be considered to have a bearing on nematode control."

The discovery of several more infested fields prompted the Ontario government to designate as a "precautionary area" the involved fields in the township of Sarnia. The government also imposed regulations designed to prevent the spread of the sugar beet nematode. All beets grown in that area were required to be shipped by railroad cars to either Chatham or Wallaceburg sugar factories. Thus it was possible to avoid some of the spread that might have occurred by means of soil scattered along the highways from trucks. Special treatment of all railway cars from the precautionary area was required and all such cars had to be thoroughly washed, at the sugar factories, before being used again. The growing of beets in the infested fields was prohibited, by a regulation incorporated into the Plant Disease Act, in 1941.[44] Those regulations were revised in Consolidated Regulations of Ontario, 1950, as regulation 477.

The discovery of nematodes parasitizing grain and sugar beets in Ontario was a major factor in the establishment of a new "unit" for research within the Dominion Department of Agriculture in 1939. It was called simply "Nematode Investigations," and its only trained investigator was the officer in charge, entomologist Alexander D. Baker (1894–1974). Baker, a veteran of the First World War, had BSA and M SC degrees from McGill University, where he was employed as an assistant in the entomology department from 1925 until he resigned in 1932, and a PH D from the University of Toronto. He was listed as a junior entomologist with the Canada Department of Agriculture in 1935, where he became involved with nematode investigations largely because he had indicated to his superiors that he was interested in nematode studies.

Laboratory facilities for nematology work in Ottawa, at the time of his appointment and for some time after, were practically non-existent. This was partly because Baker and his assistants were required to be almost fully occupied with surveys and field experiments on nematode control, and partly because a small field laboratory in Sarnia soon became available for their use. He had to wait until 1949 before permanent laboratory facilities were provided for nematode investigations, in what was then called the Science Service Building, Ottawa.[45] Baker managed to have an illustrated account of the sugar beet nematode in Ontario published in June 1945, "for the use of those who are actively concerned with the growing of crops attacked by this pest."[46]

In November 1945, a seed potato inspector in Prince Edward Island found tubers with a disease or injury that he had not seen before. Samples were sent to Baker, in Ottawa, who sent specimens to Gerald Thorne, Division of Nematology, United States Department of Agriculture. Thorne was known to be an authority on bulb nematodes, and it so happened that he had just recently described a new species of Ditylenchus from Idaho potatoes, which he named *Ditylenchus destruc-*

tor, and suggested that its common name be "potato rot nematode." Thorne reported to Baker that the nematodes found in the tubers that had originated in Prince Edward Island were potato rot nematodes.[47] That revelation, coupled with the nematode problems associated with sugar beets and oats, prompted officials of the Canada Department of Agriculture to a renewed interest in nematode investigations. One of the indirect results of this interest was more funds for Baker, who immediately looked for an assistant.

Baker's first assistant in nematology was V.E. Henderson, who, for a brief period in 1946, helped with the sugar beet nematode studies near Sarnia, Ontario. Also, Roland H. Mulvey, who was destined to become one of Canada's leading nematologists, was a summer-student assistant in the sugar beet nematode studies from 1947 to 1950. Soon after the discovery of the potato rot nematode in Prince Edward Island was confirmed, Henderson was transferred there to assist with the extensive field experiments that appeared to be so urgently needed. His paper on host relationships of the potato rot nematode did not get published until 1951.[48]

Richard R. Hurst, officer in charge of the Charlottetown Research Station, Canada Department of Agriculture, soon became involved in the investigations of the potato rot nematode. He reported that none of the ten varieties of potatoes in his trials showed any resistance to the nematodes, and that numerous wild and cultivated plants were critically examined but only field mint was found to be a host. Up to 1947, only six farms on Prince Edward Island were known to be infested with *Ditylenchus destructor*, and they, together with those adjacent to them, were put under quarantine.[49]

When H.N. Racicot (1893–1968), BA, of the Division of Botany and Plant Pathology, Ottawa, addressed members of the Canadian Phytopathological Society on the subject of nematode diseases of potatoes in Canada, he told them that the potato rot nematode was reported to have been found in ships' stores in Prince Edward Island in 1931, and that a subsequent search for the nematode in that province had failed to reveal its presence. He expressed the view that the 1945 discovery of *Ditylenchus destructor* on the Island was unrelated to the 1931 report, and stated that nematodes of a species of Aphelenchoides, considered to be parasitic, had been reported in Island potatoes and that further studies to confirm or deny this were being undertaken.[50]

It was 1948 before a chart illustrating potato tubers damaged by potato rot nematodes became available. That chart, which lacked a publication number, was a good example of the developing dichotomy of responsibility for nematodes affecting plants because it was produced jointly by three different divisions in the Department of Agriculture:

the Division of Botany and Plant Pathology, the Plant Protection Division, and the Division of Entomology. The chart advised that "specimens suspected of being infected with the potato rot nematode should be sent to the nearest Laboratory of Entomology or Plant Pathology," thus virtually necessitating the involvement of both entomologists and plant pathologists in nematological activities.

The formation of the Nematode Investigations unit in 1939, when there was no one in Canada, with the possible exception of Baker, who would have considered himself or herself to be a nematologist, *sensu* Cobb, marked the initiation of a new phase in the development of plant nematology in Canada, a modern phase which was destined to expand rapidly over the next two decades. Baker himself provided a brief résumé of the developments during that exciting period.[51] He also compiled a comprehensive list of the early records of plant parasitic nematodes in Canada, and of nematodes that had originated in Canada but which had been intercepted by American authorities as they entered the United States.[52] A dramatic expansion of nematology in Canada, much of which can be credited to the initiative and energy of Alexander D. Baker, took place in the 1950s.

7 Early Plant Disease Legislation in Canada

Many destructive plant pathogens are carried in, on, or with seeds, soil, and plants, all of which can be moved over relatively short distances by such natural means as wind, water, birds, and animals. Large bodies of water, mountain ranges, deserts, and other geographical barriers limited the spread of plant pathogens, until people began to transport them over and around the natural barriers. In doing this, people inadvertently became the chief agents for the long-distance transportation and distribution of plant pathogens, the incitants of plant diseases. As a substitute for natural barriers to the movement of these incitants, artificial ones have been created in the form of laws that control or prohibit the transportation of plant pathogens and their carriers from areas where they are known to exist to areas that are free of them.

The history of laws that restrict the freedom of Canadians, or people living in the geographical areas that are now parts of Canada, to import plants or plant products goes back to at least 1789. On 29 April of that year, Judge Ward Chapman in Saint John, New Brunswick, wrote to Judge Edward Winslow (1746–1815) in Fredericton, and commented on the recent embargo that had been placed on the importation into Saint John of certain kinds of flour suspected of harbouring insect pests.[1] Laws or decrees to prevent the export of plants or plant materials have an even longer history in Canada. On 17 September 1754, Charles Lawrence, commander-in-chief of His Majesty's Province of Nova Scotia, proclaimed that it was strictly forbidden "For any Masters of Vessels to ship on board their vessels any Corn [grain], without permission in writing, signed by myself under Penalty of Fifty Pounds

Sterling." An explanatory note accompanying the proclamation stated that its intention was to ensure that the Halifax market was supplied before corn could be shipped anywhere else.[2] In January of 1814, the governor of the Hudson's Bay Company issued a proclamation ordering that people trading in furs or provisions within the territory controlled by that company around the Red River were not permitted to take any "flesh, fish, game or vegetables procured or raised, out of the territory."[3] Because of severe food shortages, the governments of each of the three Maritime provinces prohibited the export of potatoes and grain from their respective areas of jurisdiction in 1837.[4]

Although the above examples of restrictions on the movement of plants and plant materials had very little to do with plant protection as such, they serve to illustrate the fact that the authorities could prohibit the import or the export of such materials whenever they deemed the restrictions to be in the best interest of the people concerned.

Legislation relating to plant protection was enacted under municipal and provincial authority long before national action by federal officials was contemplated. An example of early municipal legislation is "An Act for the Preservation of Apple Trees, in the parish of Montreal," apparently directed against the cankerworm, passed on 25 March 1805.[5] More than sixty-five years passed before there was similar provincial legislation aimed at plant protection. It was "An Act for the better protection of growing fruit in Kings County," enacted by the legislature of the province of Nova Scotia in 1873.[6]

The first legislative measures that were aimed specifically against plant diseases were *ad hoc* responses to what were perceived to be emergencies. For example, the Fruit Growers' Association of Ontario watched the spread of the black knot disease of plums and cherries for several years, while advocating voluntary measures to keep it under control. When it became apparent that the voluntary efforts were largely ineffective in preventing the spread of the disease, the association appealed to the government for assistance. In response to their petition, the legislative assembly passed "An Act to Protect Plum and Cherry Trees," which became law on 11 March 1879, and compelled property owners "to cut out, and immediately burn up, all the black knot found on plum and cherry trees thereon, so often in each and every year as it shall appear on such trees."[7]

Because that act made no provision for the appointment of inspectors to see that the law was being observed, it was expanded during the following year by providing that any municipal corporation in Ontario could appoint officers or inspectors to carry out the provisions of the act, which had become popularly known as the Black Knot Act. Furthermore, if no such inspector was appointed, it became the duty of the overseer of highways, upon request, to give notice to the owner or occupant to cut out and burn all black knot found on the property.[8] It may now seem inappropriate for an overseer of highways to be appointing plant disease inspectors, but the black knot disease was so easily identified that any adult with reasonably good eyesight could quickly become a competent inspector. However, it soon became obvious that those inspectors were not very efficient. Within a few years, members of the Fruit Growers' Association were complaining that "little regard is paid to the law which requires affected trees to be destroyed."[9]

When similar black knot legislation was enacted in Prince Edward Island in 1895, authority for appointing inspectors was given to the ratepayers in each school district at the annual school meeting. A unique feature of the Prince Edward Island act was that it set out fines for failure to obey the legally appointed inspectors and stipulated that the fines imposed "shall be paid into the hands of the Secretary of School Trustees for the District, ... to be used ... for the purpose of payment of the inspectors and the enforcement of the Act."[10] Thus it provided the inspectors with a financial inducement to ensure that the law was enforced. Precedent for school trustee involvement in laws pertaining to agriculture on the Island goes back to the 1883 "Act to Prevent the Spread of the Potato Bug," which authorized trustees to appoint "one or more fit or proper persons" to act as inspectors for purposes of the act.

Nova Scotia's "Act to Prevent the Spread of Diseases Affecting Fruit Trees," passed 15 February 1896, was more inclusive than a simple black knot act because it required the owner or occupier of land to "cut out and burn all the black knot or other contagious diseases found on any plum, peach, cherry, or nectarine tree on his land."[11]

While the fruit growers in Ontario were urging their government to enact legislation pertaining to the black knot of plum and cherry, they were keeping a wary eye on the spread of the yellows disease of peach trees in the adjacent state of Michigan. In 1878, A.M. Smith of Drummondville, Ontario, was warning growers that a number of trees in the orchards of his area were already infected with yellows.[12] During the next two years, many growers reported the presence of the disease in widely separated districts of Ontario, and the Fruit Growers' Association appealed to the government for legislation to help prevent its spread. Again the government responded favourably to their plea by passing "An Act to Prevent the Spread of The Yellows" in 1881.[13] That act was rewritten and expanded in the "Act to amend the Act to Prevent the spread of Noxious Weeds and of Diseases Affecting Fruit Trees," assented to on 4 May 1891. It is of interest to note that the new act included a subsection which stated that "Any person who sows any wheat or other grain knowing it to be infected by the disease known as smut without first using some proper and available remedy to destroy the germs of such disease shall, upon conviction, be liable to a fine of not more than $20.00."[14] That was the first law in Canada obliging farmers to treat seeds for disease control or be fined for not doing so.

With increased knowledge, and a greater awareness of their losses due to disease, orchardists in Ontario insisted that the laws for the protection of fruit trees were in need of further broadening and strengthening. To accomplish these two objectives, "An Act for the Better Prevention of Certain Diseases Affecting Fruit Trees," referred to as "The Yellows and Black Knot Act," became law on 27 May 1893.[15] That law was intended to prevent the spread not only of black knot and yellows but also any other destructive disease affecting fruit trees. After the provisions of that more comprehensive act came into effect, however, the fruit growers were still complaining that the law regarding black knot was being ignored. Such complaints by influential people may have been what prompted the government of Ontario to define the duties and responsibilities of inspectors more clearly through "An Act Respecting the Inspectors of Fruit Trees," on 7 April 1896.[16]

BARBERRY ERADICATION

In the 1890s, many farmers in Ontario, and a number of fruit growers, had hedges of barberry bushes because that bush "is easily propagated, sold cheaply by nurserymen, and has few insect enemies," and, being very prickly, it "is given a wide berth by animals of all kinds ... and it is a most enduring hedge-fence." The horticulturist who made those remarks at a meeting of fruit growers, in 1893, went on to state, "It is said that barberry hedges breed rust on wheat. I have never seen satisfactory evidence to that effect, and think the evil is wholly imaginary."[17] Obviously he did not know that the great German botanist Anton de Bary had, in 1865, provided conclusive evidence that the common barberry, *Berberis vulgaris*, plays a very important role in the life cycle and dissemination of the wheat rust. De Bary had shown that the rust on cereal grasses and a disease on barberry leaves were caused by the same agent, a fungus that had long been known as *Puccinia graminis*.

Before it was learned that spores of the rust fungus were being blown from warm southern regions, where they could overwinter, to the wheat-growing regions of Canada, there was a general belief among knowledgeable farmers and biologists that if susceptible barberry bushes were not at hand the disease could not complete its life cycle and therefore could not ravage wheat. As the role of the barberry became known among wheat growers, a number of them tried to eliminate the bushes, many of which were growing in parks, on city lawns, and elsewhere beyond their reach. To get at the latter, governmental support was needed in the form of legislation that would compel everyone, including towns and cities, to get rid of all barberry bushes on their property.

There was precedence for such legislative action in several countries on the continent of Europe, but not in England, where a very successful barberry eradication campaign had been conducted on a voluntary basis.[18] Knowing this, a few influential Canadians wanted stronger support for a voluntary eradication program. Some, including the Dominion entomologist and botanist, James Fletcher, had expressed the view that the rust would still be attacking wheat even if all barberry bushes were removed from the wheat-growing areas.[19] Similar statements by other influential people weakened the efforts of those who were campaigning for the compulsory removal of barberry bushes.

The province of Ontario was a Canadian pioneer in barberry legislation with its "Act Respecting the Barberry Shrub" in 1900. That act forbade the planting of barberry shrubs on or within a hundred yards of farming lands and stipulated that owners of land on which barberry shrubs already existed "may be required" to remove and destroy them.[20] In short, it was a law without very strong teeth, and with plenty of loopholes. Its revision of 1914, and the virtually new "Barberry Shrub Act" of 1927,[21] each became more restrictive, more compelling, and less easily evaded.

After the great crop loss on the prairies due to rust in 1916, almost everyone associated with the production or sale of wheat wanted the government to assume a greater role in ways and means of preventing such a loss ever again. The federal Department of Agriculture convened a conference of representatives from the prairie agricultural colleges and provincial departments of agriculture in August 1917, at Winnipeg, Manitoba. One of the results of that conference, in so far as legislation is concerned, was a general agreement that the eradication of barberry bushes should be made compulsory.[22]

As early as 1890, the province of Manitoba had among its statutes "An Act to Prevent the Spread of Noxious Weeds," the terms of which compelled owners or occupants of land to cut down or otherwise destroy all noxious weeds growing on their property. When it was decided that the elimination of barberry bushes should be compulsory, the Manitoba legislature, in 1917, amended its Noxious Weeds Act by adding "berberies [sic] vulgaris, commonly known as berbery [sic] bush" to the list of such weeds.[23] Both Saskatchewan and Alberta had "inherited" from the territorial government "An Ordinance Respecting Noxious Weeds," which they each amended and incorporated into their respective legislative acts, and they, like Manitoba, eventually included barberry in their lists of the weeds required to be destroyed. Some of the other provinces less dependent on wheat for their economic well-being waited for federal legislation to deal with the barberry problem. In 1919 the federal government did outlaw "Common or Rust Barberry (*Berberis vulgaris* L), its hybrids and horticultural varieties ... susceptible to Black-stem Rust," and made it illegal to move any of them "into any area within the Provinces of Manitoba, Saskatchewan or Alberta, throughout which Provinces they shall be exterminated without any claim for compensation."[24]

In spite of such laws, there has never been a concerted, nation-wide campaign to eliminate barberry bushes from the whole country. Consequently, one did not have to travel far in eastern Canada, up to at least 1950, to see masses of those shrubs growing within sight of the highway.

BRITISH COLUMBIA

In the early days British Columbia made its own laws with regard to plant protection, and had as severe an inspection service against the rest of Canada as against any foreign country.[25] Those laws had their beginning in the 23 April 1892 "Act to Create a Provincial Board of Horticulture," which is cited as the "Horticulture Board Act, 1892." The board was set up "For the purpose of preventing the spread of contagious diseases among fruits and fruit trees, and for the prevention, treatment, cure, and extirpation of fruit pests and the diseases of fruits and fruit trees."[26] The act authorized the board to suggest regulations for the accomplishment of those objectives, and for the appointment of "a competent person especially qualified by practical experience in horticulture, who shall be known as Inspector of Fruit Pests." It also gave the inspector, and members of the board, wide powers to inspect orchards, nurseries, fruit-packing houses, store rooms, sales rooms, etc., and authority to compel the owners or occupiers to comply with the regulations regarding the treatment, removal, or destruction of plants or plant materials harbouring pests or diseases.

After a trial period, and some changes in the act in 1893, the legislature passed "An Act to consolidate and amend the Acts respecting the Provincial Board of Horticulture," which became law 21 March 1894.[27] A supplement to that act, containing the "Horticultural Regulations, 1894," was published in the *British Columbia Gazette*, 4 October 1894. Its regulation no. 5 required that "All persons owning or having in their possession nursery stock, or trees and plants of any kind, infested with insect pests or fungous diseases, shall cause the same to be disinfected and cleansed by using the remedies herein prescribed, or such other insecticides and fungicides as may be found effective." Apparently that latter clause was seen by some people as authorization for the use of dangerous chemicals, in contravention of the Pharmacy Act. Their complaints to members of parliament, and others in authority, were eventually listened to by the government, whose response was to make the use of commonly used pesticides legal by bringing in the "Poisons Act, 1909,"[28] followed by a revision of the Pharmacy Act in 1911.

The combined effects of those acts did not satisfy the pharmacists. They continued to complain about the unlicensed sale by hardware merchants, feed houses, etc. of poisons. As a consequence of their representations to the government, the minister of agriculture issued regulations on 26 July 1921 that required dealers in agricultural poisons to have an annual licence and to keep a record of all purchases and sales of "poisonous substances used exclusively in agriculture or horticulture for the destruction of insects, fungi, or bacteria, or sheep dips or weed killers."[29]

The first specific legislation concerning chemical pesticides by the federal government was "An Act to Regulate the Sale and Inspection of Agricultural Economic Poisons" in 1927. It was replaced by "An Act to Regulate the Sale of Products Used in Controlling Agricultural Pests" in 1939. This latter act defined a pest control product as "any produce used, or represented as a means, for preventing, destroying, repelling, mitigating, or controlling, directly or indirectly, any insect, fungus, bacterial organism, virus, weed, rodent or other plant or animal pest."[30]

During the first few years of its operation, the Board of Horticulture in British Columbia required only three kinds of locally grown fruit to be inspected: apples, pears and quinces. Otherwise the work of the inspectors was centred around imported plants and plant products, which, if showing signs of infection or of harbouring insect pests, were required to be either burned or returned to the sender. The requirement did not apply to nursery stock or grain, which were handled differently. All imported nursery stock, except that which had been grown under glass, was fumigated with hydrocyanic-acid gas, and grain showing symptoms of infestation was fumigated with carbon bisulphide, at certain designated fumigation stations.[31]

After the federal government enacted the Destructive Insect and Pest Act in 1910 (see below), British Columbia used it as a guide for its own laws, and it was not long before federal employees were assisting the provincial inspectors. The inspection of potatoes for export was undertaken by the Fruit Inspection Branch of the British Columbia Department of Agriculture during the fall of 1915, in response to an embargo placed on the importation of diseased potatoes from that province by United States authorities.[32] On 17 July 1917, the federal government, through an Order in Council, reaffirmed its regulation of 25 December 1915 by declaring that potatoes offered for sale to the United States must be free from injurious diseases and insect pests. It further stipulated, with regard to imports, that all nursery stock and certain other plants destined for British Columbia and subject to fumigation or inspection must enter through the port of Vancouver. Those entering the province through any other port had to be sent in bond to Vancouver for fumigation and inspection.[33]

The federal government tried to prevent the spread of white pine blister rust into British Columbia through an Order in Council dated 4 April 1919, prohibiting the movement of five-needle pines, currants, and gooseberries from east of the Alberta-Saskatchewan border to the west of that line.[34] When that disease was found in British Columbia in 1921, the authorities there did not advocate the elimination of the currant and gooseberry bushes, because they considered the cultivated black currant industry to be more valuable than the white pines in several areas.[35] They did enact legislation, 1 March 1922, to prevent the movement of five-needle pines and all species of Ribes to the east of the Cascade Mountains, but that restriction was lifted 9 June 1925, after the disease was found there.[36]

When a certified seed-potato program was initiated in British Columbia in 1920, potato growers were encouraged to voluntarily request the required inspections of their crop, and the rules governing inspection were similar to the ones enforced by the federal authorities in eastern Canada at that time. To assist the potato growers and the new potato inspectors, Cecil Tice (1892–1944), BSA, potato specialist in the British Columbia Department of Agriculture, prepared circular no. 32, "Potato certification," and a chart of potato diseases for their use or guidance, in 1921. Tice's official title that year was soil and crop instructor, but in the annual report of the Department of Agriculture for 1923 he was referred to as chief agronomist.

In 1924 the plant quarantine and inspection services in British Columbia became consolidated by "The Agricultural Act,"[37] and, by mutual agreement with federal authorities, the provincial quarantine and inspection staff enforced not only the provincial regulations but also those under the federal Destructive Insect and Pest Act, with regard to plants imported into British Columbia, and via that province into other parts of Canada.[38] Such an arrangement avoided unnecessary duplication of officials, although all consignees were required to apply to the Destructive Insect and Pest Board, Ottawa, for permits to import plants, and copies of inspection certificates were sent to that board. It is worth noting here that the first woman to fill the position of inspector of imported plant products, or any similar or equivalent position in Canada, was Mrs Evelyn Campbell of Kingsgate, British Columbia. In writing about Mrs Campbell's appointment, in the report of the British Columbia Department of Agriculture for 1932, the chief plant quarantine Officer, W.H. Lyne, praised her work.

The cooperative arrangement between British Columbia and the federal government continued until the end of July 1933, when all plant quarantine and inspection activities then supervised by the provincial Department of Agriculture were transferred to the Division of Plant Protection, Dominion Department of Agriculture. This was followed, in 1934, by amendments to the provincial Agricultural Act to bring its provisions into line with the federal Destructive Insect and Pests Act and its regulations.[39] From that time onward British Columbia was essentially the same as the other provinces, with regard to plant protection legislation. One of the exceptions to this general statement pertains to its 1936 Act Respecting Noxious Weeds, in which the parasitic flowering plant known as dodder was included in its list of "weeds."[40]

The Revised Statutes of British Columbia for 1948 list three acts that pertain to plant pathology, namely the Natural Products Marketing Act, chapter 200; Plant Protection Act, chapter 254; and the Certified Seed-potato Act, chapter 257. These three are reviewed in the annual report of the British Columbia Department of Agriculture for 1953.

QUARANTINE AND OTHER PLANT DISEASE LEGISLATION

Much has been written about quarantine as a means of preventing the introduction of plant pathogens into an area, but "quarantine" in its original sense – detaining the carriers of pathogens for forty days – has seldom been practised anywhere in relation to plant pathogens. When plant pathologists refer to plant quarantine they mean "the legally forced stoppage of shipments against the possibility that the materials being shipped are carrying pathogens that are potentially dangerous in the area of import."[41] In that sense, the world's first quarantine legislation against a plant disease was enacted by the Dutch in 1877, when they prohibited the importation of coffee plants and seeds from Ceylon to the Netherlands East Indies (Indonesia) to prevent the spread of coffee rust.[42]

Plant quarantines have been quite effective in preventing the introduction of certain plant pathogens into Canada, and from infested into non-infested areas within the country. The first federal quarantine legislation to deal with plant diseases was "An Act to Prevent the

Hans T. Güssow (1879–1961)
Courtesy Public Archives of Canada

Evelyn (Neil) Campbell (1908–84)

Introduction or Spreading of Insects, Pests and Diseases Destructive to Vegetation," usually cited as "The Destructive Insect and Pest Act," which became law 4 May 1910.[43] The regulations issued under that act, by Order in Council of 27 February 1911, listed the destructive insects, pests, or diseases to which the act applied, including potato canker (*Chrysophlyctis endobiotica*), branch or stem canker (*Nectria ditissima*), gooseberry mildew (*Sphaerotheca mors-uvae*), and white pine blister rust (*Peridermium strobi*). The chesnut bark disease (*Diaporthe parasitica*) and both chestnut (*Castanea dentata*) and Chinquapin (*C. pumila*) were added to the act by Order in Council, 30 June 1911.[44]

The regulations specifically forbade the importation of potatoes from Newfoundland or the islands of St Pierre and Miquelon.[45] This regulation was included because the Dominion botanist, H.T. Güssow, had learned in October 1909 that a potato disease known, at that time, as canker, wart or black wart was present in the Crown colony of Newfoundland, and that it could easily reach Canadian potato fields if legislative measures were not promptly taken to prevent it. The publicity associated with the finding of the potato wart disease in Newfoundland prompted the General Assembly of that colony to pass "An Act to Prevent the Spreading of Potato Canker" on 29 March 1911. Because it had been initiated in the previous year, it is cited as "The Potato Canker Act, 1910." That act authorized the governor in Council to "make such regulations as are deemed necessary to stamp out and prevent the spread in the Colony, of Potato Canker, and prohibit the introduction into the Colony of any potatoes affected by said disease."[46]

In spite of that authority, which was continued after the Potato Canker Act was replaced by the more inclusive "Destructive Insect and Pest Act" in 1919, one of the legislators complained that he could find "nothing on record to show that anything had been done to combat the disease excepting the spreading of a few barrels of lime on infected gardens at St. John's and at Harbor Grace. A few pamphlets were distributed describing the disease, otherwise nothing has been done."[47]

When United States authorities placed an embargo on the importation of Canadian potatoes because the powdery scab disease had been found on some tubers from New Brunswick, its causal organism, *Spongospora subterranea*, was added to the list of pests in the Destructive Insect and Pest Act in 1914. Shortly after that organism was found in soils of the United States and the embargo had been lifted, it was taken from the act by Order in Council, 15 October 1915.[48]

The importance of the federal Destructive Insect and Pest Act was recognized by the various provinces, and their legislatures passed similar acts to serve the purpose of their individual needs. The first province to do so was Nova Scotia, with "An Act to Prevent the Introduction and Spread of Insects, Pests and Plant Diseases Destructive to Vegetation," 31 March 1911. This was followed by New Brunswick, 20 March 1913, and Quebec, 19 February 1914. All of the other provinces eventually followed their lead by enacting somewhat similar legislation. When the provinces revised their acts, most of them omitted a specific listing of "pests" and, using terms somewhat like Nova Scotia's 1914 revision, made the act apply to "such plant diseases, insects and pests as the Governor in Council may from time to time declare to be subject to the provisions of this Act."[49] However, the Plant Protection Act, in the Revised Statutes of the Province of Quebec, 1925, and its revision in 1941, chapter 138, named sixteen plant pests, seven of which were plant diseases.

The federal government also took this more inclusive approach when the Destructive Insect and Pest Act was revised in 1927. Specific pests and diseases were named in the regulations pertaining to the act, not in the act itself. Some of the areas of potential disagreement or conflict between that federal act and its provincial counterparts were removed by the addition of the statement "nothing herein contained shall be construed as legislation to prevent the legislature of any province from making laws in relation to any such insect pest or disease not so dealt with by the Governor in Council," in the 1932 amendment of the federal act. The rights of provinces to control pests within their respective borders were redefined by another amendment, 20 April, and an Order in Council, 9 May 1934.[50]

None of the various jurisdictions that had enacted legislation pertaining to plant diseases had really defined the term "plant disease." When Manitoba introduced its "Plant Disease Act," in 1927, there was no attempt to define the term, but the act did state that "Disease" meant the following insects and diseases in any state of development: black knot, raspberry mosaic, raspberry leaf roll, and fire blight. When that act was revised in 1940, there was no change in the designation of disease.

Some provinces used the more inclusive term "plant pests" instead of "plant diseases," as Alberta did in its 1921 "Act to Provide for the Extermination of Agricultural Pests." In the "interpretation" part of that act it states, "Pests shall include any animal, insect or disease which may, in the opinion of the Department, be likely to be destructive of, or dangerous to, grain or other crops." Ontario's "Plant Disease Act" of 1937 came somewhat closer to a modern definition by stating, "plant disease shall mean any disease caused by any insect, virus, fungus, bacterium or other organism which is designated a plant disease in the regulations."[51]

Soon after the bacterial ring rot disease of potatoes was discovered early in the 1930s, laws to prevent its spread were, in some instances, incorporated into existing legislation. Alberta did this when it amended the Agricultural Pests Act, in 1945, by adding a clause that specifically dealt with that disease. Provinces more economically dependent upon potato production created new acts to deal with the problem. This was the course followed by Prince Edward Island, which promulgated an entirely new "Act Respecting Bacterial Ring Rot," 2 May 1940.[52]

When white pine blister rust was found in the state of New York in September 1906, Canadian authorities were concerned about the possibility of its entry into this country. The Dominion botanist, Hans Güssow, had wisely insisted that white pine blister rust be included among the "pests" to be excluded from Canada through regulations under the Destructive Insect and Pest Act. He did that three years before the fungus was discovered in Canada, but he failed to stop the importation of white pine seedlings from Europe until after it was discovered in Ontario in 1914. Only then was an embargo placed on the importation of species of Ribes, Grossularia, and five-needle pines.[53] The government of Ontario took legislative action by amending its Fruit Pest Act to include the blister rust fungus within its terms, but most of the other provinces were satisfied that the federal laws were broad enough, with regard to white pine blister rust, to give them adequate protection.

Soon after the sugar beet nematode was found in several fields in the Sarnia district of Ontario, in 1939–40, a new regulation under the Plant Disease Act (Ontario), known as "The Sugar Beet Nematode Regulation," was approved by Order in Council on 12 August 1941 and expanded by another Order on 13 October 1943. This appears in the Consolidated Regulations of Ontario, 1950, volume 3, as regulation 477.

The discovery of the potato rot nematode in Prince Edward Island in 1945 prompted the general assembly of that province to expand its "Act Respecting Bacterial Ring Rot" of 1940 under the new title "An Act Respecting Bacterial Ring Rot and Other Plant Diseases and Pests." This new act, assented to 30 March 1945, defined "plant disease" as "Bacterial Ring Rot and any disease caused by an insect, virus, fungus, bacterium or other organism which is designated a plant disease in the Regulations"; thus it included nematodes without actually naming them.

Various other pieces of provincial and federal legislation were directly or indirectly aimed at plant disease control. For example, the 1922 federal "Act to Regulate the Sale and Inspection of Root Vegetables" and its 1927 revision stipulated that all carloads and cargoes of potatoes for export must be inspected and that Canada Grade A potatoes must be "practically free from ... scab, blight, soft rot or damage due to disease."[54] Thus it encouraged growers to have good disease control programs. The amended act also included grades for table turnips and celery, but excluded certified seed potatoes because rules for their production and sale came under the terms of the Destructive Insect and Pest Act.

The government of the Crown colony of Newfoundland passed "The Grading of Vegetables Act" in 1931,[55] and the several provinces of Canada each passed somewhat similar acts. The combined effects of such legislation resulted in the production of more disease-free fruits and vegetables.

In many respects, the Destructive Insect and Pest Act of 1910, and the regulations that followed it, have been the foundations upon which other Canadian laws for the protection of plants from disease have been based. Its provisions were so wide and so well phrased that no major changes were necessary prior to 1949, except for the amendments of 1932, and 1934, which provided for the inspection of exports and clarified the act in relation to provincial jurisdiction. In 1949 all regulations of the Destructive Insect and Pest Act were revoked and re-established in consolidated form.

CANADA AND INTERNATIONAL PLANT DISEASE LEGISLATION

One of the basic axioms of plant disease control is, wherever possible, to exclude the disease-inciting organisms from places where they are not known to exist. This may be from an individual farm, a county, or a geographical area as large as a country or a continent. Because disease incitants have no respect for political boundaries, it has long been recognized that although quarantine regulations by individual states or nations may tend to exclude unwanted plant pathogens, international cooperation is a highly desirable adjunct to those national laws.

Several of the plant diseases that were causing anxiety in Canada over the first few years of the twentieth century, including potato leaf roll, white pine blister rust, chestnut bark disease, and the potato wart disease, were of such nature that they did not lend themselves to control by any of the known chemicals. The realization of this fact was an important factor leading to the adoption of legislative measures against both the introduction and the spread of plant diseases.

Europeans became acutely aware of the need for international cooperation soon after the phylloxera gall louse, *Phylloxera vastatrix*, was innocently introduced into Europe from America around 1859, and devastated the vine-growing industry. That catastrophe led to a five-nation Phylloxera Convention at Berne, Switzerland, in November 1881, which set the stage for future international action for the control of plant pests. It encouraged Danish biologist F.E. Rostrup, then seventy-one years old and respected by plant scientists everywhere, to advocate international cooperative action during a congress at The Hague in 1891.[56] Other influential biologists took up the cry for some form of multinational regulations for the protection of plants, but it was not until February 1914 that an International Conference on Phytopathology was convened, in Rome. During that conference, rules for the first International Convention on plant protection were drawn up.

Unfortunately, the First World War broke out before a single country ratified the convention. But the question of the need for rules or laws relative to the movement of disease- and pest-free plants and plant products from country to country was raised again soon after the war. H.T. Güssow, Dominion botanist in the Canadian Department of Agriculture, was one of the principal advocates of such international rules or regulations. In 1923 he addressed the International Congress of Phytopathology and Economic Entomology in the Netherlands on the desirability of international plant disease legislation.[57] In 1926 he reinforced that theme during a meeting of the International Congress of Plant Science, in Ithaca, New York.[58] Such agitation by Güssow and other well-known and respected proponents resulted in an International Plant Protection Conference being convened, again in Rome, in 1929. During that conference a revised draft of the 1914 convention was discussed and a new International Convention for the Protection of Plants was accepted by twenty-six of the forty-six participating countries.

Canada was one of the nations that refused to adhere to that convention, and Güssow had to advise the members of the convention that "the Government of Canada expresses regret but must continue to request compliance with the provisions and regulations under its own Destructive Insect and Pest Act."[59] Apparently this refusal, and the reasons for not accepting the terms of that convention, were not well known in Canada at the time. Supporting evidence of this lack of knowledge is seen in a letter from F.H. Auld, deputy minister of agriculture for Saskatchewan, to J.H. Grisdale, deputy minister of agriculture in Ottawa, dated 22 April 1930. In his letter, Auld suggested that the next International Convention be held in Regina at the time of the World Grain Conference that was scheduled to be held in that city in 1932. Grisdale responded by telling him that because "Canada refused to adhere to the International Convention for Plant Protection, I do not think it would be advisable for us to attempt to get them to hold their meeting in Canada in 1932."[60]

An explanation of why Canada had not adhered to the terms of that first convention is found in a "Memo in reply to a letter from Director, of May 15th on the subject of the Rome Convention of Plant Protection." The gist of that memo is contained in its first two lines: "The United States do not adhere and it might complicate our own trade with that country were we to adhere."[61]

Güssow was not deterred by these events and continued to advocate some form of agreement between the major nations trading in plants and plant products. He even used his presidential address to the American Phytopathological Society, 3 January 1936, as a platform for his continuing crusade, commenting on some of the economically important plant disease organisms that had recently crossed oceans to invade foreign countries, and on how such unwelcome invasions might be prevented in the future through international cooperation and legislation.[62]

Once again a world war intervened, and the next International Plant Protection Conference, with Canadian participation, was convened at The Hague in April 1950; but it is too recent to be included in this essay.

8 The Early Teaching
of Plant Pathology in Canada

Teaching may be done in a variety of ways, the simplest of which occurs when one person, by the spoken word or by example, transmits some knowledge or skill to one or more others. Demonstrations, on government-sponsored farms and on owners' farms, were among the most effective early methods of teaching better farming methods and plant disease control in Canada, especially among orchardists and potato growers.

Another common method of teaching is through visual images, printed words and illustrations, such as those found in newspapers, journals, books, etc. In the early days of Canadian agriculture, most of the agricultural societies and associations were so small that few of them could afford to publish the minutes of their meetings. In the 1840s it became apparent that governmental coordination and financial aid were necessary. This happened in 1847, when an act to incorporate a Lower Canada Agricultural Society, and a similar act in Upper Canada, were passed by the governments of the day.[1] In a very real sense, those societies may be considered as precursors to present-day departments of agriculture. They sought to promote improvement through the dissemination of useful information about plant diseases and other agricultural developments.

The act to incorporate the Lower Canada society stated that one of its objectives was "the diffusion of sound and useful knowledge on all subjects connected with agriculture." Within a year that society began publishing the *Agricultural Journal and Transactions of the Lower Canada Agricultural Society,* in French and English, and it continued to do so until 1853. The journal's editor, William Evans (1786–1857), was a dedicated worker in the cause of agricultural improvement. He had privately published two numbers of the *Canadian Quarterly Agricultural and Industrial Magazine* in 1838, and the *Canadian Agricultural Journal* from 1844 until 1847.[2]

The several small agricultural societies in Nova Scotia were also faced with the problem of insufficient funds for the production of a journal. The government of that province (then a colony) began publishing, in 1865, the *Nova Scotia Journal of Agriculture* under the direction of the Central Board of Agriculture. The secretary of the board, Dr George Lawson, edited the first few issues. Somewhat similar publications, containing the occasional informative article on plant diseases, were eventually sponsored by societies and governments in several of the Canadian provinces. From the standpoint of plant pathology, the most edifying of them was the annual report of the Quebec Society for the Protection of Plants, the first of which was published privately by the society in 1909. After the society's printing plant burned in 1910, the printing of those reports was financed by the government of Quebec. For a few years that publication, with its many illustrated papers on plant diseases, was used in lieu of textbooks in schools and colleges of agriculture. Thus it was one of the most important of the early Canadian periodicals for the teaching of plant pathology. The Quebec Society for the Protection of Plants, founded at Macdonald College in 1908, largely through the efforts of Professor William Lochhead, is the oldest organization of its kind in the world still functioning under its original name.[3]

The educational aspects of the work of the early Canadian agricultural groups can scarcely be over-emphasized. Their leaders were very sensitive to the lack of educational opportunities available to them and to their children. They did the best they could to remedy that situation by inviting recognized authorities on various topics to give lectures or demonstrations at their meetings and exhibitions. One of the most indefatigable of those lecturers or teacher-demonstrators must surely have been James Fletcher (1852–1908), LLD, FLS, FRSC, Dominion botanist-entomologist for many years. The records of virtually all the major and many of the minor agricultural or horticultural organizations from British Columbia to Nova Scotia refer to at least one lecture or demonstration given by that remarkable man. Fletcher was primarily an entomologist, but many of his lectures and published articles dealt with plant diseases. He became president of section 4 of the Royal Society of Canada in 1895, and is considered to have been a noteworthy pioneer Canadian plant pathologist.[4]

Hans T. Güssow (1879–1961), Fletcher's successor, continued to focus the attention of the agricultural community on plant diseases and how to prevent or control them. Each year he gave several "addresses" to members of farmers' societies and clubs and Boy Scout groups, in addition to those given during various fruit and vegetable growers' conventions.[5] Güssow's assistants followed his example by giving lectures on plant diseases to various groups. For example, the report of the Division of Botany for 1918 shows that his assistant in charge of the field laboratory in St Catharines, Ontario, gave sixteen lectures on plant diseases throughout the fall and winter of that year to agricultural classes, farmers' clubs, and farmers' institutes.

During the latter part of the nineteenth century and the early part of the twentieth, farmers and other rural people tended to depend upon their newspapers, particularly the small-town weeklies, for news and for agricultural information. Newspapers and other privately sponsored periodicals not primarily devoted to agriculture have often been the first to draw the attention of the public to plant diseases and the resultant losses attributed to them. Thus they, through their editors and correspondents, played a significant role as early "teachers" of plant pathology. The first newspaper in what became Canada, the *Halifax Gazette*, began publishing 23 March 1752. Understandably, the newspapers published in the larger cities and towns devoted most of their contents to social affairs, industries, shipping activities, etc., whereas those of the small towns had relatively more news and articles about agriculture. Thus, *Le Glaneur*, established by J.P. Boucher-Belleville at St-Charles, Quebec, December 1836, and similar small-town newspapers were agriculturally oriented even though they may have professed to a much broader orientation.

It was not at all unusual for people with a crop disease problem to write, or send a sample of it, to the editor of a newspaper for advice. One good example, in the realm of Canadian plant pathology, occurred in the fall of 1909 when Patrick Dunphy, of Red Island, Newfoundland, sent some diseased potato tubers to the *Montreal Herald* for information about his poor crop.[6] The editor sent them to Ottawa, where Güssow identified the disease as canker or black wart, a disease that was new to North America and which soon became a topic for comment and discussion in newspapers throughout the country.

Another form of the printed word that is often omitted in the context of agricultural education and the teaching of plant pathology is the educational book. Among the earliest of the truly Canadian books on agriculture to contain a significant amount of information about plant diseases were two written by William Dawson in Nova Scotia in 1853 and 1856, and a third written in 1864 while he was principal of McGill University.[7]

A book that was widely read by Canadian agriculturists in the second decade of the twentieth century was titled *Farm economy, a cyclopedia of agriculture for the practical farmer and his family*. That book, published in the United States but distributed in Canada by the Imperial Publishing Company, Toronto, was written in the form of lessons. Lessons 19 to 21, by American plant pathologist E.C. Stakman, were devoted to plant diseases and their control.

The lesson format of such books made them similar to correspondence courses, which were gaining in popularity. An early correspondence course, dedicated to agriculture and containing references to plant diseases and their control, was provided by the School of Agricultural Extension for Home Reading, operated by the Nova Scotia School of Agriculture, as early as 1900.[8] The Manitoba Agricultural College began offering somewhat similar correspondence courses in nature study, agriculture, and horticulture in 1918.[9]

ONTARIO

Trinity College

The teaching of almost any subject, including plant pathology, is customarily thought of as taking place in a more or less formal educational situation, such as a school or college, in which a teacher and one or more pupils are involved. Taken in this restricted sense, the first formal lectures on plant diseases in Canada may well have been those given by Henry Youle Hind (1823–1908) to teacher trainees attending the Normal School in Toronto during the academic year 1848–49. Evidence of such teaching is found in the list of questions in the competitive examination of that year for the Lord Elgin prizes in agriculture. A somewhat convoluted question on that examination paper was "What is the primary cause of 'colds', and of what determination to disease which has of late years been exhibited by many vegetables, especially in the tubers of the potato?" This was followed by "What remedy, in part, would you suggest with reference to vegetables?"[10]

Some of Hind's lecture notes were published in a pamphlet titled *Two Lectures on Agricultural Chemistry* in 1850. His concept of agricultural chemistry must have included plant pathology, because on pages 64–7 of that little publication, under the heading "Diseases of Vegetables," there is a good account of rust, mildew, and smut, all of which the author asserts are "diseases caused by microscopic fungi" which are "minute vegetables." An enlarged second edition, *Lectures on Agricultural Chemistry; or Elements of the Science of Agriculture*, was published the following year. In lecture no. 8 of that edition, Hind dealt with weeds, the Hessian fly, rust of grain, mildew, and the potato disease. Some of those lectures were later expanded to include more insects and diseases, and entered in an essay competition for monetary prizes awarded by the boards of agriculture for Upper and Lower Canada. Twenty-two essays were entered in that competition and Hind's essay, which won first prize, was prefaced by a noteworthy quotation from a speech by Napoleon: "The progress of agriculture ought to be one of the objects of your constant care; for upon its improvement or decline depends the prosperity or decline of empires." That essay, titled *Essay on the Insects and Diseases Injurious to the Wheat Crops*, was published by the Bureau of Agriculture and Statistics, Toronto, in 1857.[11]

Hind was then lecturing in organic chemistry at the Medical School, Trinity College. That institution, founded as an independent university in 1851, had appointed Hind to its faculty in 1853. It also awarded him an honorary MA degree because, although he was a well educated man, he had never earned a university degree.[12] After resigning from Trinity in 1864, Hind did some exploring and writing before settling in Windsor, Nova Scotia. There, after a stint of lecturing at King's College, which awarded him a DCL in 1890, he died on 8 August 1908.[13]

Essays were popular vehicles for the presentation of ideas pertaining to agriculture throughout the nineteenth century and well into the twentieth. Many of them, especially the "prize essays," were published and widely distributed by the government of the jurisdiction in which they were written, or they were incorporated into the annual reports of the various provincial departments or bureaus of agriculture. Those reports were avidly read by progressively minded farmers and gardeners and thus were important agencies for agricultural education, which commonly included plant pathology.

University of Toronto

The earliest lectures at the University of Toronto to include a discourse on plant diseases were delivered by George Buckland (1805–1885) in 1852. (Trinity College was not affiliated with the University of Toronto until 1904.) Buckland, a farmer in England, had been invited by the Hon. Robert Baldwin to make a survey of agricultural conditions in Upper and Lower Canada in 1847. He was appointed secretary to the Board of Agriculture for Upper Canada shortly after his arrival in this country, and, in 1852, he became the first professor of agriculture in the University of Toronto.

At the university, Buckland soon proceeded to establish an experimental farm, the first such farm connected with a university in Canada. One of the topics that appeared in the calendar of the university throughout the period of his teaching was "Blight and their [sic] remedies." Buckland retained his professorship at Toronto until he died in 1885, after which the chair of agriculture was abolished.

In 1874 Robert Ramsay Wright (1852–1933) became professor of natural history and, in 1876, curator of the Museum of Natural History. He held those titles until 1887, when his designation was changed to that of professor of biology, and he lectured in botany and zoology until he retired in 1912. Wright had the distinction of being not only the first professor of biology at the University of Toronto but also the first Dean of Arts and the first vice-president. He included some plant pathology in his botany lectures at University College as early as 1876. Evidence of this is found in the question "What is the nature of ergot of rye?" in his botany examination that year. In the following year he asked his students to "Give instances of parasitic and epiphytic plants."[14]

Another man who probably included comments on plant diseases in his lectures around that time was W.R. Shaw, MD, who published a paper on the peach yellows disease in 1891. In that paper, Shaw gave his address as "Biological Laboratory of the University of Toronto," and stated that he had attempted to determine the true cause of the disease by using "the ordinary methods adopted for a bacteriological investigation."[15]

Further evidence that plant pathology was being taught at the University of Toronto in that decade is seen in the correspondence between James Fletcher and the registrar of the university. Fletcher accepted an appointment as examiner in economic botany and entomology, 1 January 1891, and he put three questions on plant diseases in his first examination paper.[16] Additional evidence to indicate that plant pathology was being taught each year may be found in the annual list of examiners. For many years the calendar of the university listed an examiner for entomology, and one for bacteriology and plant pathology. The examiner in bacteriology and plant pathology for the years 1896 to 1898 was F.C. Harrison, who later became professor of biology at Macdonald College.

The first member of the teaching staff who, in the modern sense, could be referred to as a professional botanist was Edward C. Jeffrey, a BA graduate of the University of Toronto in 1888 who earned a PH D, in botany, from Harvard in 1891. Jeffrey had been a lecturer in the Department of Biology at Toronto before going to Harvard. He returned to that position in 1891, and remained there until he resigned to accept a position at Harvard in 1902.[17] Jeffrey is noteworthy here not only because he included some plant pathology in his lectures but also because he taught, and otherwise inspired, Joseph Horace Faull

(1870–1961), to specialize in plant pathology. Faull eventually succeeded Jeffery as lecturer in botany in the biology department and was placed in charge of a sub-department of botany in 1902.

It was largely through the efforts of Faull that plant pathology became recognized as an important subject in the botany curriculum at the University of Toronto. He was a native of Michigan, but his first university studies were at Victoria College, University of Toronto, where he graduated as gold medallist with a BA degree in 1898. He went to Harvard, 1901–3, as an Austin Fellow, then returned to Toronto as a replacement for Jeffrey. Faull earned a PH D at Harvard in 1904, without interrupting his teaching. When the Department of Biology was divided, he became professor and head of the new Department of Botany and continued to teach plant pathology, with an increasing emphasis on forest pathology. This latter became necessary when the curriculum of the Faculty of Forestry, established in 1907, called for lessons on diseases of trees, which he was expected to teach. When he became president, section 4, of the Royal Society of Canada, the topic of his presidential address, in May 1920, was "Plant pathology: Its status and its outlook".[18] Later that year he was elected president of the Canadian Division of the American Phytopathological Society. Except for nearly a year of study in Europe (1909–1910), mostly with Robert Hartig (1839–1901), the "father" of forest pathology, Faull taught at Toronto until he resigned in 1928, to become professor of forest pathology at Harvard.

In 1914 there was such an urgent need for plant pathologists in Canada that Hans Güssow, the Dominion botanist, appealed to students in agricultural colleges "to devote their attention to plant pathological science, the future of which ... is most promising in the Dominion."[19] From then onward, many students did just that, particularly at the University of Toronto and McGill University.

In the academic year 1924–25, two courses in plant pathology were being offered at the University of Toronto. Course no. 22, in the botany department, was "a lecture, seminar and laboratory course of 100 hours on the diseases of plants." The second one, no. 26 in the Faculty of Forestry, was "a lecture and laboratory course of 75 hours on the diseases of plants, especially of trees."[20] In that year, Clifford G. Riley, who was destined to become one of Canada's outstanding forest pathologists, was listed with the staff of the botany department as an assistant. In the following year, Clara Fritz, who became a pioneer in the study of fungal deterioration of wood, was listed as an instructor.

Another man who taught some plant pathology at the university was George Henry Duff (1893–1958), who, according to the calendar for 1925–26, was a plant physiologist. Duff was born in China, but was well known in the botany department because he had obtained MA and PH D degrees there, the latter in 1922. His lectures in plant physiology included commentaries on plant diseases, especially those that are induced by environmental or physical stress. He was considered to be such an authority on those diseases that he was commissioned by the National Research Council of Canada, on a part time basis from 1929 until 1931, to investigate the damage done to trees and other plants resulting from the smelting of lead and copper at Trail, British Columbia.[21] Duff's plant pathological activities included the supervision of a white pine blister rust survey and some basic research on the effects of ultraviolet light on the rust spores.[22] His interest in, and increased knowledge of, tree diseases was recognized by the university in 1929, when he was promoted to the rank of associate professor. He was also given the responsibility of teaching the graduate course "Forest Pathology" until Faull's successor took over, a few months later.[23] Duff continued to teach plant physiology, with its complement of plant pathology, through the period of the Second World War and beyond.

Leslie C. Coleman (d.1954) joined the staff of the botany department in 1926, as an assistant. He had graduated from the natural science course at Toronto in 1904, and taken a PH D in Göttingen, Germany. From January 1908 until 1926, he worked in India, where he became director of agriculture in the State of Mysore, and president of the Mysore Agricultural and Experimental Union. In 1927 a cooperative arrangement was entered into between the Ontario government and the University of Toronto whereby a plant pathologist could be employed part time (summer) at the Horticultural Experiment Station, Vineland, and part time (winter) at the botany department of the university. Both institutions would contribute to salary and other costs.[24] Coleman was the first to occupy that dual position; as a consequence, he was appointed provincial plant pathologist and the university promoted him to the rank of professor of plant pathology. Thus he was the first to hold such a professorship at Toronto. Coleman did not hold either position for more than a few months because he resigned from both and went back to India later that year.

Coleman's replacement, in both positions, was Dixon Lloyd Bailey (1896–1984), who had been officer in charge of the Dominion Rust Research Laboratory, Winnipeg. Bailey had earned a BA degree in 1918 from Queen's, and an M SC from Cornell in 1919. After that he went to the University of Minnesota and studied plant pathology, physiology, and genetics for the PH D that he was awarded in 1923. In the following year he was appointed director of the Field Laboratory of Plant Pathology, and later, the Rust Research Laboratory, Winnipeg, where he was responsible for the work in cereal pathology until he went to Toronto in 1928.

At the University of Toronto, plant pathology had its major development under Bailey and his colleague Herbert Spencer Jackson (1883–1951). Jackson, who had become an authority on the rust fungi while he was chief of the botany unit at the Purdue University Agricultural Experiment Station in Indiana, was appointed professor of mycology in January 1929. At that time he had only a BA degree from Cornell, but later that year the PH D was conferred on him by the University of Wisconsin, largely for his classic paper on the evolutionary tendencies of the rust fungi.[25] Jackson enhanced his reputation as a mycologist at Toronto, chiefly for his studies on wood-destroying fungi, and he became president of the Mycological Society of America in 1934. However, he is noteworthy here for his teaching of graduate courses in forest pathology from 1930 until long after the Second World War.[26]

During much of that period, valuable assistance in the teaching of forest pathology was provided by William Robert Haddow, an employee of the Ontario Department of Lands and Forests who had studied forest pathology under Faull for the B SC F degree that he obtained in 1923. Haddow was a special lecturer in the botany department from 1931 until at least 1940. In 1936, he was awarded the PH D degree by Harvard University.[27] The university calendars from 1933 through 1935 also show that the renowned Canadian naturalist John Dearness, an authority on fungi and plant diseases, was a research associate on the staff of the Department of Botany.

In 1940, when the Department of Botany had Colin D. McKeen and Lloyd T. Richardson as class assistants and Roy F. Cain as an assistant in the mycological herbarium, Bailey added some new plant pathology courses to the curriculum, one of which was titled "Virus Diseases of Plants." Both McKeen and Richardson became highly regarded plant pathologists whose most outstanding research work was done after Second World War, as was Cain's mycological work.

There was a time in the 1940s when more than half of the plant pathologists and mycologists in governmental and university laboratories in Canada had received significant portions of their training from Jackson and Bailey. Several of those who studied plant pathology under Bailey, including Arthur N. Langford (b.1910), whose MA and PH D degrees were from Toronto, have to remain virtually unnoticed in this history of the teaching of plant pathology because they taught in institutions where they were required to teach several subjects, some of which included plant pathology although not labelled as such, and because they lacked both the time and the facilities for plant disease research.

Bailey's scientific acumen and the high quality of his teaching were soon recognized by fellow plant pathologists, who elected him president of the Canadian Division of the American Phytopathological Society in December 1928, and president of the Canadian Phytopathological Society in 1933. It does not detract from Bailey's outstanding accomplishments to state that he was very fortunate in having teaching colleagues of the high caliber of G.H. Duff, physiologist, and J.W. MacArthur, geneticist, in addition to Jackson. They all contributed to the excellent training that students in plant pathology got at the University of Toronto in the early part of the Bailey era, an era that lasted well beyond the time frame of this essay.

Ontario Agricultural College (Affiliated with the University of Toronto from 1888 until 1964)

The annual report of the Ontario School of Agriculture and Experimental Farm, the forerunner of the Ontario Agricultural College (OAC), for the year ending 31 October 1876, shows that its president, William Johnston (1848–85), BA, was delving somewhat more deeply into the general topic of plant diseases than were most of his contemporary teachers in Canadian universities and colleges. He taught physiological botany and was one of the first of the Canadian teachers to search for and comment on the causes of diseases in plants. He asked second year students, in question 11 of his examination paper that year, to give the causes of disease in plants and to enumerate the common diseases. He provided part of the answer in the twelfth question, when he asked, "Give the pathology of sweat [sic], rust, mildew and dry rot."[28] After his course became "Structural and Physiological Botany" in 1878, Johnston continued to discuss the causes, symptoms, and treatments of various plant diseases in his lectures. In 1880 the outline of the general botany course included "diseases of plants – smut, rust, mildew, etc."[29] Johnston was one of the first Canadian teachers to have field plots in which students could see living examples of diseased plants and otherwise become better informed about plant diseases.[30]

J. Hoyes Panton (1847–98), who had BA and MA degrees from the University of Toronto, was installed as professor of chemistry at the OAC in October 1878, at a salary of $1,000.00 and board. As was typical of professors of agricultural chemistry at that time, Panton included plant diseases in his lectures. In 1880 he was examiner for economic botany and asked questions on such disease-inciting fungi as *Tilletia caries, Claviceps purpurea,* and *Podophyllum peltatum.*[31] The OAC report to the commissioner of agriculture for that year lists Panton as examiner for systematic and economic botany, structural and physiological botany, geology and physical chemistry, entomology, economic botany, analytical and practical chemistry, agricultural chemistry, organic chemistry, inorganic chemistry, and five others.

It may be mentioned here that William Johnston, in addition to being president of the OAC, held the title of professor of natural history, English, and mathematics. His successor, James Mills (d.1924), would not do the work in natural history, so Panton added elements of that general field to his already heavy teaching load. Although he was relieved of much of the responsibilities for natural history in 1881, he took on a course in geology, and thus continued to deliver almost as many lectures each week.[32] It was largely because of this heavy burden of teaching that Panton resigned early in 1882 and went to Winnipeg. However, in less than three years, officials at the OAC enticed him back to a less onerous teaching program, and a salary of $1,500.00.

In 1893 Panton gave ten lectures to teachers at a summer school, the topics of which included causes of mildew, rust, and "spot" of apple. He taught virtually all of the basic sciences at the college, though his laboratory, the only one at the OAC until about 1895, had a floor space of less than 115 square feet (nearly 10.7 square metres).[33] Panton remained on the teaching staff until he died in February 1898.

The professor of biology and horticulture, J. Playfair McMurrich, BA, began teaching "Systematic and Economic Botany" in 1881 and, as Panton had done in a similar course, included diseases of plants and the life cycles of disease-inducing fungi in his lectures and in his examination questions for at least three years. McMurrich was probably the first teacher in Canada to do some thorough research on a plant disease problem. His "private investigation" in 1882 included a study of the black knot disease of plum trees and its causal fungus.[34]

William Lochhead (1864–1927), a native of Listowel, Ontario, with a BA degree from McGill University and an MS from Cornell, was appointed professor of biology and geology, 15 September 1898, as a successor to Professor Panton. Lochhead had been at the London Collegiate Institute, where he earned the reputation of being a "teacher well and favorably known through the Province."[35] At the OAC he took a more scientific approach to the teaching of plant pathology than had most of his predecessors. Lochhead's advanced ideas in this regard are seen in a commentary, in his first annual report, on the arrangement of the plant physiology and plant pathology courses. He wrote, "In the third year, plant pathology is studied after the course in physiology. Here the worker must not only study the effects of disease, but the causes which lead to the diseased condition. These causes are most frequently of a fungous nature, but sometimes purely physiological. These subjects are so important that to give them due attention the time of a separate instructor would be required."[36] A synopsis of the courses offered in 1901 shows that he also taught a laboratory and lecture course on fungi and plant pathology.[37]

When classes in nature study first began in the Macdonald Institute of the OAC, in 1903, plant diseases were included in the curriculum, probably because of Lochhead's intense interest in those topics; he was one of the lecturers and, for one year, acting director of the nature study program. In the nature study course for teachers in rural schools, the teacher trainees studied several plant diseases and were required to collect twenty specimens of diseased plants. In addition, they had to choose topics for study from a list of "Special Problems" which included parasitic plants, greenhouse pests, mushrooms, and apple scab.[38]

The 1903 annual report noted that all three departments of the Macdonald Institute – Nature Study, Home Economics, and Manual Training – were "for teachers male and female." The fact that women were not admitted to the degree program in agriculture until 1918 may have been one of the reasons for the declining enrolment, for a few years, in the agricultural classes and the steady increase in student numbers at the institute.

Lochhead resigned in April 1905 to accept a position at the newly founded Macdonald College in Quebec. However, he returned to the OAC for a few months to help with work in the Department of Botany and the program in nature study. He finally left the OAC in April 1906 to work full time at Macdonald College.

For more than a decade nature study was popular throughout the country, and the Macdonald Institute at the OAC was widely thought to be the place to study it. There, hundreds of teachers were briefly exposed to some teaching in the realm of plant pathology. They, in turn, may be presumed to have included some references to plant diseases when teaching nature study in their respective rural schools.

Francis C. Harrison (1871–1952), an OAC graduate in the class of 1892 who had helped Professor Panton with some experiments on spraying and the use of the Bordeaux mixture in 1893, is listed in the OAC report for 1894 as bacteriologist and librarian. Although dairy bacteriology was his chief concern after 1895 when he became the first professor of bacteriology at the college, Harrison maintained an interest in plant pathology. From 1895 through 1898 all of the microscopic work at the OAC, including animal and vegetable histology, cryptogamic botany, and plant pathology, was done in Harrison's laboratory.[39] In 1895 he spent his summer vacation at the University of Michigan, and in 1896 at the University of Wisconsin, studying bacteriology and its methods. Always interested in the latest work in his field of study, Harrison spent eighteen months visiting laboratories in Europe prior to November 1900.

Although Harrison had the title of bacteriologist from 1895, it was 1899 before he could devote full time to bacteriology.[40] In 1902, he published in *Science* what was to be the first of a series of papers on bacterial diseases of plants. Before resigning in 1905 to accept a new position at Macdonald College, he developed a special course of lectures at the OAC "on the relation of bacteriology to agriculture," and another course on "bacterial diseases of plants and the bacterial decomposition of fruits, vegetables, and other plant substances."[41] While at the OAC, Harrison had the help of B. Barlow, B SC, a competent laboratory demonstrator who not only helped with the teaching but also prepared notes of their experiments on bacterial diseases of plants. Barlow was joint author, with Harrison, of OAC bulletin 136, *Some bacterial diseases of plants prevalent in Ontario*, published in August 1904.

S.F. Edwards, about whose teaching little has been learned, succeeded Harrison and was the professor of bacteriology from 1905 until 1914. It may be assumed that Barlow provided some continuity between the teaching and laboratory work of Harrison and that of Edwards.

Barlow was succeeded by Dan H. Jones (1875–1954) in 1909. Jones had come from England in 1901 and almost immediately began working for a farmer who was a graduate of the OAC. It was from the farmer that Jones learned about the college and the advantages of a college education. When he was introduced to officials at the OAC, they gave him further encouragement and offered him sufficient remunerative employment to finance his first year. During the next seven years of working his way to graduation, Jones had various jobs. For a while he was dean of residence and a lecturer in English. Soon after his graduation with a BSA in 1908, he was appointed lecturer in bacteriology and did some postgraduate work in bacteriology at Cornell University before becoming head of the department in 1915. Jones wrote a number of papers on bacterial diseases of fruits and vegetables and increased the plant pathology content of his lectures prior to his retirement, due to ill health, in 1936.[42]

Following the resignations of Harrison and Lochhead, there was a reorganization of departments, staff, and responsibilities at the OAC. A Department of Forestry was added, with Edmund J. Zavitz (1875–1968), BA McMaster, M SC F Michigan, in charge. Zavitz, who was to become known as the Father of Ontario Forestry,[43] included some of the common diseases of forest and nursery trees in his lectures. In 1912 he became the first provincial forester of Ontario and the Department of Forestry at the OAC ended shortly thereafter.

In July of 1905, the biology and geology department was divided into a Department of Botany and Geology and one of Entomology and Zoology. The Department of Botany and Geology got a new professor, S.B. McCready, and two new staff members: John W. Eastham (1880–1968), a graduate of the agriculture department of Edinburgh University who had been teaching for four years in the Cheshire Agricultural College in England, and J. Eaton Howitt (1881–1966), a 1905 graduate of the OAC who had done a year of graduate work at Cornell University. Eastham left the OAC in 1911 without having been much involved with plant pathology, but Howitt, who was head of the Department of Botany from 1911 until he retired in 1948, included plant diseases in his lectures, and in his annual reports,for many years.

Although all three new staff members had some background or experiences in plant pathology, much of the plant pathological work of the Department of Botany was handled, for about two years, by Tennyson D. Jarvis, BSA, a graduate in the class of 1900 who became a lecturer in the Department of Entomology and Zoology in 1906. Jarvis also lectured on plant diseases outside the confines of the college. For example, he gave a lecture on "Fungus diseases affecting cereal crops" and exhibited examples of several important plant diseases to the Farmers' Institute in Toronto, 4 June 1906.[44] Two years later, he was lecturing on insect and fungus diseases to students in the horticulture short course.

Jarvis was relieved of most of his responsibilities for plant diseases at the OAC by Lawson Caesar (1870–1952), who joined the entomology staff shortly after graduating with a BSA degree in 1908. Caesar must have been one of the college's most outstanding graduates. He had earned a BA degree from Toronto in 1895 and done some postgraduate work at Oxford University, plus a stint of teaching as a classics master at the Port Hope High School, before deciding to study agriculture. As a new staff member, Caesar was assigned "special duties of investigating insect injuries and fungus diseases of fruit trees and plants."[45] In 1910 he gave "addresses on insect pests and plant diseases, and spray mixtures" in thirteen different towns, in addition to his normal teaching of those subjects.[46] Two years later he was appointed provincial entomologist, and, although he continued to be involved in some plant pathological work, his interest in plant diseases declined from that date. Caesar was superannuated, as professor of economic entomology and provincial entomologist, 31 May 1940.[47]

For several years plant pathology was so popular at the OAC that virtually every department having anything to do with plants was conducting research on one or more plant diseases. Presumably this was reflected in the lectures of the teaching staff of those departments. Even the president of the OAC, James Mills, and the professor of agriculture, Thomas Shaw, were, at least indirectly, involved with plant pathology. Their book, *The first principles of agriculture*, authorized by the Ontario Department of Education for use in public school agriculture, had a whole chapter on "Diseases of plants."[48]

In the Department of Field Husbandry, Charles A. Zavitz (1863–1942) was doing research on smuts of oats and winter wheat, and commenting on the diseases of field crops in his lectures and annual reports. He had been in the first OAC graduating class to receive BSA degrees from the University of Toronto in 1888. From then until he retired in 1928, Charles Zavitz served the college as an experimentalist, and from 1904, as professor in the newly established Department of Field Husbandry. He became well known as the creator of OAC 21 barley, and several other high-yielding, disease resistant crops.[49] The University of Toronto awarded him an honorary D SC in 1916.

When Eastham resigned in 1911, his position was filled by Walter A. McCubbin (b.1880), a graduate of the University of Toronto who had done some graduate studies in mycology at Harvard. McCubbin had been a lecturer in botany for one year when he resigned, in 1912, to become officer in charge of the Dominion Laboratory of Plant Pathology in St Catharines, Ontario.

McCubbin's replacement, Roland Elisha Stone (1881–1939), who had a B SC from the University of Nebraska and an MS from the Alabama Polytechnic Institute, was appointed associate professor of botany in September 1912. Stone, a charter member of the American Phytopathological Society, had been an assistant in botany at Cornell while studying for the PH D degree that he was awarded in 1913, the year in which a new laboratory at the OAC was equipped with "all the necessary apparatus to carry out research in plant pathology."[50] Stone, the most highly qualified plant pathologist on the teaching staff, taught plant pathology at the college, and he also took his knowledge of plant diseases to farmers and horticulturists through a sort of extension-teaching program and by correspondence. Roland Stone was an enthusiastic supporter of athletics at the OAC, and this support continued after he died: his will provided a gift of $1,000.00 to the college for use by the Department of Physical Education.

Another man who made a significant contribution to the plant pathology program was David Richmond Sands (1883–1966), who had taught in rural schools before enrolling for the degree courses in agriculture at the OAC in 1911. Sands is reported to have collected the first specimen of the white pine blister rust fungus in Canada when making the required collection of diseased plants for Professor Howitt's course in plant pathology.[51] After graduation in 1915, and overseas service in the army medical corps, Sands was hired as an assistant in botany in 1919, but he was soon granted leave to study plant pathology at the University of Wisconsin, where he obtained an M SC in 1922. Later that year he returned to the OAC and was promoted to the academic rank of lecturer in the Department of Botany. He became well known as a plant pathologist while serving as secretary-treasurer of the Canadian Division of the American Phytopathological Society from 1924 until 1929. When he became an assistant professor in 1931, three courses in plant pathology were being offered at the OAC: Elementary Plant Pathology, Plant Pathology, and Advanced Plant Pathology. The latter was "a study of the nature and etiology of disease and the more important diseases of field, orchard, greenhouse and garden crops." Students taking that course were required to make a collection of fifty diseased plants.[52]

There were no changes in the senior staff, nor any major change in the teaching of plant pathology at the OAC during the Second World War. One of the first additions to the teaching staff, following that war, was that of Lloyd V. Busch (b.1918), BSA, a veteran of four years in the Canadian Army, who was appointed lecturer in the Department of Botany, 7 February 1946, while studying for the M SC degree that he obtained there in 1949.

Queen's University

Queen's College, incorporated by Royal Charter in 1841, became the University of Queen's College in 1862, Queen's College and University in 1885, and Queen's University in 1921.[53] The early Queen's calendars note those changes, and name members of the teaching staff, but they reveal very little about any teaching that could be construed as being directly or indirectly related to plant pathology.

The first biologist at Queen's, and a teacher who probably commented on plant diseases to his students, was George Lawson (1827–1895). Before going to Queen's in 1858 as professor of chemistry and natural history, Lawson was a student of and assistant to the renowned Professor John Balfour in the University of Edinburgh.[54] He was also assistant secretary and curator of the Botanical Society, clerk of the Caledonian Horticultural Society, and interim lecturer on natural science at the New College, Edinburgh. Shortly before going to Queen's, Lawson acquired a PH D from Giessen in Germany; thus he was very well qualified to teach the natural sciences. Among the reasons for assuming that he included comments on plant diseases in his lectures at Queen's is the fact that he had published a paper titled "The Grape Blight (Oidium Tuckeri)" in *Chambers's Edinburgh Journal*, no. 477, in 1853. There is also the supporting fact that soon after he resigned to become professor of chemistry and mineralogy in Dalhousie University, Halifax, he published, in the first volume of the *Journal of Agriculture for Nova Scotia*, a paper on blights and diseases of plants. This was soon followed by articles on potato blight and other diseases. It is therefore almost inconceivable that a man so interested in and so knowledgeable about plant diseases would not have included comments on them in his lectures at Queen's.

Lawson is remembered mostly for his work as a botanist. It was under his enthusiastic tutelage that the short-lived Botanical Society of Canada was formed, 7 December 1860, in Kingston. To commemorate that event, the highest award that the present-day (1994) Canadian Society of Botany can bestow on an individual is the George Lawson Medal.

Lawson was succeeded by one of his former students, Robert Bell (1841–1917), who left the Dominion Geological Survey to teach chemistry and natural history at Queen's from 1863 until 1867.[55] Bell too may have included some plant pathology in his lectures, but the published copies of his examination questions provide no evidence of this.

The earliest examination question to include the name of a plant disease is no. 4 of a midwifery examination, in the academic year 1879–80. That question, posed by Nathan F. Dupuis, who had succeeded Robert Bell as professor of chemistry and natural history, and who was also a professor in the Faculty of Medicine, simply asked the student to "Give the indications for and against use of ergot." A somewhat similar question, "Give the doses of the following medicines: Ergot of rye ..." appeared in Dupuis's examination in 1881–82.[56] Such questions show that Dupuis was familiar with the uses of ergot in medicine, and it is safe to assume that he would make some comment on this disease of grasses in his lectures in botany, chemistry, or natural history. Dupuis, who as a young man had been apprenticed to a clockmaker in Kingston, was appointed a paid "observer" while still a student at Queen's. When he graduated in 1866, he became the librarian at Queen's and, in 1878, he was appointed to the chair of chemistry and natural science. After thirteen years in that position he left the natural sciences to become professor of mathematics.

The first definite evidence that plant diseases were included in the lectures of a professor at Queen's University appears in an examination paper prepared by the Reverend James Fowler (1822–1923), MA, professor of natural science from 1885 until he retired and was designated emeritus professor of botany in 1908. In an examination for second-year students in the academic year 1885–86, he asked, "Distinguish between parasitic, soprophytic [sic] and epiphytic plants and give examples of each."[57] In 1895–96, he had questions dealing with plant diseases and the fungi that cause them in both the cryptogamic botany and the physiological botany exams.[58]

James Fowler was primarily a botanist with a keen interest in fungi. This is seen in his *Preliminary List of the Plants of New Brunswick*, published by the New Brunswick Department of Agriculture in 1879 and including the names of fifty-three fungi.[59] Judging from the questions in his examination papers, especially those dealing with cryptogamic botany, it would appear that Fowler's interest in plant parasitic fungi increased as he grew older. In 1897 he changed the name of his course to "Practical Cryptogamic Botany" and began questioning his students about the host specificity of parasitic fungi. He did not actually use the term "host specificity," which was a very new concept at that time, but his use of the word "symbiotic," *sensu* de Bary, conveyed a similar meaning.[60]

Fowler was born in Bartibog, New Brunswick, and graduated from the theological college in Halifax, Nova Scotia, in 1855. He was ordained a Presbyterian minister in 1857, and it was while preaching in New Brunswick that he began to study natural history from books and through correspondence with well known naturalists, including Asa Gray. He was awarded an MA degree by the University of New Brunswick in 1872, and continued to preach until forced to stop, because of a throat infection, the following year. Fowler then taught nature study for two years at the provincial Normal School, Fredericton. He resigned from that position to accept the teaching appointment at Queen's.[61] In botanical circles, Fowler is remembered for the large collection of plants that he added to the Queen's herbarium, and for having compiled the first comprehensive lists of the flora of New Brunswick. It was, in large part, for this latter work that the University of New Brunswick gave him an honorary LL D in 1900.

William Thomas MacClement (1861–1938), or "McClement," before he learned that his father had used the former spelling,[62] was the man who really brought plant pathology to Queen's University. MacClement joined the staff in 1906, when he was forty-five years old, as a sort of helpmate for his former professor, Dr Fowler, who was about to retire. MacClement had taught in the public schools of Ontario before getting a BA degree in 1888, and an MA the following year, from Queen's. This was followed by a six-year stint of teaching in Ontario high schools, then ten years, 1896–1906, of teaching chemistry at the Armour Institute of Technology in Chicago. His teaching there must have been of superior quality, because the institute awarded him a D SC degree the year he resigned to join the teaching staff at Queen's.[63] From then until he retired in 1936, the students in the Department of Botany (Biology, after 1919) were exposed to lectures that contained increasing amounts of plant pathology.

We may never know who first aroused MacClement's interest in the fungi and plant diseases, but it is known that from 1893 until 1896 he was secretary-treasurer and then, for one year, chairman of both the Botanical Section and the Microscopical Section of the Entomological Society of Canada, at London, Ontario. The significance of this information lies in the fact that John Dearness was giving lecture-demonstrations of plant diseases in those sections at that time. Furthermore, the minute books also show that MacClement, Dearness, and William Saunders were together at several of those meetings.[64] In 1910, the course outline for the medical botany course included "Moulds, Rusts, Smuts and Bacteria," and a botany-morphology examination question that year asked the students to "Sketch the life history of two of the following parasitic fungi: Late Potato Blight, Black-knot of Cherry, Apple Scab, Potato Scab, Ergot of Rye, Timber Destroyer."[65] MacClement visited several plant pathology laboratories during his summer vacations, and otherwise kept abreast of ideas and methods in that general field of science. In 1912 he was listing such books as *Diseases of economic plants* by F.L. Stevens, *Fungous diseases of plants* by B.M. Dugger, and *Diseases of cultivated plants and trees* by G. Massee as recommended reference books for students studying plant pathology.[66]

In 1916 the course work for honours students in botany was so arranged that they could specialize in biology, paleontology, or plant pathology.[67] Although MacClement's students had been studying plant pathology as an important aspect of botany for several years, 1916 was the first year in which they could officially specialize in that subject. Even then there were no courses being offered under that name. The course that contained more plant pathology than any other was MacClement's Biology 18, Economic Fungi, which dealt with "The structure and life histories of the fungi which produce diseases in plants." It was listed for the first time in 1918,[68] and it was still on the curriculum in the 1930s. The Queen's calendar for 1932–33, in referring to graduate courses in biology, stated, "Lectures and laboratory courses are to be selected in consultation with the Head of the Department from: 110 Dendrology, 113 Plant Pathology, 115 Cytology."[69] Biology 18, Economic Fungi, was a course for undergraduate students. It consisted of "class discussions and laboratory studies of important plant diseases and the fungi producing them."[70] That course remained as part of the curriculum until MacClement retired in 1936, after which all references to plant pathology were dropped from the calendar. One year later Queen's awarded him an honorary LLD and named him emeritus professor.

Those honours were not rewards for his teaching alone but also recognition of his pioneering work in establishing the Queen's Summer School in 1910 and in being its director for nearly thirty years. MacClement took a very active role in the Canadian Division of the American Phytopathological Society and was its president in 1924–25. He was also a member of the Quebec Society for the Protection of Plants for many years, beginning in 1910.[71]

William T. MacClement's teaching at Queen's had a more profound influence on the early development of plant pathology in Canada than did that of any other native son. His influence was manifested almost entirely through his students, notably B.T. Dickson, J.G. Coulson, and D.L. Bailey, all of whom became outstanding teachers of plant pathology. For more than two decades their student graduates filled at least 90 per cent of the plant pathology positions in Canada. MacClement also taught V.W. Jackson, G.H. Berkeley, W.T. Groves, L.W. Koch, A.N. Langford, J.D. MacLaughlan, A.R. Walker, and Mildred K. Nobles, all of whom became distinguished for their research in some aspect of mycology or plant pathology in Canada.[72]

Plant pathology, or courses containing a significant amount of plant pathology, did not reappear in the calendars of Queen's until several years after the Second World War. However, the one for 1941–42 shows that the course Advanced Plant Physiology, given by Gleb Krotkov, B SC(Prague), MA, PH D(Toronto), dealt with problems associated with an aspect of post-harvest plant pathology. In 1943, he added a course called Cold Storage and Gas Storage, thus increasing the availability of lectures on post-harvest plant problems and their control. Those courses were offered until 1946, when Krotkov went on leave.

McMaster University

From 1876, when the crop of wheat at the Ontario School of Agriculture and Experimental Farm failed, "owing principally to rust,"[73] until the end of the century and beyond, there was a growing interest in plant diseases throughout the agricultural areas of southern Ontario. As a result, teachers of agriculture and nature study in schools and colleges began to include commentaries on plant diseases in their lectures to students. The teachers at McMaster University were no exception to this general phenomenon. However, it was not until 1886, nine years after the university had moved from Toronto to Hamilton, that a question about plant diseases appeared on its published examination papers. In that year A.P. Coleman, PH D, asked his students in first-year biology to "give the life history of the wheat rust."[74] Although questions about plant diseases did not appear annually on the exam papers, they were asked often enough to show that the professor of biology was teaching some plant pathology each year. In 1903 there was a question about black knot, a disease of plum and cherry trees that had become so prevalent as to prompt John Craig, the Dominion horticulturist, to publish a bulletin on the topic.[75]

Plant pathology became a more significant component of the biology courses at McMaster following the appointment of Robert W. Smith (1860–1942), BA, PH D, as lecturer in physics and biology, in May 1899. Smith was the first McMaster graduate in arts to become a member of the faculty. In 1900 his title became professor of biology and experimental physics, and after a period as professor of biology and geology, he became professor of biology in 1909, the title which he held until he retired as head of the Department of Biology in 1935.[76]

Smith asked his students for the life history of the wheat rust fungus on exam papers at least twice before 1912, the year he asked his third-year biology students to "Sketch the life history of smut (Ustilago). Make a list of the classes of fungi related to smut, and state the basis of the relationship. What are the best practical ways of combating the smuts?" The outline of his botany course that year, and for several subsequent years, included a study of "common economic fungi (a collection to be made), with further study of fungus diseases."[77]

Incidentally, students at McMaster could study for a bachelor's degree in agriculture by doing two years there and two at the Ontario Agricultural College.[78]

Smith's exam questions in 1913 included one on white rust, and the request to "Give a list of the parasitic fungi you have studied, with their scientific names and classification, and the names of their hosts." That question alone, without listing those that appeared in later exam papers, is enough to show that Smith considered plant pathology to be an important part of the biology courses he was teaching. It was Smith's enthusiasm for the subject of plant diseases and their inciting fungi that influenced H.R. McLarty, H.N. Racicot, and I.L. Conners to become plant pathologists.[79]

Smith got some welcome help with his heavy teaching load and other responsibilities when Lulu O. Gaiser (1896–1965) was appointed lecturer in biology, 14 July 1925. Gaiser was the first experienced plant pathologist to join the teaching staff of McMaster. She earned a BA from the University of Western Ontario, then, after teaching in the public schools of Ontario for two years and a period of social work in New York City, an MA from Columbia University in 1921, and, six years later, a PH D in genetics and cytology. In the interval between her two advanced degrees, she worked as an assistant in bacteriology at Columbia, and she got several months of experience in plant pathology in 1924–25 while employed as junior plant pathologist in the Bureau of Plant Industry, United States Department of Agriculture.[80]

For a few years Gaiser was chiefly interested in the genetics of fungi. During that interval she published a paper, with B.O. Dodge, on the question of nuclear fusion in the blackberry rust fungus.[81] Her background experiences in bacteriology were soon expressed in her teaching and in her examination questions. For example, in 1929 she asked the students to discuss bacteria under several headings and to describe a bacterial disease of vegetables. There were so many questions pertaining to plant diseases on that exam that one could be excused for thinking it was intended for students majoring in plant pathology.

When Smith retired, in 1936, Gaiser became professor of botany, and in the following year she was appointed acting head of the biology department. From the time of her appointment as professor of botany, Gaiser agitated for, and eventually obtained, the greenhouse she felt was needed for the development of a program in natural history. Such a program, she argued, would better prepare women students for teaching the nature study courses that were then in such demand in the public schools. She also increased the amount of plant pathology in her courses in the belief that a greater knowledge of plant diseases would be of benefit to students from the horticultural areas of the Niagara peninsula.[82]

In 1940, Gaiser incorporated some plant pathology in her course, "Biology 42, Thallophytes," which dealt with algae of fresh water, bacteria, and fungi, with special reference to those causing plant diseases. In 1942, a new Department of Botany was created with Gaiser acting as its head. When Norman W. Radforth was appointed as head of the department, in 1946, Gaiser became senior professor of botanical research, a position she held until she resigned in 1949 to become a research fellow of the Gray Herbarium, Harvard.

University of Western Ontario

The Bill to incorporate "the Western University of London, Ontario" received Royal assent on 7 March 1878, but until the Faculty of Arts was set up in 1895 very little "science" was taught at Western.[83] The only paid members of its faculty were professors of language, history, mathematics, and the classics. However, that small staff was supplemented by lecturers who volunteered their services to the university. One of these was John Dearness (1852–1954), who, without himself having a university degree at that time, gave lectures in botany.[84]

Dearness, a native of Hamilton, began his teaching career in the public schools of Ontario in 1869. After having been principal of two schools, he became a public school inspector in 1874. It was during his term as inspector, in East Middlesex, 1874–99, that Dearness acted as a volunteer lecturer in botany at Western. From 1888 until 1914 he also taught botany and zoology at the medical school of the university.[85] Although there is little direct evidence to show how much plant pathology Dearness taught, there is no doubt that he did teach some.

Dearness's early interest in plant pathology is clearly shown in the records of the Microscopical Section of the Entomological Society of Ontario. For example, the minutes of their meetings of 22 and 23 January, and again those of 27 November and 11 December 1891, refer to his lectures on parasitic fungi and his demonstrations of plant diseases. Some of the microscope slides of plant diseases that Dearness prepared for the use of his students at the university and as demonstration materials for members of the Entomological Society are in the care of Regional History, University of Western Ontario. Through extramural studies at Western, Dearness earned the degrees of BA in 1902 and of MA in 1903, and the university awarded him an honorary LL D in 1926, when he was seventy-four years old.[86]

The province of Ontario did not provide direct financial aid to the Faculty of Arts until 1914, and separate departments of Biology, Physics, and Chemistry were not established in that faculty until 1916.[87] It was then that Albert D. Robertson, BA, became professor of biology. Three years later, Nelson C. Hart (b.1888), MA, became his assistant. When the Department of Biology was divided into Zoology and Botany, in 1922, Hart became associate professor and head of the Department of Botany, with Margaret A. McIntosh as instructor.[88] The calendar for 1922–23 shows that "Phytopathology and Mycology, a course dealing with plant diseases and the classification, morphology and structure of fungi" was being offered to honours biology and science students. That was the first course at Western to focus on plant diseases. It had B.M. Duggar's book, *Fungous diseases of plants*, as a recommended text, as did the course called Ecology and Economic Botany, in which special emphasis was placed on economic fungi. Obviously, some plant pathology was included in both courses.

In the announcement for 1929–30, the course Phytopathology and Mycology is listed as one leading to the Master of Arts degree. Apparently it was also a prerequisite for the course Advanced Phytopathology, which dealt with the symptoms, causes, and control of diseases of economic plants.

In a very real sense Nelson Hart was the first professor at Western to teach a course that included a relatively large plant pathology component. Earlier teachers, including John Dearness and Albert Robertson, included plant diseases in some of their lectures, but the topic did not constitute a very significant portion of their courses. Hart was a biologist who spent a lot of time in the greenhouse, which the students jokingly referred to as "Hart House," in an oblique reference to the beautiful and stately building of that name at the University of Toronto.[89]

The increased emphasis on plant pathology within the Department of Botany, under Hart, was accompanied by additions to the teaching staff. In the context of this essay, the most relevant addition was that of Anson R. Walker (1893–1959) in 1926. Walker, who had a BA from Queen's and an MA from Toronto, had been an instructor in botany at the University of Toronto prior to his appointment at Western, which in 1923 had become the University of Western Ontario.[90] Walker, a charter member of the Canadian Phytopathological Society, was promoted to associate professor in 1930, the year in which "Plant Diseases and Fungi: a course dealing with the recognition of poisonous and edible fungi and the identification, cause and control of common plant diseases" was added to the curriculum of general course students. That was also the year in which Walker's studies on strawberry root rot were noted in the annual report of the Dominion botanist. Except for changes in course numbers, credits per course, and recommended texts, etc., the general pattern for the teaching of plant pathology at Western remained about the same until after the Second World War.

QUEBEC

McGill University

Trinity College was not the only institution of higher education in which a member of the teaching staff lectured on plant diseases in the 1850s. At McGill University, or McGill College as it was then, the principal, John William Dawson (1820–1899), who was also professor of natural history and agriculture, discussed plant diseases in his lectures, especially one that included "diseases and enemies of fruit trees, small fruits and vegetables."[91] Dawson's lectures, somewhat like those of Hind at Trinity College, were given is a course called "Agricultural Chemistry."

After Sir Humphry Davy (1778–1822) had so successfully combined the views of Priestly, Lavoisier, Senebier, and Knight in his popular *Elements of Agricultural Chemistry* in 1813, and after the subsequent impressive works on the chemistry of agriculture by Justus von Liebig (1802–1879), it was virtually a duty incumbent on all progressive agriculturists to include the word "chemistry" in the titles of their books or lectures on agriculture. Thus it was not surprising that Hind and Dawson, who were among the most progressive teachers of agriculture in Canada at that time, would give courses that had almost identical titles. As agricultural educators, Hind and Dawson were similar in several respects. They each did some teaching in a normal school, each wrote essays and published books on agriculture that included information on plant diseases, and each won an award or some form of recognition for essays he had written about plant diseases.

During his tenure as superintendent of education in Nova Scotia, Dawson entered an essay titled "The Nature, Causes, and Prevention of the Failure of the Potato Crop" in a competition for a $10,000 "reward" offered by the government of the Commonwealth of Massachusetts. That offer was to anyone within the Commonwealth who could prove that "a sure and practical remedy for the potato rot" had been discovered. Although no one received the governor's reward, Dawson's essay, albeit a "foreign communication," was published in synoptic form by the government of Massachusetts in 1852.[92] Such clear evidence of Dawson's early and informed interest in plant diseases leads one to speculate that his lectures to students of the Normal School in Nova Scotia may have included as much plant pathology as did those of Hind in Ontario, and at about the same time. Supporting evidence lies in the fact that Dawson's lectures in the McGill Normal School, in Quebec, did include the general topic of plant diseases and their control.

Dawson was professor of natural history at McGill from 1855 until 1893, and also professor of agriculture at the McGill Normal School between 1856 and 1870. His book, *First Lessons in Scientific Agriculture*, written specifically for students studying that subject and for teachers in training at the Normal School, contained nearly fifteen pages of information about late blight, rust, smut, and other plant diseases of local concern.[93]

After Dawson, plant diseases were only incidental in the lectures of botanists and mycologists at McGill University, until David Pearce Penhallow (1854–1910) joined the staff. Penhallow, a citizen of the USA with a BS degree from Massachusetts College, came to McGill, on the recommendation of Asa Gray of Harvard University, to lecture in botany during the absence of Dawson in the session of 1883–84. He was persuaded to remain at McGill to fill a newly created chair, set up through public subscription on 4 October 1884. While teaching botany and plant physiology he also studied and did research at McGill, for which he was awarded an MS degree in 1886 and a D SC in 1904.

Penhallow's early interest in plant diseases is shown by his 1883 paper on peach yellows, and papers of 1884 and 1885 on plants in relation to disease.[94] Doubtless the publication of the first two of those papers was at least partly responsible for his invitation to participate in a conference on fireblight held in Massachusetts in 1884.

Penhallow seems to have been particularly interested in and a recognized authority on the diseases of fruit trees. His first paper on a plant disease to be published in Canada was one on the spot disease of apples in 1887.[95] During the following year he became editor of the *Canadian Record of Science*, and published a paper on plant diseases in the first issue.[96] In an address to members of the Royal Society of Canada, 21 June 1897, Penhallow deplored the general lack of interest, on the part of Canadians, in their annual losses due to "the destruction of crops by the operation of disease and parasitic growths." He implied that Canada should follow the example set by the United States, where "plant pathology forms a leading feature in the work of the various experiment stations."[97]

After about 1899, when he was president of the Society of Plant Morphology and Physiology, Penhallow's interest in plant physiology, including plant pathology, became overshadowed by his increasing enthusiasm for paleobotany. Nevertheless, he continued to include examples of plant diseases in his botany and plant physiology lectures until ill health forced him to stop teaching in 1909. He died at sea, while on his way to England, 20 October 1910.[98]

During Penhallow's prolonged illness, and for a period after his death, Carrie M. Derick (1862–1941) assumed responsibility for most of his work. Derick, after several years of teaching, had enrolled in McGill's Faculty of Arts at age twenty-five, where as a brilliant student she came to Penhallow's attention. She won the Logan Gold Medal in the Natural Sciences and graduated with a BA degree in 1890. In September 1891, she became the first woman to receive a teaching appointment at McGill, albeit as a part-time demonstrator in botany for Penhallow. She was named professor of morphological botany during the academic year of 1911–12 – the year in which she delivered the Somerville Lecture. That was a public lecture which she titled "Diseases in plants with special reference to crown gall." As a plant morphologist, Derick was keenly interested in diseases that resulted in any form of plant teratology, and she included comments on them in lectures to her students.

Derick was a thesis adviser for, and otherwise took an active interest in the research of, Douglas Weir, who studied the nature and causes of certain plant malformations in the vicinity of Ste-Anne-de-Bellevue, where he was one of William Lochhead's graduate students at Macdonald College. Weir's study, for the master's degree that he was awarded in 1910, dealt almost exclusively with galls induced by insects and included only one that was fungus-induced. Regardless of their cause, Derick was interested in all kinds of plant malformations, until she retired, for health reasons, in 1929, and was made McGill's first woman emeritus professor.[99]

F.E. Lloyd (1868–1946), AB, AM, who was appointed Macdonald Professor of Botany, 12 June 1912, included plant diseases, especially physiological diseases, in his lectures on plant physiology. He became a member of the Quebec Society for the Protection of Plants, and published a paper on injury and abrasion of *Impatiens saltani* in the sixth annual report of that society. Lloyd was so highly thought of by society members that they held a special meeting in his department and elected him to honorary membership, 6 March 1930. Lloyd retired as emeritus professor at the end of the 1933–34 session because he had reached the age of sixty-five years.

To assist in the rehabilitation of soldiers returning to Canada from the First World War, McGill University created a Department of Soldiers Civil Re-Establishment. In 1920, Eric Boulden, BSA, gave a series of lectures and demonstrations on economic insects, plant diseases, and weeds to classes of returned men (as war veterans were designated at that time) registered in that department.[100]

Macdonald College of McGill University

The opening of McGill's Faculty of Agriculture in Macdonald College at Ste-Anne-de-Bellevue in 1907, and the initiation of instruction there in plant pathology, more or less coincided with the decline of interest in plant diseases by Penhallow on the Montreal campus of the university. In the beginning, practically all teaching of plant pathology, or the lectures that included plant diseases, at Macdonald College was done by two men recruited from the Ontario Agricultural College: Francis C. Harrison (1871–1952) and William Lochhead (1864–1927).

Harrison, who was born in Gibraltar, attended public school in England before coming to Canada in his late teens. He studied at the OAC, where he acquired a BSA degree in 1892, and later undertook some advanced studies at Cornell University. He also studied at universities in Michigan, Wisconsin, Berne, and Copenhagen before going to Macdonald as professor of bacteriology in 1905, when the college was still in the early planning stage.

Harrison's first course in agricultural bacteriology at Macdonald College included "the bacterial diseases of plants; kinds of microorganisms involved, methods of treatment; bacterial decomposition of fruits, vegetables, roots, etc."[101] Later, he also taught Applied Microbiology, a course that included some practical methods for the control of bacterial diseases in plants, and a course called simply "Bacterial Diseases of Plants." Harrison, who had published several papers on bacterial plant diseases and described a soft rot organism before leaving the OAC, was a recognized authority in this general field.

In addition to being professor of bacteriology and head of the department, Harrison was appointed dean of the Faculty of Agriculture and principal of the college in 1910. He held those positions concurrently for sixteen years, except for a period during the war, when, with the rank of lieutenant-colonel, he served as adjutant-general at the Petawawa Artillery Camp, and a brief period after the war, when he went overseas to be in charge of the Agricultural Section of the Khaki University.[102] Teaching and administrative duties kept Harrison from personally doing much research at Macdonald College. Nevertheless, he supervised, or shared in the supervision of, the thesis research of several graduate students who later made significant contributions to plant pathology in Canada. Perhaps the most noteworthy were Kenneth Harrison, who became a distinguished plant pathologist at the Agricultural Research Station in Kentville, Nova Scotia, Bertram T. Dickson, and Thomas C. Vanterpool, both of whom soon became well known teachers of plant pathology.

In 1926 Harrison left the Macdonald campus to become professor of bacteriology in the Faculty of Medicine on the Montreal campus of McGill University. However, the courses that he initiated on the bacterial diseases of plants were available to plant pathology students for many more years. As late as 1943, such students were required to take twelve laboratory periods on bacterial diseases of plants in the Department of Bacteriology.[103]

William Lochhead, who had a BA degree from McGill and an M SC from Cornell, plus more than seven years of teaching experience at the OAC, was invited to join the planning group for Macdonald College in 1905, at a salary of $2,000.00 per annum. This he accepted, but he went back to the OAC for a while in 1906 to help in a nature study program. However, he was on duty as professor of biology and head of the department when Macdonald College opened its doors for its first students in the fall of 1907. Lochhead was a plant protectionist with a strong leaning toward entomology that lasted for the remainder of his life. The strength of his protectionist attitude became very evident in 1908 when he, with the cooperation of several like-minded Quebec biologists, organized the Quebec Society for the Protection of Plants, the presidency of which he held for seventeen years.[104] There was a time in his career when Lochhead showed great interest in plant pathology as a major aspect of plant protection. For example, in 1906 he published five short articles on plant diseases induced by fungi in the *Canadian Horticulturist.*[105]

The first official announcement of Macdonald College shows that Lochhead taught, among others, "Course 6, Plant Diseases, a laboratory and field study of the common parasitic fungi of cultivated plants and methods of prevention and treatment; study of diseased tissues; cultural studies." To this was added, in 1912, "A collection of 50 varieties of fungi is required of each student." Lochhead also taught "Course 14, Biological theories, a discussion of the problems and factors of genetic evolution, heredity, variation, mutation, Mendelism, origin and distribution of life."[106]

In January of 1912 Lochhead hired P.I. Bryce as an assistant in biology and to help with his laboratory classes. Although there is no record of Bryce doing any lecturing in plant pathology, he was certainly involved with its teaching in the laboratory. He was appointed curator and librarian for the Quebec Society for the Protection of Plants in 1912, and published a paper on the decay of harvested apples in the fourth annual report of that society. Subsequently, he had at least one article published in each of those reports from 1915 through 1918. Bryce left Macdonald College on 31 August 1920. In the meantime, some plant pathology was also being taught in the School for Teachers, Macdonald College. The nature study course, given by John Brittain (d.1913), D SC, included a "study of the autumn stage of insects weeds and plant diseases" and in the spring, "weeds, insects and fungous diseases of the garden."[107] In 1915, D.W. Hamilton, BSA, PH D, lecturer in nature study and elementary agriculture in that school, wrote a book for his students in which he outlined two lessons on plant diseases.[108]

The teaching program within the Department of Biology underwent its first major change when William P. Fraser (1867–1943) joined the staff as a lecturer in 1912. From that time onward, Lochhead concentrated his efforts mostly on entomology and zoology, leaving the botany, mycology, and plant pathology in the capable hands of Fraser, whose work with the rust fungi in his native province, Nova Scotia, had already come to the attention of J.C. Arthur, the outstanding American authority on rusts.[109] Fraser spent the summer of 1915 in Nova Scotia studying the apple disease problems for that province's Department of Agriculture. That was also the year in which he was promoted to the rank of assistant professor, and his major course at Macdonald College became "Plant Diseases and Fungi." Fraser's teaching of that course made such a favourable impression on one of his students, Margaret Newton, that she devoted the rest of her life to plant pathology.[110] Fraser taught plant pathology until he left the college in February 1919, to take charge of a new Dominion Laboratory of Plant Pathology in Saskatoon.

Fraser's successor at Macdonald College was Bertram Thomas Dickson (1886–1982), a native of Leicestershire, England, who had earned a BA degree from Queen's University in 1915. Dickson had been an instructor of plant pathology at Cornell University before going overseas in the First World War to serve as assistant agricultural officer in the First British Army. After the armistice, he became commandant of the First Army School of Agriculture in France. In December 1918, he was appointed assistant professor of biology, responsible for the teaching of botany and plant pathology, at Macdonald College. One year later, the biology department was divided into a Department of Entomology, under Lochhead, and a Department of Botany, with Dickson as its head. In the academic year of 1919–20 a plant pathology "option" became available to undergraduate students, and the departments of Botany, Entomology, and Cereal Husbandry were authorized to offer courses for a new degree, that of Master of Science in Agriculture, designated as MSA. During the following year those departments could offer courses leading to the M SC and the PH D, in addition to the MSA.[111]

Dickson expanded the undergraduate option in plant pathology to include two plant pathology courses, plus one on research in plant pathology. This latter was of particular benefit to senior undergraduates, because they were required to do research and to write a thesis.[112] One who wrote an undergraduate thesis and went on to become a respected plant pathologist for many years and president of the Canadian Phytopathological Society was John F. Hockey (1895–1980). His thesis for the BSA that he earned in 1921 was titled "Mosaic Disease of Tobacco."

Dickson designed an entirely new set of courses to form the basis for postgraduate studies in plant pathology at the M SC and PH D levels. That new graduate program provided courses in pathologic plant histology, advanced plant physiology, research in plant pathology, history of plant pathology, and Special courses in plant pathology, in addition to systematic mycology and seminar. The special courses were in diseases of field crops (for specialists in cereal husbandry), diseases of fruit crops (for specialists in horticulture), and diseases of forest trees and timber (for specialists in plant pathology, and others interested in the topic).[113] In 1922, Dickson added "Advanced Plant Pathology" to the list of postgraduate courses, and advised candidates for the PH D that they should have a thorough knowledge of physics, chemistry, and genetics, and that a reading knowledge of Latin, French, and German was required.[114] Almost from the beginning, Dickson put emphasis on the physiological aspects of host-parasite interactions in his lectures and research in plant pathology. In this he was a North American pioneer.

In retrospect, it can be seen that Dickson was an accurate and perceptive interpreter of the level plant pathology had reached, and of the path it should take to become fully recognized as a science. He believed that the science had just about gone through its essential early stage of identifying and describing causal agents and disease symptom syndromes and should be probing the complex problems of the physiological and biochemical interrelationships of living hosts and pathogens. To strengthen his program along those lines, Dickson persuaded the authorities of McGill University in 1921 to hire John G. Coulson (1893–1974) as a lecturer in his department. Coulson had taught in the public schools of Ontario and served in the Royal Air Force before becoming a gold medallist in chemistry at Queen's University, where he got a BA degree in 1920 and an MA in 1921. At first Coulson taught botany and plant physiology to undergraduates and assisted Dickson with his graduate courses, especially those that involved physiology and biochemistry, courses which eventually became his own.

Dickson completed virtually all course work for a PH D at Cornell, but he had not liked what he considered to be the excessive emphasis on the mycological aspects of plant pathology that Professor H.H. Whetzel and his staff put into their teaching.[115] That may have been what prompted Dickson to develop a different approach, one that was such a break from the traditional that it gave the teaching of plant pathology a new purpose and outlook. His bold and imaginative program was so well received at Macdonald College that twenty-two students opted for plant pathology in 1924. In those days that was an unprecedented number.

Dickson earned a PH D degree from McGill University in 1922, largely for his studies of virus diseases. He became so well known and respected among plant pathologists that he was elected president of the Canadian Division of the American Phytopathological Society, and Canadian representative on the council of that society for 1923–24. He also became a member of the Board of Control for *Botanical Abstracts*, secretary-treasurer of the Quebec Society for the Protection of Plants, president of the Macdonald College Branch of the Canadian Society of Technical Agriculturists, and president of the Macdonald Philharmonic Society. When the Philharmonic Society produced Gilbert and Sullivan operettas, Dickson displayed his artistic talents by painting the scenery while Harrison directed the dancing.[116]

The use of the term "virology" was unknown to Dickson in 1924. Evidence for this is in the following statement in his presidential address to members of the Canadian Phytopathological Society that year: "in fact we are now accustomed to the term 'Uredinologist,' and who knows but soon we may hear of 'Virology,' ... or some other high sounding '-ology.'" That quotation is from the brief history of plant pathology in Canada that Dickson wrote in 1925.[117] In the following year, when the name of the Department of Botany at Macdonald College was changed to that of Plant Pathology – the first such departmental designation in a Canadian university – Dickson automatically became its head. He retained that position until he resigned in 1927 to become senior mycologist to the Council for Industrial Research of the Commonwealth of Australia. He sailed from Vancouver on 19 October 1927.[118]

In many respects, the teaching of plant pathology at Macdonald College reached a high point during the Dixon regime. That was when Professor Harrison, in the Department of Bacteriology, taught a course called "Bacterial Diseases of Plants," and one called "Applied Microbiology." Professor Lochhead, who considered insects to be among the causal agents of plant disease, included some plant pathology in his entomology lectures, and agronomists Robert Summerby and Emile A. Lods, who cooperated with the plant pathologists in the control of grain smuts,[119] included comments on grain diseases, and those of root and forage crops, in their lectures.

The inclusion of plant diseases in agronomy lessons, and the close cooperation with Dickson and Coulson in plant disease research at Macdonald College, began with the first professor of cereal husbandry, Leonard S. Klinck, who published a paper on the cereal smut diseases in 1910.[120] That practice was followed by Gordon P. McRostie, who, with a PH D from Cornell, came to the college as assistant professor of agronomy in 1920 to take charge of forage crop breeding. He worked closely with Dickson on studies of plant viruses, and with him published a paper on mosaic diseases.[121] For McRostie, who resigned in 1923 to become Dominion agrostologist, plant pathology was more than a mere sideline. He considered it to be a major aspect of his work as a plant breeder.

Lee C. Raymond (1889–1966), assistant professor of agronomy, continued the cooperative studies with the plant pathologists at the college. When disease symptoms in turnips assumed significant proportions in 1932, Raymond and Coulson researched that disease for several years.[122] Coulson's students also participated in that study, especially René-O Lachance (1909–1992), who made the pathological anatomy of turnips, and certain other plants, the topic of his research for both the MS and PH D degrees that he earned in 1939 and 1940 respectively.

In the 1920s the horticulturists at the college were concerned about apple canker, "the most serious disease with which we have to contend."[123] As a consequence of that concern, they included comments on this and other diseases of horticultural crops in their lectures, especially in the course called "Orcharding."

It was during the Dickson period that the Department of Chemistry first offered a course on insecticides and fungicides, and one on their analysis.[124] Some of the students in the chemistry department studied plant disease problems for their advanced degrees, as Walter DeLong did in 1924. His master's thesis dealt with the pentosan content of the apple in relation to its winter-hardiness. Throughout the Dickson era, the diploma students were also being taught plant pathology. In 1927–28 their curriculum included "Studies of important plant diseases dealing with losses, symptoms, spread and control. Preparation and testing of fungicides such as Bordeaux, lime-sulphur, etc. The uses and limits of spray calendars. Seed treatments. Quarantine regulations and inspection services."[125]

Among Dickson's students during that exciting period in the history of the Department of Plant Pathology were Thomas G. Major, M SC 1922, who stayed at the college for an additional year as an assistant in botany, and became the first secretary-treasurer of the Canadian Phytopathological Society; Gordon A. Scott, M SC 1924, who was appointed plant pathologist at the Dominion Laboratory of Plant Pathology, Saskatoon; William Popp, M SC 1926, who went from Macdonald College to the Dominion Laboratory of Plant Pathology, Winnipeg; and Thomas C. Vanterpool (1898–1984) and John E. Machacek (1902–1970), both of whom became eminent plant pathologists.

Vanterpool, a BSA graduate of Macdonald College, and winner of the F.C. Harrison Prize in plant pathology, became an assistant in the department while working for the master's degree that he obtained in 1925. The published result of his thesis research was one of the first to show that a distinctive plant disease could be due to two viruses. It became a classic in the literature of virology.[126] After a year of study with Professor A.H.R. Buller at the University of Manitoba, while holding a Hudson's Bay Research Fellowship, Vanterpool returned to Macdonald College as a lecturer in plant pathology. He resigned from that position, in 1928, to join the staff of the biology department, University of Saskatchewan.

Machacek had a master's degree from the University of Saskatchewan when he came to Macdonald College in 1926 to study plant pathology under Dickson, who employed him, at the rate of $62.50 per month, as a laboratory assistant while working for the PH D degree that he earned in 1928. His was the first doctorate to be awarded by McGill University for research in plant pathology. The results of his experimentation, supervised by Dickson, with some help from Coulson and Harrison, also became a classic publication in the realm of host-pathogen associations.[127] Machacek was promptly engaged as a lecturer in the Department of Plant Pathology, to teach mycology and to help with the teaching of plant pathology during the transitional period following the resignation of Dickson. He left that position in 1930 to become an assistant plant pathologist with the Dominion Department of Agriculture.

Machacek's successor was Dorothy E. Newton, the first woman academic to be hired by the department. She had a PH D in mycology for work done under the supervision of Professor A.H.R. Buller, at the University of Manitoba, in 1932. She and her successors, Harold J. Brodie, 1935–37, and Ivan H. Crowell, 1937–46, were primarily mycologists and their teaching was more in the realm of mycology and botany than plant pathology. However, they did teach some plant pathology and their contributions to the Department of Plant Pathology were certainly meritorious, especially those of Newton who, for a brief period, was Coulson's only professional assistant in the department.

They got welcome relief in July 1931 when Ross F. Suit, a PH D graduate of Iowa State University, joined them as an assistant professor in plant pathology. Suit's plant pathological interests, as reflected in his publications, were mostly in the realm of bacterial diseases.[128] His position within the department was officially terminated at the end of August 1936, but he was retained as an assistant until 8 November 1936.[129]

Brodie, who had earned a PH D in mycology from the University of Michigan in 1934, and done a year of postdoctoral work in mycology at the University of Toronto, was the most highly trained mycologist in McGill University. Perhaps it was because he was so highly trained that he left the department in 1937.[130] Crowell was a PH D graduate of Harvard who had studied the susceptibility of lilacs to mildew at the Arnold Arboretum. After coming to Macdonald College, he published a comprehensive study of the geographical distribution of the genus Gymnosporangium,[131] and, with the help of E. Lavalee, compiled one of the first checklists of the diseases of economic plants in Canada.[132]

Dickson was succeeded as head of the Department of Plant Pathology by Alfred H. Gilbert, M SC Wisconsin, about whom very little is known because he resigned in July 1929 and returned to the United States. Apparently Gilbert had not intended to stay very long in Quebec: he is reputed to have kept Vermont licence plates on his car all the time he was at Macdonald College.[133] He in turn was succeeded by Coulson, who, while still an assistant professor, was made head of the department in the fall of 1929.

In 1930 George W. Scarth (1881–1951), an eminent plant physiologist in the botany department of the central campus of McGill, and a close friend and golfing companion of Coulson, began teaching plant physiology in the Department of Plant Pathology.[134] Scarth was interested in physiological plant diseases, especially those resulting from low temperatures. Such diseases were important components of his lectures and the major focus of his research and that of his graduate students, several of whom did virtually all of their thesis research in the Department of Plant Pathology.

Scarth, with Coulson, supervised the studies on osmotic and water permeability relationships in the parasitism of plant pathogens carried out by Frederick S. Thatcher (b.1910) for the PH D that he was awarded in 1939.[135] Thatcher had earned a BSA in 1933, and an M SC under Coulson's supervision in 1935, and been employed as a demonstrator while working for his doctorate. After a year of postdoctoral studies at the University of Minnesota on a Royal Society of Canada Fellowship, Thatcher returned to Macdonald College in 1941 as a lecturer in the Department of Bacteriology. He became an assistant professor in 1943 and resigned in 1950 to accept an appointment in Ottawa.

Jacob Levitt, PH D 1935, and David Siminovitch, PH D 1939, both of whom became internationally known authorities on cold injury to plants, were students of Professor Scarth in that decade. They made some pioneering studies on the mechanism of plant cell freezing, freezing injury, and the resistance to freezing injury by cells and tissues. Scarth continued to work at Macdonald College throughout the Second World War, easing the pressure on Coulson and permitting him to concentrate on the improvement of his courses and the development of a new one for graduate students which he named "Principles of Plant Pathology."

Ralph Antony (Tony) Ludwig (1915–1977), who had bachelor's and master's degrees from the University of Alberta, came to Macdonald College, where he was a demonstrator in plant pathology from 1940 until he was promoted to lecturer in 1944 and then to assistant professor in 1945. Ludwig, who obtained a PH D from McGill in 1947, was somewhat more of a plant physiologist than a plant pathologist. He is included here because he assisted with the teaching of plant pathology and included diseases due to physiological causes in his lectures on plant physiology. Ludwig published a paper on the health condition of Quebec-grown oat and barley seed samples in the annual report of the Quebec Society for the Protection of Plants for 1945, and again, with Thomas Simard, in that for 1947.

Except for the continuous upgrading of courses there were no major changes in the plant pathology teaching program at Macdonald College for several decades following Dickson's resignation. That alone is mute testimony of the soundness of his philosophy and his teaching program.

Lulu O. Gaiser (1896–1965)
Courtesy Canadian Baptist
Historical Collection

Thomas N. Willing
(1858–1920)

William T. MacClement
(1861–1938)

Carrie M. Derick
(1862–1941)
Courtesy McGill University
Archives

Dixon L. Bailey (1896–1984)

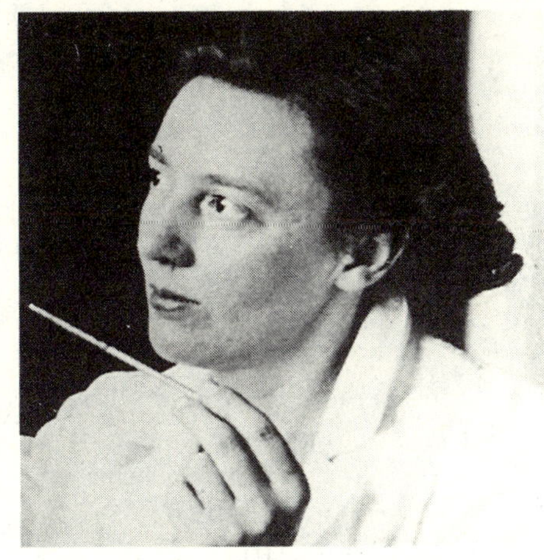

Eleanor Silver Dowding
(1901–91)
Courtesy University of Alberta
Archives

Bertram T. Dickson
(1886–1982)
Courtesy W.E. Sackston

John W. Eastham
(1880–1968)

Laval University and the University of Montreal

For a number of years in the early part of the twentieth century, two schools of agriculture were affiliated with Laval University, one at Ste-Anne-de-la-Pocatière and one at Oka. The one at Oka became affiliated with the University of Montreal when that university became independent from Laval in 1919–20. Students of agriculture in the school at Ste-Anne-de-la-Pocatière, established in 1859, were being taught some plant pathology as early as the academic year 1899–1900, in a course called "Fruit Tree Culture" that included "Different diseases of fruit trees and remedies to be applied."[136] That course, with but slight variations in its component of plant pathology, was part of the curriculum for many years, and taught by a succession of teachers, beginning with the Rev. Joseph Richard.

The teaching of fruit tree diseases at Ste-Anne-de-la-Pocatière and at Oka was a logical consequence of the establishment of several experimental fruit orchards in various parts of the province by the Quebec Department of Agriculture in 1898, and of the rising influence of the Pomological Society of Quebec, which had been organized in 1893.[137] Nevertheless, plant pathology, or vegetable pathology as it was called, did not become a significant component of the teaching at Ste-Anne-de-la-Pocatière until some time after Laval had conferred the newly created degree Bachelor in Agricultural Science (BAS) on students graduating from that school in December 1913.[138]

The first specific mention of vegetable pathology as part of the teaching responsibility of one of Laval's professors is in the annual report for the academic year 1916. In the staff list for that year, Georges Bouchard (b.1888) is designated as "Professor of Botany, Entomology, Vegetable Pathology." Bouchard, who had obtained a BA degree from Laval in 1908 and studied agriculture at Louvain, Belgium, and at the University of Angers, France, between 1912 and 1914, joined the teaching staff in 1914. He also did some graduate work in plant pathology at Cornell University in 1919, with the objective of making his teaching more effective. However, Bouchard's teaching became somewhat of a secondary occupation when he became a federal member of parliament for Kamouraska County from 1922 until 1940. He did some teaching in the academic year of 1922–23, but went to Europe before the term ended to represent the federal minister of agriculture on an exhibition train.[139]

Elzéar Campagna (1898–1987), BA, Laval, and a 1923 BAS graduate of the school, delivered some of Bouchard's lectures during his periods of absence. Beginning in February 1924, Campagna became responsible for the courses in botany, vegetable pathology, microscopy, and microbiology.[140] After nearly three years of teaching, during which there was a rising interest in plant pathology, Campagna got leave of absence and studied plant pathology at Cornell University in 1926–27.[141] When he returned from that period of advanced study, his academic title was changed to professor of botany and plant pathology, and he became chairman of the Department of Botany. Campagna was destined to become a respected teacher of plant pathology, and various other subjects, for more than forty years. He was president of the Quebec Society for the Protection of Plants, 1938–39, and Laval University recognized his distinguished services to that institution by awarding him a D SC (le diplôme de docteur ès sciences naturelles) in 1940.[142]

In 1940, Laval's School of Surveying and Forest Engineering adopted a new curriculum, and René Pomerleau (b.1904) was appointed professor of forest pathology. Pomerleau had been teaching forest pathology at the Forest Rangers' School at Duchesnay, and he continued to do so, on a part-time basis, until 1950. At Laval, he gave the course previously taught by Georges Maheux (1889–1977), who was the provincial entomologist and professor of entomology. Maheux had obtained an M SC degree from Cornell University in 1919. He continued to teach entomology and zoology until he retired in 1964. In 1937 the University of Montreal awarded him a doctorate of science *honoris causa*, and Laval University crowned his career with the award of a doctorate of forest science and named him emeritus professor. Maheux had become a member of the Royal Society of Canada in 1944, and president of its biology section the following year.[143] When the School of Surveying and Forest Engineering became Laval's Faculty of Forest Engineering in 1945, Pomerleau continued to teach forest pathology there, and, as he had been doing since 1942, also in the Department of Biology, Faculty of Science, until 1965.[144]

The "École Supérieure d'Agriculture de Sainte Anne-de-la-Pocatière," its official name as given in the report of the minister of agriculture, became the Faculty of Agriculture of Laval University in April 1940. This was followed by some alterations in the curriculum, but there was little change, for several years, in the general area of plant pathology. The teaching staff of the Ste-Anne-de-la-Pocatière school was often augmented by collaborators from the federal or provincial departments of agriculture. For example, the report of the Quebec minister of agriculture for 1942 shows that the collaborators that year included Henri Genereux (b.1911) BA, BSA, M SC, a plant pathologist with the provincial Department of Agriculture, who taught a course in plant protection to young farmers, and Champlain Perrault (1903–1988), BA, M SC, plant pathologist with the federal Department of Agriculture, who gave a course in "technical phytopathology and vegetable pathology."

Plant pathology never had a very prominent place in the curriculum of the agricultural school at Oka, although the subject did enter the lectures of more than one of the teachers there. In 1897, senior students in a two-year course were being taught "Diseases of cereals – rust, ergot, caries, smut. Destruction of the germs of caries and smut in the seed." Diseases of potatoes and beets also formed part of the course of instruction that year.[145] Those lessons were probably given by Gustave Boron, who was listed as "School Professor." In 1906, Gabriel Reynaud, who had been at the school since 1890, reported a serious fungal disease in the school orchard. He also mentioned that the students, under his direction, amputated and burned the diseased parts and thus got a practical lesson in disease control.[146]

The Oka Agricultural School became the Agricultural Institute of Oka when it officially affiliated with Laval University in March 1908. In its annual report for that year, mention is again made of Reynaud's problems with orchard tree diseases. This time he dusted some trees with Bordeaux, thus providing another object lesson for his students. Reynaud, at one time president of the Pomological Society of Quebec, and president of L'Union Expérimentale des Agriculteurs de Québec, was a well-known authority on fruit tree culture and vine growing.[147]

During the second annual meeting of the Quebec Society for the Protection of Plants, Professor G. Dimitriou told the members of the sudden appearance of mildew in the experimental plots of grain at the Agricultural Institute, Oka. Because his brief history of the disease and of methods for its control, including the selection of mildew-free plants, was published in the second annual report of that society, one may assume that his lectures to the students at Oka included a commentary on mildews and their control.

In 1912, the Rev. Father F.R. Leopold, a staff member since 1909, went to Europe for several months and continued his earlier studies in fruit arboriculture.[148] When he returned he replaced Renaud as the authority on fruit culture and fruit diseases at Oka. Father Leopold became president of the Pomological Society and, in 1914, published a pamphlet on fruit growing in the province of Quebec.[149] He was vice-president of the Quebec Society for the Protection of Plants for many years.

Plant pathology became a more important feature of the teaching at Oka after the appointment of Firmin Letourneau (b.1891), BSA, a 1914 graduate who was named "Professor of Entomology and Vegetable Pathology" in 1916. At Oka, vegetable pathology "consisted especially in identifying the most common diseases and in applying the most effective remedies to check the diseases in cultivated plants."[150]

In 1920–21, both the horticulturist and the entomologist were giving lectures and demonstrations with respect to plant diseases and their control.[151] In addition to those lectures, students at Oka could see the applications, on the farm and orchards, of methods taught in the school.

Letourneau taught plant pathology, and other subjects, until November 1921, when he resigned and a graduate of 1917, J. Romeo Cossette, BSA, was promoted to professor of agricultural botany and pathology.[152] Cossette left Oka in 1923 to join the staff of the experimental farm at Farnham, and Father Leopold once again assumed responsibility for the teaching of vegetable pathology and entomology.[153] In recognition of Father Leopold's outstanding contributions as scientist, teacher, and administrator, the University of Montreal awarded him the degree of Doctor in Science *honoris causa* on 25 May 1927.[154] Father Leopold continued to teach most of the plant pathology that was taught at Oka to the end of the period covered by this essay.

Other Agricultural Schools in Quebec

At least two agricultural schools in Quebec, other than the ones at Oka, Ste-Anne-de-la-Pocatière, and Ste-Anne-de-Bellevue, had staff members whose mandate included the teaching of plant pathology. Noteworthy in this respect is the short-lived Ste-Martine Agricultural School, which was officially opened 3 September 1933. From the beginning, its teaching staff included Fernand Godbout, MAS, professor of vegetable pathology. The outlines of courses offered at that school show that the students in their second year were getting thirty hours of lectures on entomology and vegetable pathology.[155] There were no significant changes in staff or courses at that school, relative to the teaching of plant pathology, prior to its closure during the war years.

In 1942 the Rev. Ernest Lapage, BAS, was teaching some plant pathology at the Rimouski Intermediate Agricultural School, where he was listed as "Professor of rural economy, of botanic and vegetable pathology."[156]

Doubtless some plant pathology was taught at other schools in Quebec prior to the Second World War, because the subject had been popular among agricultural educators for several years, but records of it are not known to exist.

NEW BRUNSWICK

In the field of agricultural education in general, and the teaching of plant pathology in particular, New Brunswick was slow to develop, in comparison to other Canadian provinces. (Newfoundland was not a province of Canada until 1949, and plant pathology was not taught there until after that date). This slowness extended to almost all aspects of agriculture, because the economy of New Brunswick was, for many years, based largely on lumbering and fishing.

Governor Carleton and several land-owners tried to stimulate an interest in agriculture by establishing a provincial agricultural society in 1790,[157] but the farmers did not respond as Carleton had expected, and the society accomplished very little. Throughout the following century many agricultural and natural history societies were established in New Brunswick, and several of them, at various times, advocated governmental support for methods of improving agriculture in the province, including the introduction of agricultural education in the schools.[158] Occasionally the question of an agricultural college was discussed, but this did not have the general support of farmers, at least not until near the end of the nineteenth century.

When the advocates of an agricultural college realized that each of the Maritime provinces could not individually afford an institution worthy of the name, they decided to cooperate and establish one jointly. Nova Scotia took the lead, in 1901, by allocating $50,000 for land and buildings for an agricultural college, providing it was to be located in that province.[159] In 1903 New Brunswick agreed to contribute funds for a college in Nova Scotia, with the understanding that New Brunswick students would be accepted in it on the same basis as those from Nova Scotia.[160] As a consequence of that agreement, New Brunswick has never had an agricultural college.

At the turn of the century, the farmers of New Brunswick were not at all enthusiastic about "book farming" or the thought of anyone in a school being able to teach agriculture. That attitude softened somewhat when a nature study movement, which began in the United States late in the nineteenth century, swept across the country. Nature study in eastern Canada was supplemented by the Macdonald Movement, an enthusiasm for the benefactions of Sir William Macdonald, particularly his financing of the construction, equipment, and, for a time, the operating expenses of a number of schools in eastern Canada in which manual training, home economics, and kindred subjects were taught, often for the first time. That "movement," which culminated in his founding of Macdonald College, Ste-Anne-de-Bellevue, Quebec, provided a more palatable means of introducing agriculture into the curriculum of elementary schools.[161] It became even more acceptable when the federal government provided all of the funds for teaching agriculture in the public schools, through the Agricultural Instruction Act of 1913.[162]

New Brunswick then became enthusiastic about the teaching of nature study and agriculture, making its teachers-in-training study those subjects in the provincial Normal School and adding them to the curriculum of the elementary schools. To facilitate this program, the province sponsored rural science schools in Woodstock and Sussex during the summer, for teachers who had graduated from the Normal School prior to the introduction of the new program, and for all who wanted to upgrade their knowledge of those topics. In 1915 the Sussex school was held in the new agricultural building that was officially opened on 15 July of that year.[163] In January 1916, winter short courses were inaugurated, also in Woodstock and Sussex.[164]

Teachers who took those courses must have been taught some plant pathology by Arthur G. Turney (1885–1960), B SC, the provincial horticulturist, or his assistant, Raymond P. Gorham (1885–1946), BSA, because one or the other of them was on the list of lecturers for several years during the time they were involved with the control of insect and fungal pests in demonstration orchards. In 1913, Turney also gave a series of short courses in different parts of the province, during several of which Gorham lectured on insect and fungal pests and their control; in the same year, Gorham wrote a bulletin on insecticides and fungicides.[165]

In 1924 the outline of the rural science courses specifically mentioned plant pathology as an area of emphasis.[166] The person who taught the plant pathology that summer is not identified, but R.C. Parent, BSA, M SC, taught plant physiology, which included a "study and collection of plant diseases."[167]

The New Brunswick Department of Agriculture did some "teaching" through the medium of its annual reports, many of which included references to plant diseases and their control. For example, a lecture by James Fletcher, Dominion botanist and entomologist, on "Enemies of Fruits and How Best to Fight Them" was reproduced in the report for 1892. Spraying instructions were included in many of the reports, beginning with that of 1897, and an American bulletin on "The Grain Smuts" was reproduced in 1898, to mention only a few examples of this method of "teaching" plant pathology. The Department of Agriculture also ensured that plant pathology was being taught through the medium of demonstration orchards and "Orchard Field Days," which almost invariably included demonstrations of spray materials, sprayers, and an accompanying lecture on disease and insect control. In 1915 there were sixteen illustration orchards in the province, and several demonstration plots of potatoes where the merits of dusts and sprays for the control of late blight and potato "bugs" were brought to the attention of interested farmers.[168]

In 1918, Austin C. Taylor, agricultural representative for the counties of Carleton, Victoria, and Madawaska, lectured and gave demonstrations, at ten different places in those counties, on the recommended methods of treating seed grain to control smut.[169] Taylor was just the first of the several agricultural representatives, appointed in succeeding years, who lectured on plant diseases and various other topics to groups of farmers throughout the province.

In 1923 a building that became known as the New Brunswick School of Agriculture was constructed on the grounds of the Dominion Experimental Farm, Fredericton. Its first students were studying general agriculture there in 1924, in a short course that included lectures on plant diseases by plant pathologist Donald J. MacLeod (1894–1990), BA, MA.[170] For the first few years the curriculum of that school was largely based on farm animals and their care, feeding, and management. This changed in 1934 when the teaching staff was enlarged and a more balanced curriculum, one that included more field crops and plant pathology, was developed. From that time onward MacLeod and John Lorne Howatt (1898–1959), BSA, MS, with help from Scott F. Clarkson, conducted virtually all of the work in plant pathology.[171] The New Brunswick School of Agriculture continued to offer winter courses that included plant pathology until its facilities were taken over by the military during the Second World War.

In the meantime, the Carleton County Vocational School, Woodstock – the school that the *Telegraph Journal* of 15 June 1970 referred to as "Canada's oldest Trade School" – was also offering winter courses in agriculture that included plant pathology. Robert Newton (1889–1985), BSA, the director of the school during the winter of 1914–15, was assisted by horticulturist R.P. Gorham, who taught most of the plant pathology.[172] The practice of offering courses in general agriculture that included plant diseases and insect pests was continued by Newton's successor, Robert W. Maxwell (1894–1975), BSA, until he retired sometime after the Second World War. Maxwell was an agricultural representative, as well as director of the Vocational School, and he either taught all of the plant pathology himself or had guest lecturers do part of it for him. Somewhat similar winter courses for young farmers were offered in schools in Sussex and St Leonard. All three were classed as vocational schools and under the jurisdiction of the New Brunswick Department of Education, but operated in cooperation with the provincial Department of Agriculture.

University of New Brunswick

At the University of New Brunswick, in Fredericton, some plant pathology was being taught by Philip Cox, BA, B SC, PH D, as early as 1908. His cryptogamic botany course that year included "Enemies and diseases of Economic and food Plants and their treatment."[173] Cox was professor of geology and natural science from 1907 until 1930.

The first professor of forestry at the University of New Brunswick, Robert Parcel Miller (1875–1965), BS, MF, who joined the staff on the first of October 1908, offered a botany course that autumn in which emphasis was placed on plant pathogenic fungi, "so as to provide a foundation for later work in plant diseases."[174] In 1920 that botany course was still emphasizing plant pathology by giving special attention "to the economic plant diseases of the region, such as clubroot of cabbage and turnips, late blight of potato, rusts, smuts, etc." That course, taken by forestry, engineering, and arts students, was designed to "give arts students a good elementary knowledge of plant diseases in case they might wish to specialize later in the field of plant pathology, or teach the subject in High Schools. Foresters and engineers learn the species of fungi attacking trees and timber and the methods of timber preservation."[175]

Miller resigned 15 May 1919 and was succeeded, in the autumn of that year, by Albert Van Siclen Pulling, of New Hampshire, who had a bachelor's degree from the New York State College of Forestry. He resigned in 1924 and Horace P. Webb was head of the department from 7 July 1924 until May 1929, when he was succeeded by John Myles Gibson. Gibson, who became dean of forestry in 1939, and Byron Wentworth Flieger, who had joined the staff as Webb's assistant in February 1928, taught all the forestry courses until the chair of entomology was established in 1938.[176] Throughout that period, the plant pathology content of the botany courses remained essentially as it was in the 1920s, although taught by different people.

A major change took place in 1939–40 when a new course, "Plant Pathology and Plant Physiology," was introduced by Charles William Argue, professor of biology. This was a two-term course, divided roughly into plant pathology in the fall term and plant physiology in the spring term. Almost from the beginning, the plant pathology component of the course was provided by honorary lecturer J. Lorne Howatt with the assistance of Donald J. MacLeod, both of whom were full-time members of the staff of the Dominion Department of Agriculture, at the Experimental Farm, a few kilometres downriver from Fredericton. That is where most of the laboratory work of the course was conducted, usually on Saturdays.[177]

There were no basic changes in the plant pathology offerings at the University of New Brunswick until 1946, when plant pathologist Arthur J. Skolko, B SC, MA, PH D, an employee of the Dominion Department of Agriculture, began teaching there. He taught the plant pathology course for two years, after which the course was dropped from the curriculum. Some plant pathology continued to be taught, mostly by Professor Argue, but as an incidental part of botany and biology rather than as a separate course.

NOVA SCOTIA

Nova Scotia was the first Canadian province to introduce a course of nature study into the curriculum of its public schools,[178] and, in 1885, the second one to have a School of Agriculture.[179] That latter school, established in rooms of what was then the provincial Normal School, Truro, became fully operational when Herman W. Smith (1860–1936), a B SC graduate of Cornell University, was appointed lecturer in agriculture the following year. Smith was, at the same time, lecturer in natural science on the staff of the Normal School. His duties included the teaching of agricultural chemistry and botany to the student teachers, and the conduct of a course in agricultural science to young farmers.[180]

For many years Smith was a popular guest lecturer at meetings of agricultural and horticultural groups throughout the province. When a separate building was erected for the School of Agriculture, he was placed in charge of its laboratories while continuing to teach part-time at the Normal School. He later became principal of the School of Agriculture.[181] One of the most noteworthy graduates of that school, in so far as plant pathology is concerned, was William P. Fraser, who became an internationally recognized Canadian authority on the rust fungi.[182]

When the Nova Scotia Agricultural College (NSAC) was established, early in 1905, it absorbed the School of Agriculture and Smith became its professor of biology.[183] At the NSAC his teaching included bacteriology, entomology, zoology, and botany, until 1912, when he was relieved of the entomology and zoology by Robert Matheson, provincial entomologist and professor of zoology.[184] From that time onward Smith was primarily a botanist, but with a strong interest in plant pathology. He wrote a paper titled "Some plant diseases in Nova Scotia" in 1910, and one on "Diseases of fruit plants" in 1916.[185]

Although no records of the subject matter of Smith's lectures to the students of the Normal School or to those of the School of Agriculture have been found, there is no doubt that they included references to plant diseases. Support for this assertion comes from the report of his work for 1896, in which he mentioned that his students had been conducting an investigation of turnip rot. At the same time, he requested that "farmers kindly inform me at the time when diseases or insects attack their crops and not wait until some months afterwards."[186] There is also no doubt that he taught plant pathology at the NSAC. Proof of this is in his first report as professor of biology, wherein he states that "In Botany, the first year students studied ... the more common plant diseases and their remedies ... The second year students were required to make collections of plants and plant diseases."[187] This was confirmed again when he reported, in 1912, that he gave his junior class "an elementary course in plant diseases" and his senior class a course in "anatomy, physiology, and the diseases of plants."[188]

To anyone reading the annual reports of Smith's work, it soon becomes apparent that each year he became somewhat more of a plant pathologist. In 1916 he included a discussion of the diseases of potatoes, turnips, mangolds, beets, and carrots, and in 1917 he gave an advanced course in plant diseases to his senior students.[189] In his report the following year he commented on smuts and rusts of grain, mosaic diseases, and his experiments for the control of bean anthracnose.

Smith also taught plant pathology for several years to schoolteachers attending summer school. In 1919 he wrote, "I had, as usual, my regular classes in the Rural School of Science in Plant Diseases."[190] That school was established in 1908, and the well known Canadian naturalist John Dearness, vice-principal of the Normal School in London, Ontario, did some of the teaching that summer.[191] The school, which continued to operate until 1923 in cooperation with the staff of the Normal School, was primarily for teachers in rural schools who were expected to include some elementary plant pathology in their agricultural courses.

Smith continued to lecture on plant diseases to junior and senior students of agriculture at the NSAC until he resigned in 1924. He was succeeded as professor of botany by Howe S. Cunningham, a BSA graduate of Macdonald College who had been professor of agriculture and farm mechanics at the NSAC since 1918. In the two years prior to his appointment as professor of botany, Cunningham was on part-time leave to do some postgraduate work in agronomy and plant pathology at Macdonald College for the MSA degree that he obtained in 1924. He also studied plant pathology at Cornell University. Although there was a rearrangement of course responsibilities at that time, Cunningham continued most of Smith's work in botany, including plant diseases, while also "giving instruction in Field Crops and in some branches of Farm Mechanics."[192]

Cunningham's successors, especially A.R. Prince, a charter member of the Canadian Phytopathological Society, and A.E. Roland (1910–1991), who became a distinguished botanist, included some plant pathology in their lectures, but the subject was seldom featured in any of them. Perhaps that was because they knew that their "degree students" would be going to Macdonald College, with which NSAC had become affiliated in 1926,[193] or some other institution where plant pathology would be taught by specialists in that field. They also knew that their horticultural and agronomic colleagues were including plant diseases in their lectures.

Some plant pathology had been taught at an early date in another institution in Nova Scotia. On 9 January 1894, the first School of Horticulture in Canada was opened, with thirty students in attendance, on the campus of Acadia University, Wolfville. That school, under the direction of Professor E.E. Faville, a graduate of Cornell University, made use of some of the facilities and teaching staff of Acadia University but was not formally affiliated with that institution.[194] It is not known for certain just how much plant pathology entered into Faville's lectures, even though the course of studies is known to have included lectures on fungi, and the best methods of spraying to control plant diseases and insects.[195]

After teaching for less than three years, Faville returned to the United States. His successor was F.C. Sears, MSA, who soon became active in the study of local plant diseases, especially those of apple trees in the Annapolis Valley. In 1901 he revised the courses offered by the school and outlined one called "The Spraying of Plants ... a study of the most important insect and fungous pests, with the treatment for each."[196] His interest in plant pathology is seen in his claim to have proven that the black knot disease could be controlled.[197]

The School of Horticulture, as an independent unit, closed on 1 May 1904, when it was merged with the NSAC, in which Sears became the horticulturist.[198] Sears was giving lectures on "Insect and Fungous Pests" to students in a short course that began on 7 February 1905, nine days before the college was officially opened.[199] He continued to teach some plant pathology to his regular students of horticulture, and to students in short courses, until he resigned to become professor of horticulture at the Massachusetts Agricultural College in June 1907.[200]

Sears was succeeded by Percy J. Shaw, who, within a few weeks, began a study of what he called "the hard knot disease of pears."[201] Although there is a paucity of information about his teaching, Shaw's annual reports almost invariably mentioned his work with demonstration or model orchards and tests of various spray materials for the control of apple scab. Beginning in 1916 he commented on experiments for the control of club root of cruciferous crops, and in 1921 he was trying to control potato scab.[202] Shaw acknowledged that his research was an aid to teaching and entirely subordinate to it, and for that reason alone he carried out some experimental work with fruits and vegetables.[203] Thus we have indirect confirmation that he included plant pathology in his teaching and is therefore eligible for inclusion among the early teachers of plant pathology in Nova Scotia.

The incorporation of some plant pathology in the instructional lectures of the succeeding professors of horticulture, including Charles

M. Collins, who held that position from 1937 until 1946, continued each year, and into the 1950s.

Nova Scotia Universities

Plant pathology, as a separate subject or course, was not taught in any of the universities in Nova Scotia prior to 1950. However, a question on the university graduates' botany examination in 1918, asking the students to "Write a life history of a grain rust, or the loose smut of grain, or ergot, or the black knot of the plum,"[204] shows that some plant pathology was being taught to university students in Nova Scotia at that time.

One of the early staff members of Dalhousie University, Halifax, who would have included a bit more than incidental comments on plant diseases in his lectures, was George Lawson, professor of chemistry and mineralogy from 1863, the year he resigned from Queen's University, until his death in 1895. Lawson was a charter member of the Royal Society of Canada, and in his capacity as secretary of the Central Board of Agriculture may be said to have been the head of the Nova Scotia Department of Agriculture for many years. His paper, "Blights and diseases of plants," was published in the first volume of the *Journal of Agriculture for Nova Scotia*, in 1866, and the *Transactions and Reports of the Fruit Growers' Association and International Show Society of Nova Scotia* contain several references to his lectures on plant diseases during its meetings, beginning with their spring meeting of 1888.

Another university professor who probably included plant diseases in his lectures was Hugh P. Bell (1889–1957), whose thesis for the PH D at the University of Toronto, 1922, was "Fern rusts of Abies." Bell, who had joined the Dalhousie staff in 1921, became a charter member of the Canadian Phytopathological Society in 1929. He retained an interest in the physiology of plants and in plant diseases, at least until 1941, when he published a paper on the russeting of apples due to Bordeaux sprays.[205]

PRINCE EDWARD ISLAND

There were no trained plant pathologists to teach plant pathology in the schools or colleges of Prince Edward Island prior to the Second World War. Nevertheless some plant pathology was being taught there by botanists, schoolteachers, horticulturists, newspaper editors, and others.

There was a sharp rise in interest in all potato problems following the failure of the potato crop in 1847, and the passing of "An Act to Prohibit the Exportation of Potatoes" by the Legislative Council.[206]

Such a drastic measure in a province whose economy was largely based on agriculture focused attention on potato diseases as never before. Beginning around 1846, the local newspapers, particularly their letters to the editor sections, carried reports of the appearance of blight and of whole fields of potatoes being destroyed. One account told of a farmer who "had the tops mowed down" so that fresh new stems and leaves would spring up and escape the disease.[207]

From the time of the formation of the Royal Agricultural Society of Prince Edward Island, 12 May 1849, there were organized attempts to improve the lot of farmers through education. Within the first year of its operation, that society was ordering books on agriculture from Europe, the United States, and elsewhere. Several of those books had chapters or articles on plant diseases and their control. The society ordered fifty copies of Fessenden's *Complete Farmer* and one hundred copies each of Jackson's *Outlines of Flemish Husbandry* and Johnston's *Catechism of Agricultural Chemistry and Geology*, and single copies of the works of several other agricultural writers of repute.[208] In April 1851 the society contracted for the printing of ten thousand copies of Judge Peters's pamphlet *Hints to the Farmers of Prince Edward Island*, and they arranged with the Board of Education to have them distributed through district schools for use as general class books.[209]

Judge James H. Peters was aware of the great spate of literature on agricultural improvement that was flooding the United Kingdom and the United States, as well as the older British North American colonies. He attempted to select the most appropriate features of this mass of information and condense them into one small book. The result was strange: the best scientific thought of the time, intermixed with nonsense. For example, in commenting on the use of charcoal as a means of controlling plant disease, he stated, "When charcoal is scattered over the ground, it absorbs and condenses the various gases within its pores to the amount of from 20 to over 80 times its own bulk ... It checks rust in wheat, and mildew in other crops."[210] With all its faults and weaknesses, that book had the merit of introducing the study of agriculture, including plant diseases, into the schools of the province, and it was done with the assistance of the Royal Horticultural Society and the cooperation of the Board of Education. This fact is noted in appendix M of the *Journal of the House of Assembly of Prince Edward Island for 1855*. There was no Department of Agriculture in the province at that time.

The first college on the Island was Kent College School, more popularly known as the National School, which opened in 1820. It developed into the Central Academy, in 1834, and became Prince of Wales College when Prince Edward visited Charlottetown in 1860. The provincial Normal School was amalgamated with it in 1879, while it was a junior institution, offering the first two years of high school, the training of teachers for primary schools, and the freshman year of university.[211] Agriculture and nature study became significant features of the curriculum of the college after the chief superintendent of education proposed that they be given important places on the syllabus of studies in 1888.[212]

Arthur W. Shuttleworth was appointed "Professor of Agriculture and Kindred Sciences" in 1890. He became well known throughout the province through his lectures on agriculture, including late blight of potatoes and other plant diseases, to groups of farmers. Shuttleworth was also one of the many advocates of fruit growing on the Island. His efforts, plus the enthusiastic support of Lieutenant-Governor George W. Howlan and others, resulted in the formation of the Prince Edward Island Fruit Growers' Association in 1896.[213] Shuttleworth addressed meetings of that association and the diseases of fruit-bearing plants was often his principal topic. Nevertheless, it was 1901 before the principal of the college reported that "Agriculture has been given the place it deserves in Prince Edward Island, and E.J. McMillan, BSA, an honors graduate of the Ontario Agricultural College conducts classes in this subject with students preparing for license."[214] The principal was referring to a licence to teach agriculture in the public schools. The timing of the appointments of Shuttleworth and McMillan was fortuitous in that it coincided with the rising interest in nature study and agriculture throughout the continent.

Teachers in training were taught some plant pathology in their compulsory agriculture course. This is known by the questions on their licence examinations. For example, one in 1909 asked the candidate to "Give the treatment for fungous diseases."[215] In the following year one of the questions was "Name three fungus diseases injurious to orchards and give full directions for their control."[216] Although questions on plant diseases did not always appear in the licence examinations, they were there often enough to indicate that plant pathology was being taught each year.

In 1938, Lorne C. Callbeck (1912–1979), a B SC(Agr.) graduate of Macdonald College, became professor of agriculture at Prince of Wales College. In the following year one of his examination questions was "Discuss the growing of turnips with reference to the following points only: (1) ... (2) Measures to control club-root and brown-heart diseases."[217] Thus Callbeck, who later became an authority on the epidemiology and control of late blight, continued the tradition of teaching some plant pathology at Prince of Wales College, although largely incidental to lectures on general agriculture, until the mid-1940s.

When a provincial Department of Agriculture was established in 1901, it soon began "teaching" plant pathology through orchard meetings, which usually included demonstrations of pruning and of spraying for pest control. It also arranged for the planting of five model orchards. In later years those orchards, and others, were used to teach by example, and the examples included means for the prevention or control of orchard tree diseases.

When Prince Edward Island became a province of Canada in 1872, the federal Department of Agriculture had an experimental farm at Nappan, Nova Scotia, that was intended to serve the three Maritime provinces. However, in 1909 the federal department leased land from the provincial government, established the Charlottetown Experimental Farm, and appointed J. Artimus Clark as its superintendent.[218] At that time the federal department was expected to concentrate on research and leave most of the propaganda, or extension work, to the provincial people. Consequently, in the early days, Clark and his staff did very little teaching of plant pathology. However, they did become somewhat involved in teaching by example and through demonstrations, soon after the first illustration stations were established in 1923.[219] The principal method adopted was through field days, when neighbouring farmers were invited to the stations to see demonstrations and to hear speeches delivered by officials of provincial as well as federal departments of agriculture. Many of those speeches-demonstrations dealt with the control of smut and other diseases of grain.

The first man with good training in plant pathology to work in Prince Edward Island was Paul A. Murphy, BA (1887–1938), a native of Ireland who was appointed assistant in charge of a new Plant Pathology Station in Charlottetown, in July 1915.[220] Murphy had exceptionally useful training and experiences in botany and plant pathology, which he obtained first at the Royal College of Science, Dublin, then at the Imperial College, London, and later at the Biologische Reichsanstalt, Berlin, and Cornell University. He addressed many potato growers on the subject of potato diseases, especially after he was placed in charge of a new potato inspection service for the Island and for Nova Scotia.

Murphy was expected to explain the new regulations and to ensure that the farmers had at least a rudimentary knowledge of the diseases of potatoes that would lower the inspection-grade of their crop. As a way of "teaching" or informing farmers, Murphy wrote a semi-popular article on the late blight disease that was widely distributed to potato growers.[221] This was done shortly before he resigned and left the Island in March 1920.[222]

Murphy's successor, John B. McCurry (b.1894), was a graduate of the Ontario Agricultural College who had a reasonably good background of experiences in plant pathology, gained at Dominion Laboratories in Brandon and St Catharines. He continued the program initiated by Murphy until he was transferred to Ottawa in 1923.[223]

An indication of the amount of teaching done by the plant pathologists employed by the federal government is seen in the report of the Dominion botanist for 1924. There it is noted that S.G. Peppin, the acting officer in charge of the Plant Pathology Laboratory, gave thirty-five lectures on potato diseases at short courses, plus a series of lectures on plant pathology to students at Prince of Wales College.[224]

McCurry was succeeded, in 1925, by Richard R. Hurst (1895–1961), who had worked on the Island with Murphy as a plant disease investigator before graduating from the Ontario Agricultural College in 1922. Hurst soon established himself as a respected authority on potato diseases and, in 1929, wrote a bulletin entitled *Late blight and rot of potatoes.* He too was a frequent guest speaker at meetings of farmers and the subject of his talks was usually plant diseases.[225] Between 1925 and 1929, Hurst was assisted during the summer months by J. Lorne Howatt, a student who later became an eminent plant pathologist.[226] Through the "teaching" of those and other knowledgeable men, many farmers acquired a better understanding of plant diseases. In doing so they learned why and how to rogue or remove diseased plants from their fields and to dispose of them in such a way as to prevent the disease from spreading to healthy plants.

The early teaching outlined above was very effective, and the end results were economic benefits to Island farmers, many of whom gained the enviable reputation of being producers of high quality disease-free grain and seed potatoes.

MANITOBA

Manitoba was not affected by the nature study craze to such an extent that it had to be taught in the public schools and used as a sort of priming device for the introduction of agriculture as an acceptable subject for study. Agriculture, as such, was being taught in the Normal School and in the public schools of Manitoba in the first decade of the twentieth century.

The authorized text for that study was one titled *Agriculture,* written by Charles C. James, MA, deputy minister of agriculture for Ontario, and Alexander McIntyre, BA, vice-principal of the Normal School in Winnipeg. The significance of that textbook in the present context is that it had one full chapter devoted to the diseases of plants, and it had an appendix of spraying mixtures. The text authorized for both Manitoba and Quebec in 1910, *Elementary agriculture, with practical arithmetic,* also had a chapter dealing with the protection of plants from insects and diseases.[227] Thus it is evident that some elements of plant pathology were being taught in the schools of Manitoba as early as, or earlier than, they were in the majority of similar schools elsewhere in Canada.

The University of Manitoba

The University of Manitoba was founded, by an act of the provincial legislature, in 1877. In the beginning it was patterned after the University of London, England, with its functions confined to prescribing standards of education and the curricula of students, conducting examinations, and conferring degrees. In other words it was an examining and not a teaching institution. There were no university professorships or buildings; all teaching was done in the affiliated colleges, namely St Boniface, St John's, and Manitoba College.[228] In the words of the chancellor, in an address to the governor-general of Canada, 14 April 1885, it was a "Republic of Colleges."[229]

The fire that destroyed the McIntyre block of buildings in the winter of 1897–98 consumed many of the university records prior to that date. Nevertheless, clues to what was being taught in the colleges may be found in some of the surviving examination questions posed by the university examiners. For several years the examiners in the natural sciences were Rev. Father G. Cloutier, professor of mental and moral philosophy and mathematics at St Boniface College; Rev. George Bryce, MA, LL D, professor of sciences and literature at Manitoba College; and E.B. Kendrick, BA, lecturer in natural science, St John's College.[230]

The examiners must have assumed that the teachers in the colleges were commenting on plant diseases, or at least on the parasites that could incite diseases, because in the botany examination of May 1885 they asked: "What is understood by parasitic plants? State some of their characters."[231] A year later a question was "Give an account of paler or colored (not green) parasites."[232] Here they were focusing the attention of the students on fungal parasites rather than on dodder and mistletoe, about which they asked in later examinations. In a botany test in 1895, students were asked to give some account of the death of plants due to heat, and also to cold, thus indicating that the lecturer had dealt with physiological plant diseases due to unfavourable temperatures.[233]

Although the composer of the individual questions remains anonymous, it is probable that those pertaining to plant parasites were submitted by George Bryce. He was the most avid naturalist of the three examiners, both inside and outside the university. Records of his work as a natural scientist outside the university may be found in the minutes of the Historical and Scientific Society of Manitoba, of which he became president, 6 February 1866.[234]

There were no significant changes with respect to the teaching of plant pathology at the University of Manitoba until after 1904, when A.H.R. Buller (1874–1944), B SC, PH D, became professor of botany and geology in the Faculty of Arts and Science.[235] Although Buller soon became a noted mycologist, plant pathology was a significant ingredient of his lectures to senior students in botany. Beginning in 1905, he was also one of the university examiners in natural science. Question 2 on Buller's own examination paper that year was: "What is meant by parasitism? How are the Dodder and Mistletoe adapted to their modes of life?"[236] Question 5 of his botany exam the following year asked the students to "Give a careful account of the Rust Disease of Wheat. In what important particular is it not as yet fully understood?"[237]

After 1896, when the report of the Experimental Farm for Manitoba told of the considerable damage due to rust of wheat, there was a growing interest in that disease and the controversy regarding its relation to barberry. Botanists were sometimes more involved in this latter controversy than in the disease itself. Buller, who had studied under professors Pfeffer and Hartig in Germany, and lectured in botany at the University of Birmingham, England, knew that the wheat stem rust fungus lived part of its life cycle on barberry, and he also knew that the complete life history of the fungus had not been determined.

Although mycology was Buller's first and most abiding love, he was an able botanist with a keen interest in plant diseases, especially those that lowered the yield or quality of wheat in Western Canada. His vigorous campaign for a concerted effort to combat diseases of grain, following the severe outbreaks of wheat rust during the First World War, was an important factor leading to the establishment of the Dominion Rust Research Laboratory on the campus of the University of Manitoba.[238]

Buller's mycological research, including aspects of the history of mycology, attracted a number of graduate students, several of whom eventually became internationally recognized authorities in mycology and plant pathology. Among the latter, J.H. Craigie, W.F. Hanna, and T.C. Vanterpool became well known, as did mycologists H.J. Brodie, Ruth Macrae, Irene Mounce, and the first woman to be awarded a PH D by the University of Manitoba, Dorothy (Newton) Swales. Buller retired and was made an emeritus professor in 1936. He died of a brain tumour, 3 July 1944.[239]

In 1902, the provincial legislature appointed a commission to inquire into the advisability of establishing an agricultural college, and it passed an act the following year which provided for the establishment of one. However, the inevitable interferences in such a venture delayed the actual construction until 1905. Once begun, the work proceeded well, and the Manitoba Agricultural College was formally opened, 6 November 1906.[240] The initial plan of instruction was designed as a diploma course in agriculture, to be given in two sessions of five months each, extending from October to March. Eighty-five students enrolled the first year and the first diplomas were granted in 1908. The courses available to those pioneer students indicate that they got some plant pathology from teachers in the Department of Field Husbandry. The lectures there included "the study of field crops, soils, weeds, plant diseases and farm management."[241]

In 1908 Spencer A. Bradford, who had been superintendent of the Dominion Experimental Farm at Brandon, was appointed professor of field husbandry. Because he had experimented with methods for the control of grain smut for twenty years, he was well qualified to include plant diseases in his lectures, which he did until he became provincial deputy minister of agriculture in 1912.

In 1916, when grain rust was so devastating in many areas, the diploma students were exposed to much more plant pathology. Their "Farm Botany" course included lectures on environmental effects on grain smuts, spore production, and plant diseases. The outline of their "Biology" course also included plant diseases.In addition, the students were advised that each one "should bring in such weeds, plant diseases, specimens and problems as he wishes to know more about."[242]

Pressure from the students, their parents, and others, persuaded the authorities to add sessions of advanced or degree-level courses to the curriculum. The diploma graduates who successfully completed those additional courses were qualified for the degree of Bachelor of Science in Agriculture.[243] In 1911 the first degrees in agriculture were conferred on ten students by the University of Manitoba.[244] This was possible because the Agricultural College had become affiliated with the university in 1907. It was incorporated into the university as the Faculty of Agriculture and Home Economics in 1924.[245]

After a Department of Botany and Bacteriology was established in the Manitoba Agricultural College in 1909, some plant pathology was taught there, in addition to what was being taught in the Department of Botany of the Faculty of Arts and Sciences. Charles H. Lee, BA, M SC, chairman from 1909 until 1913, included plant diseases in his agricultural bacteriology course. Lee was the first professor at the University of Manitoba, albeit in the College of Agriculture, to give a course specifically named "Plant Pathology" and to include a brief description of it in the college calendar. Plant Pathology for fourth year students in 1909–10 was described as "A course of lectures and reference work embracing a study of the origin and development of the commoner plant diseases with methods of prevention and treatment. Laboratory work on special diseases."[246] Fourth-year students also got a few lectures on diseases of wheat in course no. 14 of the field husbandry department,[247] and plant pathology for fifth-year students became "A short course of laboratory and research work on recent phases of the more destructive diseases which attack farm crops."[248]

In 1913, the Department of Botany and Bacteriology was divided and Lee was made chairman of a newly formed Department of Bacteriology. Thereafter he concentrated on various aspects of bacteria in relation to agriculture, but continued to include their role in plant as well as animal diseases in his lectures until he retired, in 1929. Bacteriology was then combined with animal pathology.[249]

The botanical work, in the Department of Botany and Biology, was taken over by Vincent W. Jackson, BA Queen's, who continued to teach the plant pathology very much as Lee had taught it until 1915, when Jackson changed the outline of the course to read "Plant Pathology – microscopic study of moulds and mildews and the rusts and smuts of grain; origin and development; methods of prevention and treatment; artificial germination of spores; life history of smuts, rust and ergot; classification of fungi."[250]

Jackson was a plant pathologist who, after a few years of teaching in Ontario and New Zealand, and at the University of Manitoba, did his graduate work at Minnesota where he obtained his master's degree in 1923. At the University of Manitoba, his objective was to provide a well-rounded program in plant pathology. He prepared his students for the course work in plant pathology by including a study of the "relation of fungi to other plants and to plant diseases" in his cryptogamic botany course. He also tried to tailor his courses to fit the anticipated needs of the students. This is seen in the plant pathology course outline for fifth year students taking the horticulture and forestry option. In 1915 it was: "Fungoid diseases of garden and greenhouse plants; diseases of stored vegetables; fungoid growths and diseases of trees."[251] In that outline we see one of the very early Canadian references to a study of post-harvest pathology, or as Jackson called it, "diseases of stored vegetables."

Jackson had the reputation of being a good teacher, and Bryce J. Sallans, who in later years became head of the Plant Pathology Section, Canada Department of Agriculture Research Station, Saskatoon, gave Jackson credit for arousing his interest in plant pathology while he was an undergraduate student in the University of Manitoba.[252]

Although Jackson continued to include some plant pathology in his lectures, until he retired in 1933, much of the work in plant pathology was taken over by Guy R. Bisby (1889–1958) when he joined the department in June 1920. Bisby, a native of South Dakota, USA, with a B SC from that state college, an AM from Columbia University, and a PH D from Minnesota, was appointed specifically as professor of plant pathology.[253] That was an historic event, because it was the first time any Canadian educational institution had hired a professional plant pathologist and named him as such in its staff list. Regardless of his title, however, Bisby was more of a mycologist than a plant pathologist. He went on leave for a while, during the academic year 1921–22, to learn more about the relation of fungi to plant diseases at the Imperial Bureau of Plant Pathology, London, England. After that he took somewhat more interest in the teaching of plant pathology.

Soon after the department was recognized as being equipped and qualified to give graduate courses, Bisby began offering a course called "Advanced Plant Pathology," in cooperation with the Dominion Rust Research Laboratory.[254] The mycology courses, including the mycological component of the cryptogamic botany course, became Bisby's responsibility, and he taught them as if they were prerequisites to plant pathology.[255] Although Bisby had been hired as a plant pathologist in the College of Agriculture, he was transferred to the Department of Botany, Faculty of Arts and Science, and demoted to the academic rank of assistant professor of botany, in 1933.[256]

There was no approved program for PH D degrees at the University of Manitoba until sometime in the early 1940s. Nevertheless, it was possible for particular students to proceed to that degree. One such student was John H. Craigie, who was awarded the PH D degree in 1930, largely for research, supervised by Buller, that led to his discovery of the function of the pycnia of rust fungi.[257]

Until Bisby resigned, at the end of 1936, there were no significant changes in the outlines of his plant pathology courses. Perhaps this was because he was so busy performing the duties of a provincial plant pathologist, and working with others, that he had little time to revamp his courses. Bisby collaborated with Buller and John Dearness in the preparation of a comprehensive list of the fungi of Manitoba,[258] and with D.L. Bailey, officer in charge of the Dominion Field Laboratory of Plant Pathology that had been established and housed in the Manitoba Agricultural College in 1923,in cereal smut investigations.[259]

Shortly after the retirement of Buller and the resignation of Bisby, the Board of Governors announced the appointment of William Leach, M SC, PH D, D SC, of Birmingham University, England, to the chair of botany, and Harold J. Brodie, M SC, PH D, as assistant professor of botany.[260] Brodie assumed the responsibility for much of the teaching of mycology and plant pathology, including the course "Advanced Plant Pathology," for which he provided a new outline.[261] He was fortunate in having the continuing cooperation of staff members of the

Rust Research Laboratory, who gave many lectures in plant pathology to his students. They also permitted selected graduate students to do research for advanced degrees in their well-equipped laboratories.

Brodie continued to teach plant pathology at the University of Manitoba until he resigned at the end of the Second World War. Apparently he was not replaced until after 1951, because the president's report for that year commented on the urgent need of a plant pathologist.

SASKATCHEWAN

The University of Saskatchewan

The first University of Saskatchewan, at Prince Albert, was little more than a university on paper – the paper on which the federal parliament granted its charter, in 1883. The only college to be officially connected with it was Emanuel, a college established at Prince Albert by Rev. John McLean, first bishop of the Anglican Diocese of Saskatchewan, in 1879.[262]

For a brief period there were two Universities of Saskatchewan. Soon after Saskatchewan became a province, with jurisdiction over education, its legislature established a provincial University of Saskatchewan. The act of establishment, passed in April 1907, stipulated that no other institution in the province could grant degrees, except in theology, and that "no other University having corporate powers ... shall be known by the same name."[263] When Saskatoon was chosen as the site of the new university, Emanuel College became affiliated with it under the name of the University of Emanuel College.[264]

There is no conclusive evidence to indicate that any plant pathology was taught at Emanuel College, but there may have been some. The college owned two hundred acres of good land and agriculture was taught there for more than one season.[265] A similar lack of evidence applies to St John's College, an Anglican theological college at Qu'Appelle that was established in 1885 and closed in 1894. It too had a farm, and agriculture was part of its training program.[266] One may reasonably assume that the boys who took advantage of the training and agricultural experiences provided by those two colleges were made aware of plant diseases. They would have had to contend with smut of grain, which was prevalent at that time, and they probably learned about the current theories and methods of smut control.

William Henry Coard, LL D, in the North-Western Agricultural College and Experiment Station, Regina, was probably the first person to teach plant pathology in Saskatchewan. Coard had resigned from the federal Department of Agriculture to become principal of that college, and its professor of agriculture and entomology. His review of the work of the college for the first term, in bulletin no 1, dated 31 December 1903, included an advertisement that outlined the curriculum of the college. It stated, among other items, that "Bacteriology will be confined to bacteria in their relation to agriculture, dairying, plant and animal diseases," and that "The Entomological section will deal with noxious insects and insecticides, fungi and fungicides."[267] In a second document, accompanying the bulletin, headed "Courses of Instruction for a Diploma," there is this statement: "Botany ... also economic Botany, including smut, rust, mildew, etc., will be taught during the thirteen weeks of the course."[268]

When the present University of Saskatchewan opened its doors to receive students in September 1909, there were no scientists or dedicated naturalists on its teaching staff. Thus it may be assumed that there were few if any references to plant diseases in the university lectures until the following year, when John Bracken (1883–1969) began his teaching.

Bracken was a 1906 graduate of the Ontario Agricultural College, where he had won the Governor-General's medal for general proficiency in his second year of the agriculture course. He had been working for the federal Department of Agriculture before being employed by the Saskatchewan Department of Agriculture as superintendent of institutes, in charge of extension work. He was transferred to the staff of the fledgling College of Agriculture, as professor of field husbandry, in 1909. However, he spent a year doing advanced studies at the University of Illinois before becoming involved in teaching at the college.[269]

The College of Agriculture, which from the beginning was an integral part of the university, did not open for teaching until the farm and its buildings were ready, in October 1912.[270] The early appointees to its faculty did their teaching in the rooms and laboratories of the Faculty of Arts and Science.[271]

As professor of field husbandry, Bracken initiated the first research on crop plants at the university, and it included a study of wheat rusts. He was also responsible for a course called "Horticulture and Arboriculture" in which methods for the protection of small fruits from insects, weeds, fungus diseases, and unfavourable climatic conditions were described.[272] One of Bracken's students was Arthur W. Henry, who helped with the rust research and, in 1919–20, was an instructor in his department.[273] Bracken resigned in 1920 to become president of the Manitoba Agricultural College.

Another one of the first group of appointees to the staff of the college was Thomas N. Willing (1858–1920), assistant professor of natural history. Willing, a native of Toronto, Ontario, had travelled to the

North-West Territories as a surveyor in the late 1870s, and settled in Calgary in 1881. Little is known about his early education, but, while working as a surveyor, and during his eight years in Calgary, Willing became a self-taught naturalist and widely known as such. In 1889 he was appointed chief weed inspector and chief game guardian in what is now Saskatchewan. His knowledge of weeds and animal life so impressed the authorities of the College of Agriculture that they hired him as a teacher.[274]

Willing's first course in natural history, one on plant life, included lectures on smuts affecting cereal grasses and fungi that parasitize crop plants in general.[275] It was 1914 before the published outlines of his courses included plant diseases, even though he had included grain diseases in his lectures before that. Evidence for this is in his report to the president of the university for 1913, wherein he wrote that his "Spring term work" included a study of "weeds, poisonous plants, smuts, rusts, insects, birds, gophers, etc. in their relation to agriculture." He also mentioned that fifty-four agriculture and five arts students had taken his course.[276] By 1915, when he was made a full professor, Willing, like so many natural history teachers in Canada in that decade, was devoting a significant portion of his time to the teaching of "nature studies," including some plant pathology, to schoolteachers of the province. He served the college from 1911 until he retired in 1920.[277]

Students studying for degrees at the University of Saskatchewan could take appropriate courses in either the College of Agriculture or the College of Arts and Science. Those working for the MSA degree in 1915 were required to include a major in one of agricultural engineering, animal husbandry, or field husbandry, and two minors from bacteriology, chemistry, physics, or economics. In addition, they had to have a reading knowledge of French or German.[278] PH D degrees were not authorized at the University of Saskatchewan until 1949.[279]

In 1913, when Walter P. Thompson (b.1889) became professor of biology, and the only member of that department in the College of Arts and Science, some plant pathology was being taught in both colleges. Thompson had grown up on a farm in Ontario, attended the University of Toronto for a BA, and Harvard University for MA and PH D degrees. He taught a course called "Seedless Plants," the outline of which, in 1916, included rust, smut, and related organisms. That course was offered at the same time as Willing was dealing with "smuts and rusts affecting crops" in his "Natural History" course.[280] Thompson's course, "Bacteriology and Microbiology," comprised "the essentials of bacteriology and a study of related microorganisms, chiefly parasitic and pathogenic bacteria, fungi, including rusts, smuts, moulds, yeasts, etc."[281]

For several years after 1915 Thompson taught the seedless plants course in the College of Arts and Science and a bacteriology course in the College of Agriculture, both of which contained elements of plant pathology. In his departmental report for 1916–17, he stated that a change in his seedless plants course had been made from the program as published in the calendar, and that "Agriculture students were permitted to devote nearly all of the second term to the fungi, particularly the disease producing forms." He also commented that "Both fungous diseases and plant physiology are subjects which no specialist in field husbandry should miss."[282]

Thompson, who headed the Department of Biology until 1949, was a pioneer in research on the chromosomes of wheat and the hybrids between related genera, and on the genetics of the grain rusts. The high quality of both his research and his teaching stimulated several students to study plant pathology at the graduate level. Among the most noteworthy of those was Arthur W. Henry (1896–1989), a graduate of the class of 1917 who went on to become the first person to be awarded the MSA by the university.[283] Henry's research on the wheat rust problem began under the aegis of Thompson, with some help from Bracken with whom he worked, but was concluded under the supervision of W.P. Fraser in 1920.[284]

The University of Saskatchewan has been singularly fortunate in its relation to fine research facilities sponsored by federal and provincial governments. In 1919 a Laboratory of Plant Pathology was established on the university campus by the federal government, with W.P. Fraser as officer in charge.[285] Thompson was quick to see this new development as an opportunity to strengthen the plant pathology aspects of the course offerings of his department. Within a relatively short time, Fraser was appointed, on a part-time basis, as a lecturer in biology. Because he had taught plant pathology at Macdonald College for the previous four years, Fraser was well prepared to teach the new course, "Fungi and Plant Diseases," in the spring of 1920.[286]

Fraser held the dual appointments with the federal government and the university until 1925, when he resigned from the former to become a full-time professor of biology on the staff of the College of Arts and Science. In the academic year of 1920–21 he taught a "half class" in "Fungi and Plant Pathology," which was "a systematic study of the fungi with particular reference to those causing plant diseases, plus a short study of other cryptogamic groups." That course was listed by both the College of Arts and Science and the College of Agriculture.[287]

Shortly after Fraser became a part-time teacher at the university, Margaret Newton (1887–1971), who had been one of Fraser's students at Macdonald College and a colleague at the Dominion Laboratory of Plant Pathology, joined the staff of the Department of Biology as an instructor. At first she assisted Thompson with his research on the genetics of the grain rusts, but soon became involved in his laboratory classes and the teaching program of the department. After obtaining a PH D from Minnesota, in 1922, Newton was promoted to the rank of assistant professor. She resigned in 1925 to accept a position in Winnipeg with the Dominion Department of Agriculture.[288]

The 1923 calendar of the university lists, for the first time, a course specifically named "Plant Pathology," offered in the College of Agriculture, and gave notice of a course called "Pathology of Plants" being offered in the College of Arts and Science. Those two courses, each with the same brief outline indicating that they dealt with "The diseases of plants and their eradication and control," were offered until 1931, when "Pathology of Plants" became the name of the course offered in both colleges of the university.

Apparently Fraser taught plant pathology for both colleges regardless of the names of the courses. While doing this he showed that he was keeping up to date in his field of teaching by recommending the latest, or what he considered to be the best, textbooks to his students. At first he recommended one by B.M. Duggar, *Fungous diseases of plants,* published in 1909, which was the first plant pathology text to be written by an American. Because it was already somewhat out of date, Fraser welcomed the *Manual of plant diseases* by F.D. Heald soon after it was published in 1926. This in turn was followed by *Principles of plant pathology* by C.E. Owens in 1928. Fraser let his students know that the text by Owens was available but continued to recommend the one by Heald.[289] When Fraser retired in 1937, the university awarded him an honorary LL D and made him an emeritus professor.[290]

Several of Fraser's students for the master's degree went on to make noteworthy contributions to plant pathology. They include such notables as A.W. Henry, G.A. Ledingham, H.W. Mead, R.C. Russell, Mable Ruttle, and B.J. Sallans.

After Margaret Newton left, in 1925, Fraser taught practically all of the plant pathology, mycology, and plant physiology offered in the university. He got some gratifying relief when Thomas Clifford Vanterpool (1898–1984) joined the staff of the Department of Biology and took over the teaching of plant physiology in 1928.[291]

Vanterpool, a charter member of the Canadian Phytopathological Society, was born in the Netherlands West Indies but got most of his early education at Harrison College in Barbados, British West Indies. When he left that school in 1916, he had earned the Agricultural Science Diploma of Barbados, and the Oxford and Cambridge Higher Certificate. After a two-year stint as an overseer of a sugar plantation,

Vanterpool entered McGill University as a student of agriculture at Macdonald College. There, the stimulating influence of Professor B.T. Dickson persuaded him to become a plant pathologist.[292] He stayed at Macdonald College until he had earned a BSA, plus the Governor-General's Medal and the F.C. Harrison Prize in plant pathology in 1923 and an M SC in 1925. In 1925–26 he studied under Professor A.H.R. Buller at the University of Manitoba, on a Hudson's Bay Research Scholarship, and then returned to Macdonald College as an assistant, 1926–27, and lecturer, 1927–28.

In 1928 Vanterpool accepted President Murray's invitation to join the staff of the biology department, University of Saskatchewan, as an assistant professor. In the academic year 1935–36 he went to Britain and studied in the laboratory of Professor W. Brown, the dean of British plant pathologists, at the Imperial College, London, and mycology for a few weeks at the Imperial Mycological Institute, Kew.[293] Those studies, and the exposure to the work and facilities of others, were intended to prepare Vanterpool for the additional teaching and research responsibilities he would be having when Fraser retired, as he did in 1937.

Vanterpool soon realized that students studying for advanced degrees in phytopathology at the University of Saskatchewan were not being offered as many courses in plant pathology as they needed. To correct that situation, he developed the course "Advanced Plant Pathology and Mycology" especially for them. It was largely composed of assigned readings, with students being required to make written reports of their reading. Thus designed, it could be readily tailored to suit the individual needs of the students.[294] The university recognized Vanterpool's outstanding work in plant pathology, especially the teaching of that subject, by naming him professor of plant pathology in 1939.[295] He became president of the Canadian Phytopathological Society for the year 1944–45.

The university calendar for 1945–46 shows that Ralph C. Russell (1896–1964), a plant pathologist at the Dominion Laboratory of Plant Pathology, Saskatoon, who had earned a PH D from the University of Toronto, is listed as one of the lecturers. Except for this, there were few, if any, significant developments in the teaching of plant pathology at the University of Saskatchewan during, and for a while after, the Second World War.

ALBERTA

As in most of the other Canadian provinces in the early decades of the twentieth century, educators in Alberta were caught up in the general fervour for a return to nature and the consequent desire to teach nature study and agriculture in the public schools. In 1907 the agricultural component of the program of studies included "A general knowledge of ... Crops; their growth, management, rotation and *diseases*" (the emphasis is mine).[296] The general enthusiasm for the teaching of agriculture was intensified following the federal Agricultural Instruction Act of 1913, which provided money to the provinces to supplement and extend agricultural education. In response to increasing popular demand, a summer school for teachers, sponsored by the Alberta Department of Education, was authorized in the spring of 1913, and the first session was held at the University of Alberta from 7 July to 8 August of that year.[297]

The summer school offered a program for teachers who wished to increase their knowledge and efficiency in any one or more of several subject areas, including nature study, school gardening, and agriculture. In 1919 the courses dealt with high school agriculture at two levels, both of which involved some instruction in plant pathology under such headings as agricultural bacteriology, studies of seeds affected by smut, etc.[298] Although no record has been found to indicate who gave the instruction on plant diseases to the student teachers, he or she was probably the first to teach at least some plant pathology at the University of Alberta.

To assist the graduates of those summer schools with their teaching of agriculture in the elementary schools, the minister of education authorized a textbook for their use which had a full chapter on "plant enemies," including smut, rust, ergot, blight, etc.[299] A few years later a similar textbook was authorized for high schools, and it too had a chapter devoted to plant diseases.[300] Thus the Alberta Department of Education encouraged the teaching of plant pathology in the rural schools under its jurisdiction. Proof that plant pathology was actually taught in those schools may be found in their examination papers. For example, a Grade 8 examination, in 1912, asked the students to "Describe any common disease of plants and tell what steps should be taken to prevent its recurrence."[301]

In the meantime, the Honourable Duncan Marshall, minister of agriculture for Alberta, had decided that the way to present the best agricultural practices to farmers and their children was through the medium of demonstration farms and schools of agriculture located in several parts of the province. In 1911 seven farms were purchased for this purpose; within a year school buildings were being constructed on the farms at Claresholm, Olds, and Vermilion, and all three were officially opened in the fall of 1913.[302] At Olds, and for a three-year interval at Vermilion, F.S. Grisdale taught field husbandry and probably included grain diseases and methods for their control in his lectures, until he entered politics in 1930.[303] Doubtless somewhat similar instruction was given at the other schools, even though their original staff did not include anyone listed as a specialist in field husbandry.

University of Alberta

Those three schools of agriculture, and others which followed them, were feeder schools for the Faculty of Agriculture that was inaugurated at the University of Alberta in September 1915.[304] George Harcourt, former deputy minister of agriculture, was appointed as a special assistant to the dean of the new faculty and taught field husbandry and horticulture until 1917, when Garnet H. Cutler arrived to head a new department of field husbandry.[305] Cutler, a BSA graduate of the University of Toronto, had been assistant, then lecturer in cereal husbandry at Macdonald College, 1909–13, and professor of cereal husbandry, University of Saskatchewan, 1913–17, before going to the University of Alberta.[306] Harcourt continued as lecturer in horticulture until he retired in 1934; he was succeeded by James S. Shoemaker, BSA, MS, PH D.[307] Doubtless both Harcourt and Shoemaker included some plant pathology in their lectures because that subject had become so popular by 1920 that it was being taught in three faculties of the university, including the Faculty of Agriculture.

In the Faculty of Arts and Science, Francis J. Lewis, D SC, of the Department of Biology, was teaching agricultural botany to second-year students. His course consisted of "General vegetable physiology, diseases of plants, Mendelism and genetics."[308] In the Faculty of Medicine, one or more of the doctors in the Department of Bacteriology and Hygiene were teaching "Course 52, Agricultural Bacteriology," the outline of which specifically mentioned plant diseases, plus soil bacteriology and dairy bacteriology.[309] The calendar of the university for 1921–22 shows that Cutler added plant diseases to the outlines of two courses, "Forage, Roots and Potato Crops" and "Cereal Crops," both of which included diseases and insect pests peculiar to the crops dealt with in his lectures.[310] It was in 1920 that Cutler roused the interest of Guthrie B. Sanford (1890–1977) in the scab disease of potatoes, which led to the latter's pioneering studies in antibiosis.[311]

For a relatively brief period, beginning in 1919, Cutler was aided by Robert Newton, who was appointed assistant professor of field husbandry. Within two years Newton left the department to take up graduate studies at the University of Minnesota, where he earned an M SC degree and completed most of the work for a PH D, before returning to the university, in 1922, to become professor of plant biochemistry.

He was awarded the PH D in 1923 and, following Cutler's resignation, became head of the Department of Field Husbandry in 1924.[312]

Newton's research interests were mainly in the realm of biochemistry and plant physiology, and he applied his knowledge of those subjects to research into the reactions of plants to disease, cold, and other environmental adversities.[313] In teaching, he offered an honours course in plant biochemistry, which included the biochemistry of plant diseases and cold resistance,[314] in addition to his responsibility for the courses that Cutler had taught.

Shortly before Cutler resigned, the president of the university, Dr H.M. Tory, asked him to detail the salient points in support of research in plant pathology. Cutler's response, in a letter dated 21 May 1923, emphasized the need for a plant pathologist on the staff "who should be available at the University on full time," and added, "I urgently hope that steps may be taken whereby the services of a competent man ... may be secured."[315] Shortly after the name of the Department of Field Husbandry became that of Field Crops, Newton was authorized to search for a plant pathologist to add to the staff of his department. Eventually he chose a former classmate of his Minnesota days, Arthur W. Henry, who was appointed assistant professor of plant pathology in 1927.

Henry, a native of Fredericton, New Brunswick, had BSA and MSA degrees from the University of Saskatchewan, and a PH D from Minnesota, where he studied plant pathology under E.C. Stakman, an internationally recognized authority on the grain rust diseases. Henry had continued doing research and some teaching at Minnesota until 1926, when he went to Europe for a year of study with a fellowship from the Rockefeller Foundation. It was while in Europe that he accepted Newton's offer of a position in his Department of Field Crops. Laboratory and office space for Henry were made available on the ground floor of the "North Lab," in a room formerly occupied by the soils department. His staff there consisted of "lab man" E.A. Thompson and graduate student assistant W.R. Foster.[316]

At first, Henry taught agronomy because the curriculum of the Faculty of Agriculture did not include a special course in plant pathology, but Newton was largely instrumental in changing that. The minutes of the faculty meeting of 6 December 1927 show it was he who proposed that the lecture time in the course "Field Crops 50" be increased from two to three hours per week in order to give more instruction on their diseases. He further proposed that a graduate course in plant pathology be authorized.

As a consequence of Newton's initiative, Henry's field crops course included additional lectures on diseases and their control, and he began teaching an advanced course in plant pathology in 1928, in a newly constructed building that became known as the "West Lab." That was the year in which W.R. Foster became the first graduate student to register for a master's degree in plant pathology.[317]

After Newton resigned in 1927, Olaf S. Aamodt, BS, MS, PH D, who had been associate professor of genetics and plant breeding, became head of the Department of Field Crops.[318] Although primarily a plant breeder, Aamodt was interested in grain diseases, especially the smuts,[319] and he included discussions of them in his lectures until he resigned in 1935 and was replaced by Kenneth W. Neatby (1900–1958), BSA, MSA, PH D.[320] Neatby did not need to put as much emphasis on the diseases of field crops as did his predecessors because within a year of his appointment Henry was teaching a course devoted to diseases of those crops.

In the meantime, diseases of plants continued to be included in the course "Agricultural Botany," now being taught by Ezra Henry Moss (1892–1963). Moss, who had University of Toronto MA and PH D degrees, the thesis research for both of which was based on rusts on coniferous trees, was an assistant professor of botany in the Faculty of Arts and Science. In 1933, when he was promoted to the rank of associate professor, Moss increased his plant pathology offerings with "Diseases of Plants," a course for third-year agriculture students. Moss retained his interest in the rust fungi and became one of the few scientists of that era to study the parasitic behaviour of dodder on non-agricultural plants.[321]

In 1936, Henry, now associate professor of plant pathology, was teaching three plant pathology courses: Diseases of Field Crops, Diseases of Horticultural Crops, and a graduate course, Advanced Plant Pathology, and Moss put "diseases of plants" into his course, Elementary Mycology.[322] That situation pertained until 1939–40 when Henry added a fourth course, one called "Elements of Plant Pathology," to his repertoire of plant pathology courses.[323] He was promoted to the rank of full professor in 1947, and there were no significant changes in the plant pathology offerings to students at the University of Alberta until several years after the Second World War.

BRITISH COLUMBIA

British Columbia responded to the nationwide movement for the introduction of agricultural education and nature study into the curriculum of the public schools by using a portion of its share of the funds provided by the Canadian government through the Agricultural Instruction Act of 1913 to hire John W. Gibson, MA, as director of elementary agricultural education in 1914.[324] To assist in the implementation of its program, the Department of Education prescribed John Brittain's book, *Elementary agriculture and nature study*, as the textbook for school use throughout the province. Brittain was the professor of nature study at Macdonald College of McGill University, and his book contained several pages on the diseases of fruit trees, thus making it an appropriate text for students in the fruit-growing areas of the province.[325]

Like other provinces, British Columbia provided short courses and a summer school for teachers of agriculture, and during each of them the teachers were exposed to a few lectures on plant diseases, and other aspects of plant pathology. Probably the first series of lectures on plant diseases and their control in the province, delivered by a man with training in plant pathology and experience in teaching, were those of provincial plant pathologist John William Eastham (1880–1969) to students attending a short course on agriculture at Kelowna in 1915.[326] Eastham also "gave a short course of lectures and demonstrations on bacteriology and plant pathology" to a summer school for teachers of agriculture, held in Victoria in July 1919.[327]

University of British Columbia

The first lectures on plant pathology in the classrooms of the University of British Columbia were also those given by Eastham. As a special lecturer, for three winter terms beginning in 1920, Eastham gave a course of lectures and demonstrations in plant pathology to the agricultural classes in the Department of Soldiers' Reestablishment.[328] A native of Liverpool, England, with a B SC degree from the agricultural department, University of Edinburgh, Eastham had taught chemistry and biology in England for seven years, four of which were at the Cheshire Agricultural College at Holmes Chapel, before coming to Canada in 1906. He had also been a teacher at the Ontario Agricultural College for five years, and a graduate student in plant pathology at Cornell University. Thus he was an experienced and knowledgeable teacher of agriculture, including plant pathology, prior to becoming provincial plant pathologist in 1914. He was also a member of the Quebec Society for the Protection of Plants and a charter member of both the Canadian and American Phytopathological Societies, and served one year as president of the Pacific Division of the American society.[329] He seldom missed an opportunity to teach or give lecture-demonstrations on plant pathology, and the above examples illustrate the wide range of groups he attempted to serve, in addition to his regular work as provincial plant pathologist.

Eastham participated in horticultural short courses for several years,[330] and the calendar of the University of British Columbia for 1921–22 shows that he was a lecturer in plant pathology in the

Department of Botany. It also shows that plant pathology was an integral part of at least four courses available at the university: Economic Botany, Plant Pathology, General Plant Pathology, and Forest Pathology.

After 1923, virtually all of the lectures on plant diseases at the university were given by Frank Dickson (1891–1969), who was appointed assistant professor of botany that year. Dickson, a native of England, graduated from the University of London and came to Canada in 1912. He served in the Canadian Army, and then taught school while working toward the BA degree that he earned at Queen's University in 1920. From Queen's, Dickson went to Cornell University, where, for the first year, he was an assistant in plant pathology, then an instructor, until he obtained a PH D degree in that subject in 1923. Like so many other budding plant pathologists, Dickson accepted the appointment at the University of British Columbia before completing the writing of his thesis, which he finished before the end of the year.

The plant pathology offerings at the university were soon increased following Dickson's appointment. This is shown in the calendar for 1924–25, which lists courses in elementary plant pathology, advanced plant pathology, history of plant pathology, and mycology.

Although Dickson had a heavy teaching load, he conducted a study of the decays of Alpine fir trees in the upper Fraser River region on behalf on the provincial forest service in 1927 and in 1928, and supervised the studies and research of several students in mycology and plant pathology who became well known in these fields. Former students Cecil E. Yarwood, Ewart Woolliams, and Norman S. Wright became well-known plant pathologists. Dickson is fondly remembered by many as having been a good teacher and an avid basketball fan. His laboratory courses were uniquely programmed, with the required work outlines and their accompanying instructions conveniently placed in numbered boxes.[331]

The calendars of the university show that an undergraduate option, or major, in plant pathology was introduced in 1928, and that Dickson was the only teacher in that general field for many years. His course in mycology put emphasis on the fungi of economic importance in British Columbia, and the advanced plant pathology course, designed for honours and graduate students, put emphasis on methods for the isolation, culture, and identification of fungi from infected plants. It was given in two lectures and four hours of laboratory work each week. Identical outlines of the three plant pathology courses appeared in the calendars of the university for both the Faculty of Arts and Science and the Faculty of Applied Science until sometime after the Second World War.

According to the calendar for 1949–50, Donald C. Buckland joined the staff of the Department of Biology and Botany that year as an associate professor of forest pathology. He took over the course in forest pathology that Dickson had taught, and three additional courses, namely: Applied Forest Pathology, Problems in Forest Pathology, and Advanced Forest Pathology. There were no major changes in the teaching of plant pathology at the University of British Columbia until a Faculty of Forestry was established there early in the 1950s.

SUMMARY

In the early days of plant pathology in Canada, much of the teaching of that subject was done through reports of agricultural societies and the medium of newspapers and magazines, especially those that catered to the agricultural community. The acquisition of useful knowledge for its members and their children was a major objective of virtually all of the early agricultural societies in Canada. In their attempts to reach that objective, some information about the protection of crops from insects and diseases was often provided by guest speakers at their meetings, or through whatever books and other forms of published material they were able to acquire. Eventually, in every province, there were agricultural societies, editors of farm-oriented periodicals, and others interested in the advancement of agriculture and agricultural education, advocating the teaching of plant pathology as an aspect of agriculture in public schools. To do this effectively, several provinces included agriculture, with its component of plant pathology, in the course of studies for their teachers-in-training, and then made the subject compulsory in the curriculum of their rural schools. After the fungal nature of many plant diseases became generally known, the topic of plant diseases commonly entered into such courses as biology, botany, and mycology, wherever those subjects were taught.

When agriculture became a subject for study in colleges and universities, first in Ontario, Quebec, and Nova Scotia, some aspects of plant pathology were invariably included in the course of studies. It was an important component of the curriculum of the agricultural colleges and of institutions having departments or faculties of agriculture. In several such institutions of higher learning, beginning in the 1920s, plant pathology was offered as a separate course, and students could specialize in that general field for advanced degrees. By 1949, plant pathology was being taught, at various levels in schools and colleges, in every province of Canada except Newfoundland, which joined the rest of Canada that year.

9 A History of Early Mycology in Canada

People living in the geographical area that is now Canada were collecting mushrooms and other fungi as a hobby in the first half of the nineteenth century. However, their pursuit of mycology (the branch of biology that deals with mushrooms, mildews, and other fungi) during most of that period has to be viewed in the Victorian tradition of natural history, and their study of fungi, mostly the fleshy forms, mainly as an appreciation of nature. Those early "mycologists" were, almost without exception, clergymen, doctors, or businessmen, many of whom developed a scientific spirit that has seldom been seen since their time. Thus in the beginning, mycology in Canada was dominated by enthusiastic amateurs who did little more than collect fleshy fungi, identify them or have them identified by someone else, usually a European mycologist, and then publish a list of what they had collected and identified.

Around the middle of the nineteenth century a change took place. Until then European authorities and institutions were looked upon as guides and models. That was logical, because European mycologists had set up classification schemes and worked out the life cycles of many fungi long before any original mycological work was done in Canada, and until mid-century virtually all of the books dealing with fungi were written by Europeans. But after about 1850, Canadians began to join American associations, and they were as likely to exchange information and specimens with Americans as with Europeans. Such activities would have been unthinkable just a few decades earlier, because of the anti-American feeling that had persisted since the war of 1812–14 when American armed forces invaded Canada. Doubtless the change in attitude was speeded up by the friendly greetings and exchanges of information that took place during the meetings of the American Association for the Advancement of Science that were held in Montreal in 1857.

Beginning in the 1850s, great changes and improvements in methods and aims of mycological work began to take shape in Europe, changes that were destined to have a profound influence on the development of mycology in the United States and Canada. Conspicuous individuals in this regard were the German biologist Heinrich Anton de Bary (1831–1888), the French brothers Louis-René (1815–1885) and Charles Tulasne (1816–1884), and the English clergyman Miles Joseph Berkeley (1803–1889), all of whom gave some attention to the physiology of fungi and the pathological aspects of their subjects. Thus it is understandable that Canadian scientists, including botanists and mycologists, considered themselves to be British scientists as well as Canadian. Many of them were members of the British Association for the Advancement of Science, which held its first meeting outside Britain in Montreal in 1884.

When it became widely recognized that the great majority of plant diseases were caused by fungi, there was a natural tendency for students of the fungi to become "economic mycologists," or "applied mycologists," both of which were transitional toward, or precursors of, "plant pathologists." Until physiological disorders and virus diseases were recognized as important factors in plant diseases, applied mycology and plant pathology were virtually synonymous. For example, beginning in 1922 the Commonwealth Mycological Institute surveyed the literature on plant diseases in its *Review of Applied Mycology*. The title was not changed to *Review of Plant Pathology* until 1970.

Mycology became the chief cornerstone of the emerging science of phytopathology, and a growing number of plant pathologists recognized this by getting a good grounding in mycology. They did this largely because a knowledge of the life cycle of a fungus pathogen was more important to a plant pathologist looking for weak or vulnerable links in the cycle than it was to the mycologist who had only an incidental interest, or none at all, in controlling or killing the organism. Consequently, many plant pathologists, of necessity, became "acting mycologists," or they published works that were mycological in orientation. In other words, the dividing line between the publications of mycologists and plant pathologists became blurred or even non-existent in many instances.

Such a situation makes the task of the historian of either science a difficult one because the histories of mycology and plant pathology often overlap. Nevertheless, this brief history of mycology in Canada, the coverage of which is from its beginnings to shortly after the end of the Second World War, arbitrarily excludes most plant pathologists, medical mycologists, and mushroom growers, unless they published a description of a fungus or a list of the fungi they collected, or otherwise made a significant contribution to the advancement of mycology in Canada.

NOVA SCOTIA

Nova Scotians were pioneers in the study of mycology in Canada, and the earliest published list of fungi collected by one of them is in the book by Thomas Chandler Haliburton (1796–1865) titled *An Historical and Statistical Account of Nova Scotia*, published in 1829. Haliburton, author of the humorous "clockmaker" series of fictional stories featuring Sam Slick, was referring to a list of plants, including ten fleshy fungi, collected by the Rev. William Cochran (1757–1833), one-time president of King's College, Windsor, Nova Scotia.[1]

In 1831, Titus Smith (1768–1850), of Halifax, prefaced his list of the principal indigenous native plants with a discussion of "the operation of fungi in disintegrating vegetable substances."[2] Smith was a remarkable individual – one who has been referred to as farmer, surveyor, botanist, author, philosopher, and journalist. Among his endeavours were experiments, within plots in his garden, to acclimatize plants, grown from seeds purchased in England, to the environmental conditions of his farm near Halifax. In this, and some of his other work with vegetation, Titus Smith was a pioneer plant ecologist in North America, and one who also theorized on the physiology of fungi.[3]

Nearly half a century passed before another Nova Scotian, Newfoundland-born John Somers, MD, began a series of talks on fungi to members of the Nova Scotia Institute of Natural Sciences. The first, on 26 January 1880, included "a very short list of some common species of our mycological flora, the result of three month's study of the local botanical region." Many of the thirty-three species of fleshy fungi in his list were classified on the basis of spore colour. On 10 December the same year, he provided a list of nearly eighty species, some of which had been sent to him by his colleague Alexander H. MacKay. For at least eight years Somers collected fungi, not all of which were fleshy forms, and kept adding to his published lists. There was a slime mould in his second list, and the one in 1886 included a Sphaeropsis and a Cladosporium.[4]

Alexander Howard MacKay (1848–1929) was a native of Pictou County, Nova Scotia, where he attended Pictou Academy, the school of which he later became principal for sixteen years. MacKay earned a BA degree at Dalhousie University, which later awarded him an honorary LLD, as did St Francis Xavier University. He won an enviable reputation as an outstanding and progressive educator while serving as principal of the Pictou Academy, 1873–89, as principal of the Halifax Academy for one year, and as Nova Scotia's superintendent of education for thirty-six years. During that time MacKay also served as secretary of the Canadian Botanical Society for twenty years, president of the Canadian Educational Association for four years, and president of the Nova Scotia Institute of Science for one year. He became a Fellow of the Royal Society of Canada in 1889 and president of its Geological and Biological Section in 1900.[5]

MacKay's published works give no indication when his interest in the fungi was first aroused, but it is known, from John Somers's lists, that he was collecting fleshy fungi and lichens in the early 1880s. His own "provisional list" was presented to members of the Nova Scotia Institute of Science, 8 December 1902, but was not published until 1906. In that paper he mentioned three additional amateur mycologists in the region: R.R. Gates of Middleton, J.M. Swaine of Truro, and C. Stanley Bruce of Shelburne.[6]

Reginald R. Gates, MA, who became a teacher at Mount Allison University in Sackville, New Brunswick, had a remarkably good knowledge of mycology for a man of his time. In a paper read before the institute on 8 September 1902, he referred to mycorrhizal and other relationships between fungi and higher plants. He commented on wound parasites and fungi on trees, and stressed the importance of mycology to the science of forestry.[7] From Sackville, Gates also contributed to the list of New Brunswick fungi compiled by G.U. Hay (see below).

MacKay, who was so interested in fungal-induced plant diseases that he became a member of the American Phytopathological Society, had some of his fungi identified by Charles H. Peck, state botanist of New York, and he followed the general classification scheme of the English mycologist M.C. Cooke. His lichen collection, of approximately one hundred species, is now (1993) in the herbarium of the Nova Scotia Museum of Science, Halifax. In his "First Supplementary List," MacKay acknowledged contributions from Minnie C. Hewitt, science teacher in the Lunenburg Academy, and from Clarence L. Moore and William P. Fraser, teachers in the Pictou Academy.[8] (Clarence Moore was particularly interested in the slime moulds, but he also studied and published on aquatic fungi and the rusts.[9]) Those teachers, and MacKay himself, had a far-reaching influence on the development of mycology and plant pathology in Canada. They and their students were among the first in Canada to specialize on the slime moulds, the micro-fungi, and fungi that induce diseases in plants.

William Pollock Fraser (1867–1943), destined to become Canada's outstanding native-born authority on the rust fungi, took his students on field trips to collect specific groups of fungi. For a while they concentrated on the powdery mildews, and seven of his students wrote brief papers on their collection. Annetta Bishop, 1909 school gold medallist, wrote on Uncinula; Emily Spicer and Emeline Mackenzie, a 1909 silver medallist, wrote on Microsphaera; Mabel McKay and Jean Henry on Phyllactinia; John Cameron on Podosphaera; and John Craigie on Erysiphe. (This is the same John Craigie who, while working in Manitoba in 1927, was acclaimed for his discovery of the function of the pycnia of the rust fungi.) Fraser encouraged his students by having their work, together with his own, published in the first *Bulletin of the Pictou Academy of Science Association* in 1909.[10] Further evidence of the emphasis on powdery mildews at that time is seen in the notes on those fungi by H.H. Mussels and E.T. Parker in the same issue of that bulletin.

In July 1909, Fraser became the first person in North America to discover the aecial stage of the blueberry rust fungus, a fact that was confirmed by the American authority, J.C. Arthur, with whom he had been corresponding.[11] In 1910, Fraser published a description of ninety-two species of rust fungi, several of which had not previously been reported on this continent.[12]

The extraordinary nature of Fraser's early work on the rust fungi was formally recognized by Dalhousie University when it conferred on him the MA degree in 1910. Fraser had earned an AB degree from Cornell University in 1906, when he was thirty-nine years of age. His education had been interrupted because the early death of his father forced him to support his mother and sister until he had saved enough money to attend Cornell University in 1905.[13]

A.R. Prince, who taught botany and bacteriology to students of agriculture in the Nova Scotia Agricultural College, was an avid collector of various kinds of fungi and slime moulds in the late 1920s. Specimens of Lycogala, Hemitrichia, Albugo, Pilobolus, Dibotryon, and Hypoxylon, collected by Prince in Nova Scotia, are in the herbarium of the University of Alberta. Notes accompanying several specimens indicate that they had been identified by L.E. Wehmeyer, a mycologist from the USA with whom Prince went on several mycological forays in Nova Scotia.

Many foreigners have published lists of the fungi they collected in Canada, and Lewis E. Wehmeyer(b.1897), a nativeof Illinois whose wife's parents lived in Nova Scotia and who collected fungi in the

Maritime provinces over a period of years, is one of the most historically important of them. Beginning in 1935, he published a series of "Contributions to a study of fungi in Nova Scotia," the sixth of which is in the 1942 volume of the *Canadian Journal of Botany.*

Wehmeyer's outstanding work on the mycology of the Maritime provinces appeared in 1950, when he produced *The fungi of New Brunswick, Nova Scotia and Prince Edward Island.* In that publication he reviewed the early history of mycology in the Maritimes and referred to a number of individuals whose names have not been seen elsewhere as collectors of fungi in Nova Scotia. These include J.A. Boyle (pp. 15 and 29), G.W. Hope (p. 28), R.M. Lewis (pp. 18 and 32), D.E. McCuish (p. 23), and R.W. Ward (p. 106). Two or three well-known Canadian botanists-mycologists who collected fungi in Nova Scotia are also mentioned. These include Ibra L. Conners (pp. 22 and 23), John Dearness (p. 111), and Albert E. Roland, a long-time teacher of botany in the Nova Scotia Agricultural College whose name appears at least a dozen times. Wehmeyer also mentions (p. 28) that Grant D. Darker, a native of Streetsville, Ontario, who became an authority on the Hypodermataceae of conifers while working in the USA, has specimens collected in Nova Scotia.[14]

More recent students of the fungi in Nova Scotia, also mentioned by Wehmeyer, include plant pathologists John Frederick DeWitt Hockey (1895–1980), B SC, officer in charge of the Dominion Laboratory of Plant Pathology, Kentville, from 1924 until he retired in 1960, and his long-time colleague, Kenneth A. Harrison (1901–1991). Although no published list of Hockey's collection has been found outside of Wehmeyer's notations, dozens of specimens bearing his name, and those of Harrison, are in the herbarium of the Nova Scotia Museum of Science.

Harrison, a native of New Brunswick who graduated from the Nova Scotia Agricultural College, earned a B SC for studies in horticulture at the Ontario Agricultural College in 1924, and studied plant pathology at Macdonald College of McGill University for the M SC that he earned in 1925, was publishing on the fleshy fungi of Kentville as early as 1927.[15] His interest in those fungi intensified after his service in the Second World War. Harrison established a mycological herbarium at the research station in Kentville, where he was employed for many years, to which Hockey and others contributed. Most of Harrison's collections, other than those in the Nova Scotia Museum of Science and nearly two thousand duplicates donated to Acadia University, are now in the National Mycological Collection in Ottawa.

The six-volume "Index of the Mycological Writings of C.G. Lloyd" has references to many fungi having been sent to American mycologist

Curtis G. Lloyd (1859–1926) by Canadians. Two of these from Nova Scotia were E.D. Lordly of Chester, whose contributions are noted in letters 3, 4, 17, 29, 49, and 60, and R.R. Gates of Middleton, in letter 4. The addresses of many Canadian contributers are so vague as to be almost worthless. For example, Rev. Joseph Mignault (vol. 3, p. 29, and vol. 4, letter 39) is merely listed as being from Canada, as are more than twenty others.

NEW BRUNSWICK

The earliest account of fungi in New Brunswick appeared in "a preliminary list of the plants of New Brunswick" compiled by the Rev. James Fowler (1829–1923), MA, in 1878, and published in the report of the secretary of agriculture the following year. That list included fifty-three fungi collected in the counties of York, Kent, and Restigouche.[16] Fowler was appointed to teach natural science in the Normal School, Fredericton, in 1878. Two years later he moved to Ontario and became professor of biology, Queen's University.[17]

Fowler's list of New Brunswick plants was prepared too early to have been included in the *Bulletin of the Natural History Society of New Brunswick,* the first number of which was published in Saint John in 1882. Although there are one or two oblique references to fungi in the early minutes of that society, the first noteworthy published account is the Rev. George U. Hay's "A Preliminary List of New Brunswick Fungi" in 1901.[18]

George Upham Hay (1843–1912) was born in the village of Norton, Kings County, New Brunswick, where he got his early education prior to obtaining a B PHIL degree from the Illinois Wesleyan University and an MA from Acadia, the university that eventually awarded him an honorary PH D. Hay won for himself an enviable reputation in journalism and as a teacher in the public schools of New Brunswick for thirty-four years. During that period he served, at various times, as editor of the *New Brunswick Journal of Education,* president of the Botanical Club of Canada, president of the New Brunswick Historical Society, and president of the Natural Sciences Section of the Royal Society of Canada, to which he had been elected Fellow in 1894.[19]

Hay's several published lists include the names of more than 320 species of fungi from divers parts of his native province. He acknowledged that his lists included fungi collected by others, including John Moser, who contributed many specimens from Kings and Queens counties accompanied by his field notes; J. Vroom of St Stephen; Reginald R. Gates of Sackville; and Mary and Adaline van Horne, who sent specimens from the general area of St Andrews.[20]

Mary van Horne talked about her collection and displayed preserved specimens, spore prints, photographs, and paintings of fungi during the 2 April 1902 meeting of the Natural History Society of Montreal. In her talk she said that her brother, Sir William van Horne, president of the Canadian Pacific Railway, had photographed and "made some very fine water-color drawings of a number of the plants." Mary's niece, Adaline van Horne (Sir William's daughter), was also an amateur mycologist, and she too gave an illustrated talk about her collection of fungi to members of that society, 26 February 1912.[21]

Both of the van Horne ladies took a very professional attitude toward their hobby and sent duplicate specimens of many fungi to well-known authorities for identification. They received belated national recognition when the federal Department of Marine and Fisheries published a list of 108 fleshy fungi collected by them, in a supplement to its forty-seventh annual report in 1915. There is a sad note associated with that publication, the manuscript of which was read by D.P. Penhallow of McGill University: both Mary van Horne and Penhallow died without seeing the paper in print.

Lewis E. Wehmeyer, in his *The fungi of New Brunswick, Nova Scotia, and Prince Edward Island*, indicates that Miss G.E. Fisher, S.F. Clarkson (p. 24), J.L. Howatt (p. 101), and John Ehrlich (p. 103), of New Brunswick, each contributed to his list of fungi. Except for these there have been few, if any, published accounts of fungi collected in the province between 1915 and the late 1940s.

PRINCE EDWARD ISLAND

John MacSwain, who presented three papers that dealt with fungi to members of the Natural History and Antiquarian Society of Prince Edward Island in 1899, may well have been the first man to be recognized for his collection of fungi there. The first paper was an unpublished commentary on fungi in general, whereas his talk on 24 January, titled "Some of Our Fungi," and the one on 7 February, "Rust of Wheat," were published in the bulletin of that society. Later in the same year MacSwain also published "Notes on Fungi collected by Miss Pippy, 1899."[22] Apparently Miss Pippy was so well known in Charlottetown that neither MacSwain nor the secretary of the society felt it necessary to use her given name. She was elected vice-president during one meeting, and the minutes show that a meeting was held at her home.[23] June (Pippy) Middleton, of Hazelbrook, Prince Edward Island, the present-day historian of the Pippy family, provided evidence to indicate that Eleanor Sarah Pippy (b.1851), who studied briefly in France and Germany, was the Miss Pippy noted above.

Paul A. Murphy (1887–1938), BA, officer in charge of the Plant Pathology Field Laboratory, Charlottetown, for nearly four years from July 1915, and a widely recognized authority on diseases of potatoes, was well qualified to appear in this history, but he neither collected nor described fungi while he was in Prince Edward Island, even though he worked with many of them in relation to plant diseases.

Richard R. Hurst (1895–1961), BSA, who joined the staff of the Charlottetown laboratory in 1925 and became officer in charge in 1927, was somewhat more of a mycologist. His reports of the presence of fungi and of plant diseases in the province appear in many issues of the annual report of the Canadian Plant Disease Survey between 1925 and 1940, and he is mentioned more than a dozen times by Lewis E. Wehmeyer in *The fungi of New Brunswick, Nova Scotia and Prince Edward Island*. Wehmeyer also refers to collections made by Blythe Hurst, G.C. Warren, and J.H. Eastham, on pages 26, 27, and 90 respectively. Collections by B.V. Baxter are mentioned at least three times (pp. 62, 70, and 71). Wehmeyer, who did some collecting on the Island, also names sixteen or more fungi that were collected there by Canadian naturalist John Macoun (1831–1920).

QUEBEC

Philip H. Goss (1810–1888), an American-born naturalist who farmed for a while near Compton, Quebec, mentioned mushrooms in his book *The Canadian Naturalist*, published in 1840, but the earliest published list of fungi in Quebec is in "A Provisional Catalogue of Canadian Cryptogams," prepared by D.A. Watt in 1865.[24] Scottish-born David A. Watt (1831–1898) came to Canada in 1846 and soon established himself in the business life of Montreal, where he helped organize a corn exchange. He became editor of the *Canadian Naturalist* and a life member of the Natural History Society of Montreal, which had been founded in 1827. In his leisure time, Watt was a botanist, with mycology as a major hobby interest. One of his unaccomplished botanical objectives was the compilation of an "annotated catalogue of Canadian plants."

Watt's list of fungi included the collection of Dr W.P. Maclaggan, whose species had been determined by British mycologist the Rev. M.J. Berkeley, and whose collection of *Claviceps purpurea* is listed in Berkeley's book, *Notices of North American Fungi*. Watt said his own collection had "passed under the eye of Rev. M.A. Curtis of North Carolina." The remarkable feature of both collections is that they were composed largely of microscopic fungi listed under headings that appear to have been proposed by Berkeley. There are 152 species under the general heading "Mycetales," in which the genera are listed alphabetically. One section begins with 18 species of "Aecidium" and concludes with "Uredo"; thus the rust fungi are well represented. The list also includes approximately 75 species of lichens.[25]

Another noteworthy amateur mycologist in the Montreal area around that time was the Rev. Robert Campbell (1835–1921), who had BA, MA, and DD degrees from Queen's University. In 1903 he published a paper titled "Canadian fungi," in which he referred to 129 species collected from three sites in Quebec. Campbell discussed fungi in relation to their environment in a 1910 paper, and noted that certain trees are the "favorite prey" of certain fungi.[26]

Thirty years later, John Adams was collecting fungi on Anticosti Island and the Gaspé Peninsula of Quebec. In his published account of that work, in 1935, reference is made to Joseph Schmitt's 1904 "Monographie de l'Ile d'Anticosti," in which eleven species of fungi and several genera are noted.[27] Adams was helped with his identifications by Ibra L. Conners (1894–1989), MA, curator of the National Mycological Collection, who added a number of fungi to Adams's list of flora from that area in 1937.[28]

Virtually everyone associated with the National Herbarium in Ottawa did some collecting on the Quebec side of the Ottawa River and added specimens to the herbarium, but few of them published lists of their collections.

In September 1922, J.W. Grenier, about whom little has been learned, collected *Gloeophyllum saepiarium* and *Polyporus abietinus* on the Gaspé peninsula. These specimens bear accession numbers 3995 and 3994 respectively in the herbarium of the University of Alberta.

Apparently mycology had little or no appeal to early French-Canadian naturalists. Abbé Leon Provancher (1820–1892) had a chapter devoted to fungi in his 1862 book, *Flore Canadienne ou Description de toutes les plantes des Forêts, Champs, Jardins et Eaux du Canada*. He also published, in 1886–87, two articles on fungi in volume 16 of *Le Naturaliste Canadien*, of which he was founder and long-time editor, and deplored the paucity of mycological information in the libraries of Quebec. Provancher attempted to rectify that situation by encouraging others to become interested in fungi. He set an example by publishing a brief article on mushrooms and the names of twenty-five species that had been collected in Quebec and identified by Baron Felix de Thümen of Austria.[29]

No one seems to have followed the abbé's example; consequently there is no list of Quebec fungi in any of the French-language publications prior to 1950 that is comparable to Watt's, for example, in the English. There are brief unsigned comments about fleshy fungi in *Le*

Naturaliste Canadien, and a well-referenced commentary on mycorrhizae of the Canadian yew (*Taxus canadiensis*) by Henri Prat, D SC,[30] but the authors did not publish lists of any fungi that they may have collected. There are abstracts of papers pertaining to mycology by such outstanding Quebec plant pathologists-mycologists as René Pomerleau, Elzéar Campagna, Champlain Perrault, and Emile Jacques in the first volume of *Annales l'ACFAS* (Association Canadienne-Française pour l'Avancement des Sciences), which was published in 1937. Abstracts of papers by Omer Caron and Abbé Ernest Lapage are in volume 4, and similar brief abstracts by various authors appear in succeeding volumes.

As early as 1897 a course was being offered in the Agricultural School at Oka, Quebec, on diseases of cereal grains that included rust, smut, and ergot, but there was no publication devoted to fungi, or the description of a fungus (except those mentioned above) from any French-language school in Québec prior to the Second World War.

The most noteworthy native-born mycologist in Quebec is René Pomerleau (b.1904), B SC, M SC, D SC, who made the first systematic study of any group of fungi in Quebec. He did this at the suggestion of his supervisor, Professor Bertram T. Dickson, while studying for a master's degree in the Department of Plant Pathology at Macdonald College, McGill University. His master's thesis, on Pyrenomycetes of Quebec, was published in 1927.[31] His 140-page thesis for the D SC, University of Montreal, in 1938, was titled "Recherches sur le *Gnomonia ulmea.*"

Pomerleau and Jules Brunel, of the University of Montreal, provided a descriptive inventory of Quebec fungi, but not necessarily of ones collected by them, in *Le Naturaliste Canadien* from 1938 until 1940. They may have been inspired to do that as a result of their active participation in the Quebec foray of the Mycological Society of America, in August 1938. The names of the participants, and most of the 836 species and varieties of fungi collected, were listed in *Mycologia* the following year.

Outside the ranks of mycologists, Pomerleau is best known for his contributions to forest pathology, the field in which he was to work for more than forty years. Nevertheless, throughout his working life, especially after studying mycology in France, 1927–30, he has been a widely recognized authority on the higher fungi of eastern Canada. Pomerleau's major publishing achievements in this regard, which are among the best in Canada, came after 1950.

Pomerleau's friend and companion on a number of mycological forays, Henry A.C. Jackson (1877–1961), was a competent amateur mycologist and a well-known commercial artist from Montreal. He collected fungi mostly for their beauty, because the shapes and colours of the fleshy forms appealed to his aesthetic sense, and many of them became models for his artistic work. After he retired from commercial life in 1941, Jackson spent the summers in his country home where he had his own private laboratory and painting studio, plus a library that was largely devoted to birds and the higher fungi. His paintings of fleshy fungi were a unique combination of artistic beauty and accuracy. In addition to his beautiful paintings, Jackson published at least two papers on fungi and in one, with Irene Mounce, he mentioned that Mrs Charles A. Lewis had collected an interesting specimen in a dry spruce woods at Metis Beach.[32]

At McGill University, American-born botanist David Pearce Penhallow (1854–1910) wrote several papers on plant diseases, but only one that dealt specifically with a fungus.[33] In 1918, Professor R.J. Blair of McGill University sent several fungi to the editor of *Mycologia* for identification. The names of those fungi are given in a "Notes and brief articles" section of volume 10, page 290.

William Lochhead (1864–1927), BA, MA, professor and head of the Department of Biology at the newly constructed Macdonald College of McGill University, in Ste-Anne-de-Bellevue, published several papers on plant diseases, but not on the causal fungi alone. Nevertheless, he is included here because he taught mycology and, beginning in 1908, required his students to make an identified collection of fifty fungi; thus he advanced the science of mycology in Canada.[34]

Lochhead's course, with its collection requirement, was taken over by William P. Fraser when he joined the staff as a lecturer in 1912. Fraser taught a course designated as "Plant Diseases and Fungi" until he resigned in February 1919 to work on grain diseases in western Canada. It is noteworthy that one of Fraser's undergraduate students, Margaret Newton, while working at Macdonald College, became the first Canadian to discover pathogenic strains in a rust fungus.[35]

Fraser's successor, Bertrum T. Dickson (1886–1982), who taught "Systematic Mycology" and "Advanced Mycology," wrote at least two papers on fungi, one that described a fungus growing on a cow's horn and hoof that he found in Morgan's Woods, near Macdonald College, and one that dealt with the then relatively new topic of "saltation" in a fungus.[36]

Others who taught mycology at Macdonald College prior to 1950 were John G. Coulson (1893–1974), who taught "Physiology of the Fungi" to graduate students for several years beginning in 1925, Thomas C. Vanterpool (1898–1984), John E. Machecek (1902–1970), Dorothy Newton (b.1901), and Harold J. Brodie (1907–1989), each of whom taught or assisted in the teaching of mycology for relatively brief periods but did not publish, during that time, on mycological work performed in Quebec.

Ivan H. Crowell (b.1904), B SC F, AM, PH D, who was an authority on Gymnosporangium before joining the Macdonald staff in 1937, published two papers on that genus in 1940 and one on new Canadian smut fungi in 1942.[37] Crowell, who liked to work with his hands, published two papers on mycological methods in volume 33 of *Mycologia*, and became so interested in handicrafts, while at Macdonald College, that he resigned, in 1946, to devote himself full time to handicraft work. Crowell was succeeded, in 1947, by Eric O. Callen (1912–1970), PH D, who taught mycology but did not publish on mycological work done in Canada prior to 1950.

Robert Hagelstein (1870–1945), of New York City, collected slime moulds in the Morgan Arboretum of Macdonald College, Quebec, while attending the foray of the Mycological Society of America that was based at that college 25–28 August 1941. He donated named duplicates to the McGill University Herbarium, many of which are mentioned in his 1944 book, *The Mycetozoa of North America*.

In the summer of 1936, Dr Nicholas Polunin collected fungi in the arctic regions around Wolstenholm, Sukluk, and Cape Smith in northern Quebec, and in the subarctic region around Port Harrison. The record of those collections was not published until after the Second World War, when Polunin was visiting professor of botany, McGill University.[38]

NEWFOUNDLAND AND LABRADOR

There were very few, if any, native amateur or professional mycologists in Newfoundland or Labrador prior to 1950, but a number of "outsiders" collected there. Arthur C. Waghorne (1851–1900), an English missionary who worked in Newfoundland from 1875 to 1899, collected fungi, bryophytes, and vascular plants from several areas on that island. In a very real sense, Waghorne was Newfoundland's first resident botanist-mycologist, even though he was entirely self-taught. Except for correspondence with knowledgeable individuals, he had no personal contacts with mycologists. In part 3 of his *Flora of Newfoundland, Labrador and St. Pierre and Miquelon*, published in 1896 by the Nova Scotia Institute of Science, Waghorne mentioned the fact that he had not included fungi and expressed the hope that he would be able to do so in a following publication. His failing health may have been the factor that prevented him from fulfilling that desire.

For whatever reason, Waghorne did not name a fungus, nor did he publish a list of the fungi that he collected. However, dozens of fungi collected by him were named by others to whom he sent specimens.

Noteworthy among these is J.B. Ellis, who named and deposited many of Waghorne's collections in the herbarium of the New York Botanic Garden. Ellis sent some duplicates of those fungi to Hans Güssow in Ottawa, who, in 1909, was purchasing exsiccati sets and acquiring duplicate specimens from well-known mycologists for the collection which, in 1932, became the National Mycological Herbarium, the Index Herbariorum code for which is DAOM.[39]

Waghorne seems to have known the leading American mycologists of his day, and he sent specimens to a number of them. C.H. Peck acknowledged that several of the fungi in his "New Species of Fungi" had been collected by Waghorne in Newfoundland and Labrador, and W.C. Coker, in his *The Clavarias of the United States and Canada*, stated that *Clavaria cristata* had been collected by Waghorne.[40]

Others who collected in Newfoundland or Labrador include Robert Bell, whose list of ten fungi is in the report of the Geology and Natural History Survey of Canada, 1882–84, appendix DD; Canadian naturalist John Macoun, who sent at least twenty-six species of fungi collected there to J.B. Ellis for identification in 1885;[41] J.D. Soper, who collected there in the 1920s,[42] and Nicholas Polunin, who went on a number of northern expeditions sponsored by the Canadian government, and collected a few fungi around Burwell, Labrador, in September 1936.[43]

F.A. Gilbert found *Hypoxylon multiforme* on St John's Island, and *Cintractia caricia* near Raleigh, Newfoundland, in August 1925, and contributed them to the herbarium of the University of Alberta, where they bear accession numbers 445 and 707, respectively. Records, if they exist, of fungi described or collected in Newfoundland and Labrador between 1940 and 1950 have not been seen.

ONTARIO

The earliest published reference to a paper on fungi in Ontario is one read by James Hubbard of Knox's College during a meeting of the Canadian Institute, in Toronto, 13 February 1863.[44] Several years passed before there was another published account of fungi in that province, and it was Daniel Knode Winder's twenty-four-page booklet, *The Mushrooms of Canada with engravings, and Catalogue of the Fungi of Canada*, published in Toronto in 1871. Winder's book or booklet has the distinction of being the first one published in Canada that was devoted exclusively to fungi. Up to about that time, the majority of Canadian collectors of fungi were accepting European mycologists' names and descriptions of fungi and applying them to local species.

Among those who departed from that tradition was Cephas Guillet, who sent fungi that he collected to C.H. Peck, New York State botanist, for identification.[45] Perhaps Guillet was following the example of his friend John Dearness (1852–1954), a largely self-taught naturalist-mycologist living in London, Ontario, who had been getting more help from American than European mycologists. This is exemplified in an early work titled simply "New Species of Canadian Fungi," published as a joint paper with J.B. Ellis, an American mycologist who was also largely self-taught.[46] Many of those so-called new species were given "Canadian" names, but because they lacked Latin descriptions they were not accepted by the majority of mycologists. It had been decided, by an overwhelming majority of those attending the Fifth International Botanical Congress, Cambridge, England, in 1930, that the published names of new groups would be valid only when accompanied by a Latin diagnosis, after 1 January 1932. Dearness persisted in his practice of not using Latin and gave his reasons in an addendum to a paper that he published in 1941.[47] His reasons are more valid today than they were in 1941.

John Dearness was a remarkable man, not only because he lived 102 years, but because from the time he began teaching school, at age seventeen, until a short time before his death he maintained an active interest in several aspects of what he would have called nature study. One can see Dearness's early interest in mycology in the minutes of the meetings of the Microscopical Section of the Entomological Society of Ontario. For example, it is recorded that Dearness gave a lecture-demonstration on the fungal nature of the black knot disease of stone fruits during the meetings of 23–24 January 1891. Minutes of the 27 November 1891 meeting show that he gave a talk on the nature and habits of "a class of fungi known as the Erysiphaceae." During that meeting the members examined leaves and, by means of large diagrams provided by Dearness, identified an Uncinula and a Phyllactinia. During the following meeting, the members, with guidance from Dearness, identified six more powdery mildews. Over the next ten years Dearness's name appeared in the minutes of many meetings of that society, and several of these appearances are in relation to some aspect of mycology.

In his presidential address to the Entomological Society of Ontario in 1897, Dearness talked about fungi and the possibility of using selected species to control insects, thus indicating that he was keeping abreast of the latest advances in mycology. In 1916 he began a series of papers in *Mycologia* on new or noteworthy fungi, and in the following decade he (in a few instances with others) named and described more than 175 species, varieties, or combinations. During that period he also published "An annotated list of Anthracnoses."[48]

Dearness and American mycologist Homer D. House, of the New York State Museum, Albany, named a number of rust fungi, many of which were noted by J.C. Arthur in his *Manual of the rusts of United States and Canada.* They also published a list of more than a dozen new species in New York State Museum bulletins 205 and 206, in 1919.

Incidentally, after Dearness had been teaching at the University of Western Ontario for fourteen years, and was fifty years of age, he earned his first degree, a BA. This he did through extramural studies at Western, which granted him an MA the following year and awarded him an honorary LLD in 1926, when he was seventy-four.

Dearness, long-time dean of Canadian mycologists, was for many years an outstanding contributor to foreign exsiccati. His first such specimen appeared in Century 3 of Ellis's *North American Fungi* of February 1880, and the last in Fascicle 11 of Anna Jenkins and A.A. Bitancourt's *Myringales Selecti Exsiccati.* That final one was collected on 6 August 1951, when he was ninety-nine years of age, making a span of seventy-one years of mycological activity.[49]

In 1937, at age eighty-five, Dearness became the first Canadian-born president of the Mycological Society of America. Four years later the Montreal Botanic Garden named the Dearness Research Laboratory in his honour, and on his one hundredth birthday the American Association for the Advancement of Science sent him an address and made him a life honorary member. Dearness's home town, London, Ontario, named a home for elder citizens and a public school in his honour.[50] When he died in 1954, his mycological collection was given to the Montreal Botanic Garden.[51]

The Ottawa Field Naturalists' Club had a Microscopical Section similar to that of the Entomological Society of Ontario and it was during a "microscopical soirée" in Ottawa, 25 February 1892, that Adolph Lehmann read a paper on parasitic fungi. In presenting that paper Lehmann became one of the earliest of the Canadian amateur mycologists to put emphasis on the vast number of microscopic forms which, he suggested, were being neglected by virtually all naturalists in the country.[52]

John Macoun (1831–1920), an Irish immigrant and self-taught plant collector who became botanist to the Geological Survey of Canada in 1882, was the first federally employed botanist to publish on fungi. Although not widely known for his mycological work, Macoun collected fungi in the vicinity of Belleville as early as 1878, when he was professor of natural history at Albert College, and during his lifetime he collected several thousand fungi from Labrador to Vancouver Island. In 1884 he published a paper on edible and poisonous fungi in which he mentioned that Mrs C. Chamberlin (1833–1913) had collected mushrooms in the Rideau Hall Woods, Ottawa, and made watercolour paintings of several choice specimens. It was also mentioned, as a footnote, that Mrs Chamberlin's aunt, the well-known Canadian naturalist Catharine Parr Traill (1802–1899), was in the audience and that she had also collected many of the species illustrated by Mrs Chamberlin.[53]

Macoun collected hundreds of fungi around Ottawa in the autumn of 1898, the year in which Percy Saunders made a special study of the fungi there.[54] That was also the year in which Macoun's classic work on lichens of the Ottawa area began to appear in the *Ottawa Naturalist.* Apparently his interest in lichens was one of long standing, because, according to his autobiography, he made an extensive collection of lichens on the Gaspé Peninsula of Québec in 1922. Although he was an avid collector, he never really learned how to properly identify fungi and his publications often consisted of little more than lists of names, usually provided by someone else, that soon became hopelessly out of date. However, many of his specimens are in herbaria, especially DAOM and that of the University of British Columbia (the international code for which is simply UBC), and may be useful for comparing the past with present conditions in the areas of his collections.

Federal employee James Fletcher (1852–1908), in recognizing the importance of fungi as plant pathogens, may be justifiably credited with having set the stage for mycological studies in the Dominion Department of Agriculture. He was a member and probably the founder of the Ottawa Mutual Research Society and a co-founder of the Ottawa Field Naturalists' Club, besides being a member of several other scientific societies. Some of the fungi collected by him are now in DAOM. Fletcher was particularly interested in, and, for his own enlightenment, made a study of, the rust and smut fungi, and a number of others that cause plant diseases.[55]

Those studies were continued by Fletcher's successor, Dominion botanist Hans Theodor Güssow (1879–1961), a native of what was then Prussia, who had studied in the German Universities at Breslau, Leipzig, and Berlin. Whenever he had the time, Güssow, often with his wife, would join with fellow members of the Ottawa Field Naturalists' Club on their mushroom forays. One of his colleagues at the Central Experimental Farm, forest pathologist Alan W. McCallum (1893–1967), MA, who also collected fungi during those outings, reported that Güssow's wife and her companion Mrs R.A. Inglis were the first to collect *Bulgaria platydiscus* in Canada.[56] Although Güssow considered himself to be an economic botanist, his papers on the nature of parasitic fungi and on edible, poisonous, and other fungi in 1911, plus the fact that he named at least two species of Actinomycetes (then considered to be fungi) and published a book, *Mushrooms and toadstools,* written jointly with amateur mycologist W.S. Odell, in 1927, would have been sufficient to stamp him as a botanist-mycologist.[57]

Güssow wrote the text of Dominion Department of Agriculture circular no. 45, *Mushroom culture,* in 1926, which he revised in 1938, and it was he who initiated the collection of fungi that, in 1932, became the National Mycological Herbarium in Ottawa. In 1934 he was the Canadian representative to what was then the Imperial Mycological Institute (IMI), an institution at Kew, England, which became functional in 1920, and which has been of inestimable aid to the development of mycology in Canada. Many Canadians have gone there for periods of study or to compare fungi with their herbarium specimens or their many named cultures of living fungi. For years the IMI, which became the Commonwealth Mycological Institute (CMI), was a source of pure cultures of fungi for use in the teaching of mycology, or for research purposes in Canada, and many Canadians sent fungi there for identification or to have a tentative identification confirmed.

W.S. Odell was recruited, in 1917, by James M. Macoun, chief of the Division of Biology in the Museum Branch of the Department of Mines, Ottawa, to survey the area around Ottawa for fleshy fungi. This he did until at least June 1926, when he placed on record "nearly 400 species and varieties," most of which had been identified by John Dearness and incorporated into his collection.[58] Additions to that list were published, by Odell, in the *Canadian Field Naturalist*, in 1931, and still more were added by J.W. Groves in 1938 and published in the same journal. Groves acknowledged that Michael Timonin had collected some of the ones in his list, and identified several that were collected by Groves. Another of Güssow's colleagues in the Department of Agriculture who went on mycological forays with the Ottawa group was Dominion chemist Frank T. Shutt (1859–1940), the first Canadian authority to publish on the food values of mushrooms, in 1904 and in 1905.[59]

In 1907, John William Eastham (1880–1968) of the Ontario Agricultural College (OAC) spent a month with Professor G.F. Atkinson, mycologist at Cornell University, for a special study of fungi, and used his vacation time to collect specimens of the Agaricaceae and Polyporaceae that became the nucleus of a herbarium collection at the OAC.[60] He resigned in May 1911 to accept a position with the Dominion botanist in Ottawa.

Eastham was succeeded, in September of that year, by Walter A. McCubbin (b.1880), who had studied mycology for a year at Harvard after having graduated from the University of Toronto where he was Professor J.H. Faull's first student for the MA degree. McCubbin added to the fungus collection at the OAC and made a list of the eighty-eight species of fleshy fungi and nineteen myxomycetes that were in the herbarium before he also resigned to accept a position with the Dominion botanist, but at St Catharines rather than Ottawa.[61]

In 1912, Eastham published on the myxomycetes of the Ottawa district,[62] and McCubbin, working in the Niagara Peninsula, published a bulletin on what he called an "edible toadstool." During the following year, McCubbin produced a bulletin titled *The morel,* wrote an article on the function of toadstools in nature, and prepared a display of *Amanita phalloides,* together with edible mushrooms, in the window of a local shop at a time when people were disturbed by reports of a fatal case of mushroom poisoning.[63]

Thomas Langton, MA, QC, who collected a wide range of fleshy Ascomycetes and Basidiomycetes, chiefly in the neighbourhood of Toronto and in the general areas of Muskoka and Parry Sound, gave an account of his mycological activities during a meeting of the Canadian Institute, of which he was a member, in Toronto, 15 April 1911. However, a list of the fungi he collected, together with his general classification of them, was not published until 1913.[64]

T.H. Bissonnette, MA, of Queen's University, Kingston, collected fungi in the vicinity of Georgian Bay in August and September of 1912. In reporting on that work he acknowledged that collections made by a Miss Penson and a Mr Woodhouse were in his list, which included (using his terminology), Agarics, Polyporaceae, Hydnaceae, Thelephoraceae, Clavariaceae, Tremellini, Ascomycetes, Nidulariaceae, Basidiomycetes, Sphaeriaceae, and Myxomycetes.[65]

William G. Evans (b.1885), who identified himself as demonstrator in botany at the OAC, collected species of Puccinia and Colletotrichum in the Port Stanley and Guelph areas in 1920. Specimens of those collections were contributed, in an exchange arrangement, to the herbarium of the University of Alberta.

From around 1912, and for the next twenty-six years, Nebraska-born Roland Elisha Stone (1881–1939), who had a B SC from the University of Nebraska, an MS from the Alabama Polytechnic Institute, and a PH D from Cornell University, was involved in a research and extension program of plant pathology and mycology at the OAC. His first mycological paper was published in 1916, and his last one, on the mushrooms of Ontario, in 1938.[66] Specimens of Peronospora, Phragmidium, and Macrosporium, collected by Stone in Ontario in 1920 and 1921, are, like those of Evans, now in the herbarium of the University of Alberta, although the major part of his mycological collection remained at Guelph.

Alfred Brooker Klugh (1882–1932), a graduate of the OAC, where he was an instructor in nature study before going to Queen's University for the MA degree that he obtained in 1910, was the first to record the presence of *Morchella bispora* in Canada.[67] Klugh, who manifested many of the traditions of the early naturalists who took all nature as their field, earned a PH D from Cornell in 1926. He rose to the rank of associate professor at Queen's, where he was employed from 1910 until the time of his tragic death in a car-train accident, 1 June 1932.

In the decade after about 1912, and largely because of international import and export restrictions on the movement of fungal-infected plant materials, there was a great demand for plant pathologists with a good knowledge of mycology. One of the agricultural students who responded to that demand was Frank Lisle Drayton (1892–1970), a native of Barbados and a 1914 graduate of Macdonald College, Quebec. Drayton had been in Ottawa only a few months, as an assistant plant pathologist in the Department of Agriculture, when he joined the Canadian Army and served overseas, where he was severely wounded. Following his discharge from the armed forces in 1919, Drayton returned to Ottawa as a plant pathologist and became an authority on diseases of bulbous ornamental plants before taking leave to study for the PH D that he obtained at Cornell University in 1932, for research under Professor H.H. Whetzel.

In the next two decades, Drayton, with some help from J. Walton Groves, elucidated the sexual mechanisms of a number of fungi, and learned how to induce them to develop fruiting structures in pure culture. That work has been considered to be among the more noteworthy contributions to mycology.[68] Drayton was a charter member of the Mycological Society of America, and he became an elected Fellow of the Royal Society of Canada, and of the Agricultural Institute of Canada, before he retired, in 1957.[69]

Garnet S. Bell, who at one time was president of the Mycological Society of Ontario, collected fleshy fungi in and around Toronto for more than twenty years. He searched the literature that was available to him and found what seemed to be satisfactory descriptions and names of at least 427 of the species that he saw within a range of fifty miles (80.46 kilometres) of Toronto. His list, which he compared with Thomas Langton's earlier record of fungi from the same general area, was published in 1933.[70]

Michigan-born Joseph Horace Faull (1870–1961), the "Father of Forest Pathology in Canada," was a teacher in the Department of Botany, University of Toronto, from 1902 until 1928. He had earned a BA degree at Toronto in 1898 and studied mycology under Roland Thaxter for the PH D degree that he was awarded by Harvard in 1904. The breadth of Faull's botanical and mycological interests is reflected in his 1913 publication, *The natural history of the Toronto region*, which included sixty-one species and varieties of slime moulds.

Faull's graduate student, Mary E. Currie, MA, made a study of the slime moulds of the whole province of Ontario, and was an authority for the identification of many slime moulds collected by F.B. Adamstone in the Lake Nipigon district in 1921.[71]

Present-day natural scientists recognize that the slime moulds are not true fungi, but the early students of mycology thought that they were, and many modern mycologists continue to collect them. Around the turn of the century, Canadians who collected slime moulds commonly had them identified by Arthur Lister or his daughter Gulielma in England. Consequently, it was not unexpected to find thirteen species that had been collected in Ontario, two from Quebec, two from New Brunswick, and one from Newfoundland, included in their *Monograph of the Mycetozoa.*

Lists of slime moulds are sometimes found in unexpected places. For example, a list of ten previously unpublished names of species that had been collected around Toronto and Montreal and identified by Miss Lister are in W.N. Cheesman's paper on slime moulds from the Rocky Mountains,[72] and several slime moulds, and filamentous fungi, from the Kenora region of Ontario are in Bisby, Buller, and Dearness's list of Manitoba fungi.[73]

Faull saw great economic importance in the large stands of spruce and fir trees in eastern Canada, and his interest in these, plus his background in mycology, can be seen in the thesis research of his graduate students, three of whom, namely Lillian M. Hunter (1892–1970), Hugh P. Bell (1889–1957), and Ezra H. Moss (1892–1963), did thesis research on the rust fungi that parasitize those trees.

Lillian Hunter's thesis for the MA degree, "Rusts of Abies," was completed in 1924. A few years later she went to Radcliffe College, where she earned a PH D in biology in 1935. Because of the economic depression, she could not find an academic position in a university or college, so she taught in various schools from 1936 until sometime after the Second World War. In 1948 she published "A study of the mycelium and haustoria of the rusts of Abies" in the *Canadian Journal of Research.* That paper dealt with nineteen rusts on the leaves of fir trees, most of which were described for the first time on their aecidial hosts.

Bell, who had been wounded while serving as a captain in the Canadian army in France during the First World War, studied the balsam rusts as a thesis project for the MA degree that he was awarded in 1920. Although he was appointed assistant professor of botany, Dalhousie University, Halifax, that year, Bell continued his rust studies under a thesis titled "Studies on the fern rusts of Abies" for the PH D that he earned, under Faull's supervision, in 1922. Two years later that paper, which outlined the life histories of various rust disease fungi of Nova Scotia and described three new species, was published in the *Botanical Gazette.* While Moss was studying for the three degrees that he obtained at Toronto, his mycological interest was not confined to the fungi of trees. In 1920 and 1921 he collected the specimens of Albugo, Cystopus, and Synchytrium that he took with him when he joined the staff of the University of Alberta in 1921 and placed in that university's herbarium.

Faull's graduate student Ibra L. Conners, who was awarded a master's degree in 1920, went to western Canada for several years before becoming curator, in 1929, of what was soon to become the National Mycological Herbarium, Ottawa. Conners, at the same time, headed the Canadian Plant Disease Survey.[74] For a while he was assisted in that dual task by Eleanor Silver Dowding, (1901–91) a graduate of the University of Alberta who earned a PH D for research on fungi under Professor A.H.R. Buller at Manitoba. While in Ottawa, Dowding continued her studies, begun in Manitoba, on coprophilous fungi, and described Gelasinospora, a new genus.[75] Soon after that she turned her attention to medical mycology, in Alberta.

Faull's interest in the fungal destruction of trees and wood can be seen in his published papers on *Fomes officinalis* and other wood destroyers. It is also seen in his guidance of the research by James Herbert White (1875–1957) for his PH D, which dealt with the biology of a tree fungus, and in the master's research of Alan W. McCallum, who became the first forest pathologist in the federal Division of Botany. McCallum became an avid collector of fungi that grow on trees and wood, specimens of which are in the DAOM herbarium, Ottawa.

Faull's interest is even more evident in the doctoral research of Clara Winnifred Fritz (1889–1974), an M SC graduate of McGill University who studied cultural criteria for the identification of wood destroying fungi for the PH D that she was awarded in 1924. Her published thesis, a basic reference for many years, brought her recognition as one of the outstanding Canadian pioneers in research on fungal deterioration of wood. She was the first woman in Canada to qualify in the field of forest pathology and was largely responsible for making the first collection of cultures of wood-destroying fungi in this country.[76]

Fritz was followed, in that general area of study, by another of Faull's outstanding graduate students, Irene Mounce (1894–1987), whose research for her 1929 PH D was on the biology of *Fomes pinicola*. Mounce had joined the staff of the Department of Agriculture, Ottawa, in 1924, where, except for leave periods, and periods of research in other parts of Canada, she remained until 1941, helping to lay the foundation for important work on sexuality in fungi and studies of wood-destroying species. Those latter studies, in which she was a leader for many years, took her to British Columbia, where she collected and cultured forest fungi during the summer of 1938. An account of that work was published, a year later, in the *Proceedings of the Canadian Phytopathological Society*. In 1941 Mounce was transferred to her native province, where she studied fungi in seeds and diseases of vegetables at Saanichton, British Columbia, until she resigned in 1945.[77]

When Faull left Toronto, in 1928, he was succeeded by Herbert Spencer Jackson (1883–1951), who was appointed professor of mycology in 1929, the first such appointment in an Ontario university.

Jackson became head of the Department of Botany in 1941 and retained that position until he died in 1951. During that period he served a term as president of the Mycological Society of America, and as president of the Biological Section (section 5) of the Royal Society of Canada. His presidential address to members of the Royal Society of Canada in 1944 was titled "Life cycles and physiology in the higher fungi." For many years he taught a general, two-year undergraduate course in mycology and a course on the taxonomy of fungi to graduate students. Jackson was born in the state of New York and earned a BA degree from Cornell in 1905, where he was a student of Professor G.F. Atkinson. The year he went to Toronto, a PH D was conferred on him by the University of Wisconsin on presentation of his classic paper, "Present evolutionary tendencies and the origin of life cycles in the Uredinales."[78]

Jackson's early work was mostly on the rust fungi, but, following the lead of Faull, he established a second reputation in Toronto with his studies on wood-destroying fungi. However, his major contribution to mycology in Canada was through his teaching of graduate courses in that subject, and in the guidance he provided to such students as Mildred Nobles, Ruth Macrae, Roy F. Cain, and J. Walton Groves, to name only four of those who became noteworthy mycologists. Harold J. Brodie (1907–1989) spent a postdoctoral year of study with Jackson and, while there, made a critical analysis of the mutual aversions between fungi.[79] After that, most of Brodie's contributions to mycology were made in western Canada.

Ontario-born Mildred K. Nobles (1903–1993) received her early education in the schools of Regina, Saskatchewan, but went to Queen's University for a BA degree in biology and chemistry. She transferred to the University of Toronto to work under Jackson and was awarded an MA degree in 1931. That was around the time Jackson was becoming interested in the Thelephoraceae, so it is understandable that Nobles's thesis research for the PH D, which she obtained in 1935, would deal with those fungi.[80] She joined the staff of the Canada Department of Agriculture where, beginning in 1936, she began research on the improvement of methods for the identification of wood-destroying fungi, based on their cultural characteristics. The manual that she published in 1948 established her as an internationally recognized authority for the identification of those fungi.[81]

Ruth Macrae (b.1903), a student from Montreal who had won the Hiram Mills Gold Medal for biology and the Penhallow Prize for botany before graduating from McGill University with a BA degree in 1924 and an M SC in 1926, worked for three years as Buller's research assistant in Manitoba before joining the staff of the Department of Agriculture,

Ottawa, as a research officer. She assisted Irene Mounce in research on wood-destroying fungi and in maintaining that part of the herbarium containing wood-inhabiting Hymenomycetes. Eventually Macrae became fully responsible for the latter work, while continuing her studies on the taxonomy of the Polyporaceae. Her genetic and sexuality studies, by herself and with Irene Mounce, on some of the higher Basidiomycetes,[82] were precursors to thesis studies for a PH D from the University of Toronto, obtained under Jackson's supervision, in 1941. She continued those studies, after returning to the Ottawa laboratory, where she became almost fully occupied with the isolation and identification of the fungus causing the Dutch elm disease in Canada.

Roy F. Cain (b.1906) graduated from the University of Toronto with a bachelor's degree in 1930, a master's in 1931, and a PH D in 1933 – not a good year for a young person to find employment in Canada, because of the economic depression. Jackson got him to sort the mycological collection in the department and from that beginning he became largely instrumental in organizing the fungal component of the University of Toronto Cryptogamic Herbarium (TRTC), which, with lichens and bryophytes, became one of the major herbaria on the continent. In the university calendar for 1938–39 Roy Cain is listed as a technical assistant in the Department of Botany, but he was destined to rise through the ranks to became professor of mycology. His 126-page *Studies on Coprophilous Sphaeriales in Ontario*, published in 1934 as University of Toronto study no. 38, was followed by a wide-ranging series of papers, the publication of most of which occurred after the period covered by this essay, as did his presidency of the Mycological Society of America.[83]

Following a stint of teaching in the public schools, James Walton Groves (1906–1970), a native of Kinburn, Ontario, entered Queen's University, where he earned a BA in 1930. He then transferred to the University of Toronto for graduate studies in mycology under Jackson, and plant pathology under Dixon Lloyd Bailey (1896–1984), for which he was awarded an MA degree in 1932 and a PH D in 1935. In June 1936 Groves joined the Division of Botany and Plant Pathology, Department of Agriculture, Ottawa, where, for several years, his publications indicated that he was a graduate assistant.

While in Toronto, and at the suggestion of Jackson, Groves had undertaken in 1932 a cultural study of the life histories of fungi in the family Dermateaceae. Those studies were continued in Ottawa, where he published at least six papers on work done, or initiated, under the direction of Jackson,[84] and then went of to publish many others. During the Second World War there was need for much more information about the fungal deterioration of seeds and of seed-borne patho-

John Dearness (1852–1954)
Courtesy D.B. Weldon Library,
University of Western Ontario

Henry S. Jackson (1883–1951)

genic fungi, and Groves was called upon to supervise the mycological aspects of a large seed-testing program. In this latter work, he was assisted by Arthur J. Skolko (1912–89), B SC, MA, PH D, who had been one of Jackson's students at Toronto.[85]

William R. Haddow, B SC F, LL D, who was a special lecturer in the Department of Botany, University of Toronto, between 1931 and 1940, investigated the classification, nomenclature, hosts, and geographical range of *Trametes pini* in 1938, and decided that the various forms that he saw represented a single species. He published on the history and diagnosis of *Polyporus tomentosus, P. circinatus,* and *P. dualis* in 1941.[86]

In 1931 the Elizabeth Ann Wintercorbyn Award of the University of Toronto, for research likely to prove of most benefit to agriculture, went to George Aleck Ledingham (1903–1962), who had B SC and MS degrees from the University of Saskatchewan. His study of a grass root disease under D.L. Bailey became a mycological problem under the guidance of Jackson. Ledingham became widely known among mycologists for his discovery of *Polymixa graminis,* the study of which was the major feature of his research leading to the PH D that he was awarded in 1932. Despite his unique ability in discovering those elusive organisms, there was no opening for him in the general field of mycology in Canada, so he accepted a teaching fellowship at Harvard University.

After a brief sojourn at Harvard, Ledingham returned to Canada as an editor of the *Canadian Journals of Research*. In 1934 he transferred to the Division of Applied Biology of the National Research Council of Canada, in Ottawa, after which his work became more biochemical than mycological.[87]

John J. Miller (b.1918), who earned a BA degree at the University of Toronto in 1941, stayed there and studied, under Bailey, species of Fusarium involved in the muskmelon wilt disease as the topic of research for the PH D degree that he was awarded in 1944. Miller continued to study fusarium diseases and the taxonomy of Fusarium species while employed as a plant pathologist for three years with the Canada Department of Agriculture in Harrow and in St Catharines. He became the mycologist (officially, assistant professor of botany) at McMaster University, Hamilton, Ontario, in September 1947, the year before he published notes on the maintenance of fungus cultures.[88]

Another plant pathologist working in the Niagara Peninsula who made noteworthy contributions to mycology was L. Ward Koch, who studied *Didmella applanta*, the cause of spur blight of raspberries, and in doing so corrected some misconceptions concerning its life history. This general topic formed part of both his MA (1928) and PH D (1931) theses at the University of Toronto. Koch discovered the tobacco blue mould fungus in Canada, and he studied *Dibotryon morbosum* and the overwintering of certain other fungi associated with fruit trees before he and others studied *Macrophomina* Phaseoli.[89]

The fleshy fungi figured prominently in one of the early waves of enthusiasm for nature study, and in response to this fervour, Hugh H. Halliday, Hans T. Güssow, and Henry J. Scoggan produced popular illustrated articles for *Canadian Nature*,[90] the leading Canadian journal in that "back to nature" movement.

For several years the *Revue Canadienne de Biologie* listed the titles, and sometimes the abstracts, of papers presented to members of the Ontario Society of Biologists. For example, volume 3, page 279 (1944), notes that Rosemary Biggs gave a talk, 3 January 1936, on "An interesting undescribed Ascomycete," and page 223 notes that H. Berdan's paper on "A revision of the genus Ancylists of the Phycomycetes" was presented 3 May 1941.

MANITOBA
AND THE NORTHWEST TERRITORIES

The first record of fungi being collected in Manitoba is in a botanical appendix to Captain (later Sir) John Franklin's *Narrative of a Journey to the shores of the Polar Sea in the years 1819, 20, 21, and 22*, published in

1823. John Richardson, the expedition's surgeon and author of that appendix, lists some nineteen fungi collected in the wooded area that is between latitude 50° and 64° north. He also commented that the fungi were examined by Professor Hooker, referring to Sir W.J. Hooker (1785–1865), director of the Royal Botanic Gardens, Kew, England, into whose herbarium many of Richardson's specimens were eventually placed.

Most of Richardson's collecting area is within what is now the province of Manitoba, the northern boundary of which lies along latitude 60° north. Thus, all of the fungi collected north of that line were from the geographical area now known as the Yukon or the Northwest Territories of Canada.

John Richardson also collected fungi during Franklin's second overland expedition across the Northwest Territories to the Arctic Ocean, which he referred to as the Polar Sea. The names of those fungi are somewhat scattered in M.J. Berkeley's account of exotic fungi in the collection of Sir W.J. Hooker, published in 1839, and in his 1841 supplement to that account.[91]

Members of the Canadian Arctic Expedition of 1913–18 brought back more than a hundred species of fungi that were identified by John Dearness of London, Ontario.[92] And Hugh S. Spence, who was associated with the Dominion Department of Mines, wrote about the abundance of mushrooms that he saw, and collected, near the southeast corner of Great Bear Lake, Northwest Territories, in August 1931.[93] Five years later Nicholas Polunin, visiting professor of botany, McGill University, collected fungi from Baffin Island, from Devon and Ellesmere Islands in Hudson Bay, and from the Chesterfield district on the west coast of that bay. Polunin was indebted to David H. Linder, curator, Farlow Herbarium, Harvard University, for writing the chapter on fungi in *Botany of the Canadian Eastern Arctic*, of which Polunin was editor. Linder also listed seven fungi that O.E. Jennings had collected on Southampton Island, Hudson Bay, and the eleven listed by John Dearness as having been collected on Baffin Island in 1926.[94]

In 1947 Polunin, through the cooperation of the Royal Canadian Air Force, collected fungal spores from the air over northern Canada. The general story of this cooperative enterprise, in which the botany department of McGill University, Montreal, and the Dominion Laboratory of Plant Pathology, Winnipeg, became involved, has been told by Stuart M. Pady, associate professor of botany at McGill. Pady also reviewed the work of earlier collectors of fungi from arctic air, one of whom was Colonel Charles A. Lindbergh, who collected spores during his historic flight to England via Greenland in 1933.[95]

Apparently there were no accounts of edible mushrooms collected by citizens of Manitoba earlier than 1889. Charles N. Bell, in his inaugural address as president of the Manitoba Historical and Scientific Society, commented on the number of members who were enthusiastic mushroom hunters, and asked, "Will one of them not give us a paper on the edible fungi of the Province?" He went on to state that "So little is known of the value of this form of food, in a country producing spontaneously such a large number of varieties, that really good services would be rendered by an article plainly describing the forms and their usual places of growth."[96] In spite of Bell's plea, however, it was two years before anyone published a paper on Manitoba fungi, and that was by John Dearness, a man from Ontario who seems to have ignored the edible mushrooms. During his visit there, in 1891, Dearness collected several rust fungi, plus *Trametes carnea*, *Phragmidium potentillae*, and *Didymosphaeria manitobiensis*.[97] That last one was a new species, named by J.B. Ellis and B.M. Everhart and listed for the first time in their book, *North American Pyrenomycetes*, in 1892.

The first record of mycological activity in Manitoba by a native of that province is that of Norman Criddle (1875–1953), who collected and made a number of coloured illustrations of fleshy fungi in the first decade of the twentieth century. His earliest publication concerning the fleshy fungi dealt with an Amanita and the effect it had on cattle that had eaten it.[98]

The Criddle family lived at Treesbank, and three of the boys, Norman, Evelyn, and Stuart, became widely known naturalists. All three collected a broad range of biological specimens, including fungi, but Evelyn seems to have specialized in the rust fungi. He contributed several rusts, including *Melampsoridium betulae*, *Puccinia apocrypta*, *P. grindeliae*, and *P. rubellu*, for inclusion in Bisby, Buller, and Dearness's book, *The fungi of Manitoba*. Norman became a very competent entomologist employed by the Dominion Department of Agriculture. He was also an artist, and as such provided many of the illustrations for departmental publications on various topics, especially weeds and fodder plants. Stuart's specialty, if he had one, is not known, but he discovered hoards of fungi that had been collected by squirrels and sent at least two large lots of them to A.H.R. Buller for identification and comment.[99] Stuart is also known to have planted in sand a sclerotium from which two large Polyporus-type fruit bodies developed, and he is reported to have collected *Melampsoridium betulinum* at Treesbank, 6 October 1922.[100]

A.H.R. Buller (1874–1944), who became the best known of the early mycologists in Canada, was born in Birmingham, England. In 1904 he accepted a position as professor of botany at the University of Manitoba, in Winnipeg, and was not there very long before becoming aware of the abundance of *Puccinia graminis* on the grain in that area. This fact is mentioned several times in his later accounts of the rusts of grain. Most of his Canadian work with fungi (he did some work with fungi before coming to Canada) is well documented in his *Researches on fungi*, six volumes of which were published in his lifetime; a seventh, the manuscript of which he had nearly completed, was published by friends several years after his death.

In his research, and in the expression of ideas, Buller coined new terms and made a number of discoveries, or paved the way for others to do so. He discovered nuclear migration through septal pores in the hyphae of fungi, and in describing it became the first to draw attention to the discoid pad of material that is now known to be part of the dolipore type of septum. He was the first to demonstrate wound healing in the hyphae of a fungus, and he made such a thorough study of hyphal fusions that the dikaryotization (a term that he coined) of a monokaryon by a dikaryon became known as the "Buller phenomenon." Because he was one of the first to critically examine and describe the discharge of basidiospores, the associated drop (bubble) became known as "Buller's drop." Buller retired in 1936 and died in 1944.[101]

Buller's reputation as an excellent teacher, and his many publications on the fungi, attracted students, particularly graduate students, from other universities. The best known of these are Harold Brodie, John H. Craigie, William F. Hanna, Irene Mounce, Dorothy Newton, and Thomas Vanterpool, each of whom became a well-known mycologist or plant pathologist in Canada.

Harold Brodie (1907–1989), a native of Winnipeg, attended the University of Manitoba for the B SC degree that he obtained in 1929 and the M SC a year later. Buller advised him to go to the University of Michigan to study under Professor C.H. Kauffman for a doctorate, but Kauffman died before Brodie had the opportunity of meeting him. Nevertheless, Brodie did go to Michigan, where he was a student of L.E. Wehmeyer while studying for the PH D that he was awarded in 1934. After that, he had a year of work at the University of Toronto and a stint of teaching at Macdonald College before returning to Winnipeg, where he was an assistant professor in the Department of Botany from 1937 until 1946. For a while, in Winnipeg, Brodie made some important contributions to our knowledge of the powdery mildews,[102] but from the autumn of 1941, when Buller first aroused his interest in the bird's nest fungi, he began a concerted study of those fascinating organisms, and eventually became the world authority on them.[103]

John H. Craigie (1887–1989), who became Officer in charge of the Dominion Laboratory of Plant Pathology, Winnipeg, and who is remembered, and excessively honoured, for his discovery of the function of the pycnia of the rust fungi,[104] also wrote a paper on the epidemiology of stem rust that is a modern classic on that topic.[105]

William F. Hanna (1892–1972), a Nova Scotian by birth, had a BA from Dalhousie that was earned just before he volunteered for overseas service in the First World War. After that war, Hanna earned B SC and M SC degrees from the University of Alberta, and was awarded the first scholarship to be offered by the Canadian Society of Technical Agriculturists in 1923. That scholarship was to enable him to study mycology under recognized leaders in the field. He studied with Buller in 1923–24, and, after brief study periods at the University of Minnesota and the Imperial College, London, England, returned to the University of Manitoba. There, under Buller's tutelage, he obtained the first earned PH D degree awarded in western Canada, 17 May 1928.

Hanna's thesis consisted of four published papers on sex and the inheritance of spore size in mushrooms. By a technique that he and Buller devised, involving the pressure of a cover slip against the surface of a young gill of a Coprinus in such a way that the spore tetrads adhering to the cover slip could be removed individually by means of a dry needle, he was able to transfer the spores to a suitable culture medium and test the resulting mycelia for mating type.[106]

Hanna was almost immediately hired as a plant pathologist at the Dominion Laboratory of Plant Pathology, in Winnipeg, to study the smut fungi. Although his work for the next few years was primarily related to plant diseases, Hanna had an abiding love for mycology. He became a member of the Mycological Society of America in 1924 and published papers on *Clitocybe illudens*, *Coprinus urticaecola*, and the genus Entoloma. It was Hanna who provided the excellent photographs of species of Coprinus that are in *The fungi of Manitoba and Saskatchewan*, by Bisby, Buller, Dearness, Fraser, and Russell, and he was the authority for confirming the names of many others listed in that book. Hanna found species of Sporobolomyces in Manitoba, and he described and named *Sporobolomyces albus*. He also cooperated with W. Popp in a study of the distribution of the cereal smut fungi before becoming involved in the Second World War as an officer in the Royal Canadian Air Force.[107]

British Columbia–born Irene Mounce had BA and MA degrees from the University of British Columbia before becoming one of Buller's graduate students for the M SC that she was awarded in 1922. At Manitoba, where she held a Hudson's Bay Company Research Fellowship, and a studentship from the Honorary Advisory Council for Scientific and Industrial Research, Mounce studied and published on homothallism and heterothallism in Coprinus.[108] She was an enthusiastic colleague of Buller on a number of fungus collecting trips and it was she who contributed the *Aleurodiscus acerinus* and *Corticum vellereum* that are listed in *The fungi of Manitoba and Saskatchewan*. Aftergoing to Toronto for the PH D that she earned there in 1929 under the supervision of J.H. Faull, Mounce made no additional contributions to mycology in Manitoba.

Dorothy Newton had obtained BSA and M SC degrees from McGill University through work done at Macdonald College in 1921 and 1922. In Winnipeg, she took up Buller's continuing interest in the sexuality of species of Coprinus by making further studies on that topic, and on the differentiation of species in the higher fungi.[109] In 1932 she became the first woman to be awarded the PH D degree by the University of Manitoba, and when Charles W. Lowe, M SC, a lecturer in the department, went on leave to study at McGill University for a year, Newton was his temporary replacement. Shortly after that she returned to Quebec.

Thomas C. Vanterpool spent the year 1925–26 in Manitoba as the holder of a Hudson's Bay Company Research Fellowship, studying with Buller. While there, he and Buller showed that *Tilletia tritici*, the cause of a smut in wheat, although germinating and infecting wheat seedlings underground, discharges its basidiospores by the "drop-excretion method."[110] Vanterpool returned to his position at Macdonald College, from which he had obtained leave, without continuing his studies for a doctorate in Manitoba.

Charles Lowe, in addition to being a popular lecturer at the University of Manitoba, was also a collector of fungi. He is listed in *The fungi of Manitoba and Saskatchewan* as having collected *Gymnosporangium clavipes* near Winnipeg and *Disciseda subterranea* near Melita. He, with Buller, also studied the micro-organisms in the air of Winnipeg.[111]

The eminent British mycologist P.H. Gregory, who became a widely recognized authority on airborne fungi, "went to Winnipeg to work under Buller's direction."[112] While there he joined in the local enthusiasm for mycological forays and the collection of fungi, and is credited in *The fungi of Manitoba and Saskatchewan* with having found *Puccinia uliginosa, Melampsora bigelowii,* and *Phyllosticta brunnea*.

Another man at the University of Manitoba who made a major contribution to the mycology of that province was Guy Richards Bisby (1889–1958), a native of South Dakota, USA, where he got his bachelor's degree from the state college in 1912. He earned an AM degree from Columbia University before attending the University of Minnesota for the PH D that he was awarded in 1918. Bisby did some teaching at Minnesota, and had a period of employment at Purdue University where he worked with J.C. Arthur, North America's leading authority on the classification of the rust fungi, before going to Manitoba to become the first professor of plant pathology in a Canadian University. Except for leave periods, one of which was for nearly a year (1921–22) to study mycology and plant pathology in England, Bisby remained at the University of Manitoba until he resigned in 1937 to accept a position in the Commonwealth Mycological Institute, Kew, England, where he had a distinguished career.

Bisby cooperated with Buller to produce a *Preliminary list of Manitoba fungi* in 1922. In 1924 he published an account of the fungi he collected north of Lake Winnipeg, and, with plant pathologists I.L. Conners and D.L. Bailey, produced *The parasitic fungi found in Manitoba.*[113] However, from a mycological standpoint, Bisby is perhaps best remembered for his leadership in the preparation and publication of two books on fungi: *The fungi of Manitoba*, in cooperation with Buller and veteran mycologist John Dearness in 1929, plus its two supplements, the second of which contained 289 additions and was included in the thirteenth annual report of the Canadian Plant Disease Survey in 1934; and their extension, *The fungi of Manitoba and Saskatchewan* in 1938, for which he had the additional assistance of W.P. Fraser and R.C. Russell. Several thousand fungi collected by Bisby, including a few by Buller and others mentioned in those books, were placed in the herbarium of the University of Manitoba, from which they were eventually transferred to DAOM, in Ottawa.

In the preface of *The fungi of Manitoba*, a book which lists many of Bisby's early publications, he was able to state, "Manitoba ranks mycologically amongst the better known areas of the world." At the time it was written, that was a very true statement, thanks largely to Buller and Bisby, but also to the plant pathologists and others working in federal laboratories in that province.

Virtually everyone working in the Winnipeg "Rust Lab" went on one or more of the mycological forays led by Buller or Bisby. A number of these people have been credited, in *The fungi of Manitoba*, as having been the first to find certain fungi in that region. For example, I.L. Conners and W.L. Gordon are each mentioned nine times; D.L. Bailey three times; W. Popp, T. Johnson, and W.F. Hanna each twice; and W. Hagborg at least once.

Others not employed at the rust laboratory, but who collected fungi in Manitoba and are named in that book, include W.P. Fraser, mentioned nine times, and E. Criddle, six times. V.W. Jackson, head of the Department of Biology at the university, who with J.F. Higham and

Herbert Groh (b.1883), BSA (a temporary appointee as plant disease investigator in Manitoba for the Dominion Department of Agriculture), prepared a *Check list of Manitoba flora* for the Manitoba Department of Agriculture in 1922, is mentioned three times. Incidentally, W.P. Fraser, while on summerleave from Macdonald College to make a field survey of the rusts that attack cereals in western Canada, found a number of grass and sedge rust fungi in Manitoba, several of which he cultured "in a well lighted room in the Dominion Laboratory at Brandon, Manitoba."[114]

Students and others, some of whom did most of their professional work outside of Manitoba, also contributed to the mycology of that province. They include Dorothy Newton, T.C. Vanterpool, A.W. Henry, R.C. Russell, N. Criddle, S. Criddle, C.W. Lowe, J.B. Wallis, W.N. Denike, and Mrs Kirk Scott Wright, all of whom are acknowledged as collector-contributors to *The fungi of Manitoba*. Mrs Wright, who, as Ida Kirk Scott, BA, B SC, was a demonstrator in Buller's department for several years, contributed the chapter on "The lichens of Manitoba."

Other fungus collectors in Manitoba, though not mentioned in the book, include J.E. Machacek, who collected *Pseudopeziza medicaginis*; B. Peturson, who collected *Acremoniella atra*; William Güssow, who collected *Coniochaeta discospora* and *Pucciniastrum sparsum*; A.M. Brown, who collected *Didymium anellus* and *Lentinus underwoodii*; F. Johansen, who collected *Puccinia drabae*; and George Mayer, who collected *Polyporus tuckahoe*. They are given credit for these collections in *The fungi of Manitoba and Saskatchewan*, published in 1938.

The versatile mycologist bacteriologist plant pathologist Michael I. Timonin (1900–91) was a student of both Buller and Bisby for his BSA degree in 1932, whereas his research for the M SC in 1934, was done largely under the supervision of Bisby. Timonin was born in Russia in 1900 and had obtained a diploma from an agricultural college in Czechoslovakia before arriving in Manitoba. He was one of the early investigators to isolate fungi from butter, and from the soils of that province, which he did, in cooperation with Bisby and others, between 1933 and 1935.[115] Following stints of employment with the Division of Botany and Plant Pathology, and the Division of Bacteriology and Dairy Research, in Ottawa, Timonin went to Rutgers University for the PH D that he was awarded in 1939. His mycological activities after that were largely in eastern Canada.

The study of soil fungi in Manitoba was a popular occupation in the 1930s and 1940s. Plant pathologist J.E. Machacek, working out of the Rust Laboratory in Winnipeg, began a project on soil fungi in 1936 that was terminated in 1942. His major paper on that work, dealing with the prevalence of *Helminthosporium sativum*, *Fusarium culmorum*, and certain other fungi in experimental plots, was not published until 1957.[116] Machacek's colleague in the study of soil microflora, W.L. Gordon, became a noted authority on species of Fusarium and published papers on that genus over a period of several years.[117] Gordon also published on physiological forms of *Puccinia graminis avenae*.[118]

Margaret Newton (1887–1971), BSA, M SC, PH D, joined the research group of the Rust Laboratory in 1925. At that time she was one of the most highly trained rust specialists in Canada, and had more research experience with the rust fungi than any other member of the staff. Before she retired in 1945, Newton was the author or co-author of more than fifty publications on various aspects of the rusts of grasses.[119] In her search for rusts in Manitoba and elsewhere, she collected *Pucciniastrum sparsum*, and other fungi, as far north as Churchill, in 1936. She also collected fungi in the Northwest Territories.[120]

Newton's colleague, and co-author of several publications, Thorvaldur Johnson (1897–1979), B SC, BSA, M SC, PH D, made some pioneering studies on the genetics of cereal rust fungi and their physiologic specialization, including the effect of environmental factors on the variability of physiologic forms. Johnson continued to do research on the rust fungi throughout the Second World War and until he retired in 1962.[121]

A.M. Brown, who had been appointed as an assistant to Newton and Johnson in 1926, made a number of independent contributions to mycology. For example, he had two papers on fungi in the *Proceedings of the Canadian Phytopathological Society* for 1937, and two again in 1946. He also published on the sexual behaviour of several plant rusts in the *Canadian Journal of Research* in 1940.

SASKATCHEWAN

Saskatchewan was not as well researched, mycologically speaking, as the provinces to the east of it. However, some very good amateur and professional work in mycology was done there prior to 1950.

T.N. Willing, the first professor of natural history in the University of Saskatchewan, included lectures on the smuts and rusts that affect crops in his natural history course, and he collected fungi for classroom use and for exhibitions. A few of the fungi collected by Willing are in the Fraser Herbarium at the university. In 1906 he and G.A. Charlton wrote Saskatchewan Department of Agriculture bulletin no. 2, a bulletin that was designed to provide farmers with information about smut of wheat and methods for its prevention. Willing made a number of incidental collections of fungi between 1910 and 1920, and is credited with having found *Melampsorella cerastii* near Rosetown.[122] In 1917

Willing was requiring his students to make a collection, during the summer, of the weeds, fungi, etc. that they had studied.[123]

When Walter P. Thompson (b.1889), BA, MA, PH D, became professor of biology, and the only member of that new department, in 1913, there was less need for Willing to put emphasis on the mycological aspects of his natural history course, because Thompson introduced a course called "Seedless Plants" in 1914 and one called "Bacteriology and Microbiology" the following year; both had significant mycological components. In 1916, the seedless plants course was divided into two parts, one for each term. In the first term, Thompson dealt with bacteria and fungi, and the second term was devoted to algae, ferns, and other green but seedless plants.

In the context of mycology, Thompson is best remembered for his pioneering research on the grain rust fungi. Understandably, his mycological activities decreased as his research and administrative responsibilities increased, as they did when he became dean of Arts and Science in 1939, and president of the university in 1949.

It was Thompson who induced William P. Fraser, AB, MA, to teach mycology in the University of Saskatchewan, in 1919, on a part-time basis, January through March, while he was head of a new Dominion Laboratory of Plant Pathology in Saskatoon. This arrangement continued until 1925 when Fraser left the government service and became a full-time professor of biology at the university. His first teaching at the university included a course called "Fungi and Plant Diseases." That course was, according to the university calendar, "A systematic study of the fungi, with particular reference to those causing plant diseases. A short study of other cryptogamic groups." When he became a full-time member of the academic staff, Fraser was teaching two courses that were largely mycological: "Diseases of Plants" and "Mycology, a systematic study of the fungi." These he continued to teach, with variations, until he retired and was named emeritus professor of biology in 1937.

Most of Fraser's students for the master's degree became plant pathologists, and their mycological work, except for the thesis research by some of them, was either relatively insignificant or done outside of Saskatchewan. Exceptions to this general statement have to be made for Arthur W. Henry (1896–1989), Ralph C. Russell (1896–1964), and George Aleck Ledingham (1903–1962), all of whom were largely mycologists at heart, regardless of their official titles in later years.

Henry, a native of Fredericton, New Brunswick, had a BSA degree from the University of Saskatchewan, and in 1920 he was the first student to receive an MSA degree from that university. In his master's thesis, Henry dealt with a species of Helminthosporium that had not been previously reported. Thus he was the first Canadian to bring the presence of that fungus on barley and wheat to the attention of the scientific community. After that, Henry's mycological work was done in the USA and in Alberta.

Russell, with BSA and MS degrees from Saskatchewan, was an employee at the Dominion Department of Agriculture Research Station in Saskatoon from 1925, when Fraser left, until he retired in 1961. He took time off to study at the University of Toronto on a T. Eaton scholarship that was awarded through the Canadian Society of Technical Agriculturists (forerunner of the Agricultural Institute of Canada) for the PH D that he was awarded in 1934. His thesis research, on the fungus *Ophiobolus graminis*, was based on work that he had been doing in Saskatoon, and which had been initiated by Fraser. The calendar of the University of Saskatchewan for 1945–46 shows that Russell was a part-time lecturer in the Department of Biology that year. He collected fungi widely in Saskatchewan and was one of several Canadians who sent specimens of their collections to Franz Petrak for inclusion in his exsiccata, *Mycotheca Generalis*.[124] Eventually, the herbarium at the research station where he worked for many years was named the Ralph C. Russell Herbarium in his honour.

Russell provided a thumbnail sketch of the early history of mycology in Saskatchewan when he contributed a chapter to the book *The fungi of Manitoba and Saskatchewan*. In that chapter he listed the names of more than twenty people who, between 1917 and 1936, contributed to the mycological herbarium in the Dominion Laboratory of Plant Pathology, or to the one in the biology department of the University of Saskatchewan, both of which were in Saskatoon.

Understandably, the principal contributor to those herbaria for many years was W.P. Fraser, who was to Saskatchewan what Bisby was to Manitoba. They each taught plant pathology, and they each made extensive mycological surveys and collected many fungi in their respective provinces. As a mycologist, Fraser was primarily interested in the rust fungi, in which he became a widely recognized authority. He exchanged specimens with the noted American rust specialist J.C. Arthur, who acknowledged their cooperative work several times in his *Manual of the rusts of United States and Canada*, first published in 1934. Fraser, together with Russell, collaborated with Bisby, Buller, and Dearness in their production of *The fungi of Manitoba and Saskatchewan*, published in 1938.

In Saskatchewan, Fraser continued the cultural studies of rusts that he had started earlier in Nova Scotia.[125] He and Ibra Conners provided a comprehensive report of the Uredinales of the prairie provinces in 1925, to which Fraser made additions in 1931.[126] With G.A. Scott, he studied smut of western rye grass,[127] before he and G.A. Ledingham published their joint papers on sedge rust, in which the contributions of Mrs Bernhard Nabel (née Mabel Ruttle) were acknowledged, and on crown rust.[128] There are a number of fungi, in addition to rusts, in the Fraser Herbarium that were collected by Mabel Ruttle. She and Arthur Brown, at one time head of the science department of the provincial Normal School and a good friend of Fraser, went on fungus collecting trips with the latter, and gave him some of their best specimens.

Ledingham, who was born in Ontario but from an early age grew up in Saskatchewan, attended Normal School and was granted a licence to teach, thus enabling him to become principal of a school. By intermittently teaching, farming, and attending university, Ledingham earned a B SC in biology when he was twenty-four years old. He was awarded the M SC a year later, in 1928, the thesis research of which was done largely under Fraser's supervision. He stayed on for an additional year of post-master's research with Vanterpool before going to Toronto for the PH D in plant pathology, under D.L. Bailey, that was awarded him in 1932. Most of his mycological work after that, and prior to the end of the Second World War, was done in Ontario or British Columbia. Thomas C. Vanterpool, who had been on the staff of Macdonald College for two years, was hired in 1928 to teach plant physiology and botany and to do research in the Department of Biology with Thompson and Fraser. In 1935–36 he went on leave to England and studied mycology and plant pathology at the Imperial College, London, and at the Imperial Mycological Institute, Kew, to better prepare him to take over from Fraser. When the latter retired the following year, Vanterpool took on the teaching of plant pathology and mycology, and the guidance of candidates for advanced degrees in those fields.

The first student to earn a master's degree under Vanterpool's supervision was Robert J. Ledingham (1912–1983), whose thesis topic was "A Study of some fungi isolated from wheat roots with special reference to their pathogenicity." That thesis was largely a continuation and a refinement of research on which Ledingham and P.M. Simmonds had published in *Scientific Agriculture* in 1937. After obtaining a master's degree in 1938, Ledingham was almost immediately promoted to agricultural assistant, at the Dominion Laboratory of Plant Pathology, Saskatoon, where he had been employed part time as a plant disease investigator. Although he became a specialist on root diseases, he made no further significant contributions to mycology before 1949.

Vanterpool was, primarily, a plant pathologist, but he made important contributions to mycology in Saskatchewan, other than through his teaching. His paper on *Asterocystis radicis* in 1930[129] was soon followed by one on cultural and inoculation methods with species of Tilletia.[130] In the meantime, he and G.A. Ledingham described a new genus and species while studying the browning root rot of cereals.[131]

For a while Vanterpool, like so many mycologists at that time, studied the sexuality of some of the fungi in which he was interested. In doing this he showed that at least five species of Pythium were homothallic.[132] He and J.H.L. Truscott studied the relationships between certain parasitic species of Pythium and a plant disease,[133] and, while investigating a disease of linseed flax during the Second World War, Vanterpool discovered the causal fungus which he eventually described and named.[134] These and several more of his published papers were largely mycological or contained a significant amount of mycological material.[135] Part of Vanterpool's private collection of fungi is stored with the Fraser Herbarium.

Howard W. Mead (1899–1973), whose first two degrees were from the University of Saskatchewan, studied methods for the isolation of fungi from seeds and roots of wheat,[136] and did some plant disease research at the Dominion Laboratory of Plant Pathology, Saskatoon, before going to the University of Toronto for the PH D in plant pathology that he obtained in 1942. After that his research was somewhat more phytopathological than mycological.

Another plant pathologist at the Dominion Laboratory of Plant Pathology, Saskatoon, and a former graduate student of W.P. Fraser who made a contribution to mycology in that province was Argentina-born Bryce J. Sallans (1901–1974). Sallans had obtained two bachelor's degrees – a BSA in general agriculture from the University of Manitoba and a BA from McMaster University – before going to the University of Saskatchewan for the M SC that he earned in 1929 with a thesis titled "Studies of Alternaria occurring on wheat in Western Canada." He took some time off from his work as assistant plant pathologist, 1928–45, to earn a PH D from Wisconsin in 1939. Sallans often accompanied Fraser on his mycological forays and is credited with having found *Diaporthe crataegi*, *Puccinia sporoboli*, and *Uromyces intricatus*.[137]

ALBERTA

Alberta has never had the equivalent of a Buller, a Bisby, or a Fraser, nor did it have a native son or daughter, prior to the end of the Second World War, who could be considered to be more than an amateur mycologist. Consequently, the early mycological history of Alberta has been formulated almost entirely by hobby-mycologists, botanists, plant pathologists, and people from outside the province. One of the noteworthy "outsiders" was Edward Willet Dorland Holway (1853–1923), who made so many trips to the Rocky Mountains and wrote so much about the vegetation and scenery there that he was honoured by having a mountain named after him. Some of the fungi collected by Holway near Laggan, Alberta, are mentioned by D.S. Hone, who also collected a few fungi in the same area, near the beautiful Lake Louise.[138]

G.R. Bisby may have done some collecting in Alberta; no fewer than thirty species of fungi, mostly rusts, are mentioned in *The fungi of Manitoba and Saskatchewan* as having been found in Alberta, especially around Edmonton, Beaver Lodge, and the Peace River district. Except for the collections of *Puccinia granulispora* and *Uromyces zygadena*, which were made by A.H. Brinkman, and *Plenodomas meliloti*, by G.B. Sanford, the names of the collectors in Alberta are not given, leaving one with the impression that some of those thirty fungi may have been collected by Bisby. There is no doubt that the first collection of stripe rust in Canada was made in Alberta by W.P. Fraser in 1918.[139] There is a specimen of *Puccinia agrimoneae* in the DAOM that was collected near Edmonton by Fraser, and specimens of *Phragmidium rosae-ark* that were collected there by R. Peace and Margaret Newton.

An Alberta Natural History Society, which developed from the Territorial Natural History Society that broke up when the provinces of Alberta and Saskatchewan became a reality, was formed sometime in 1905 or early in 1906. The annual report of the Alberta Department of Agriculture for 1907 shows that its first annual meeting was held in Innisfail, 28 November 1906. When first formed, the society was composed largely of birdwatchers; when the interests of its members expanded it became the Natural History Association in 1910. It did not survive the First World War. A scan of the few documents pertaining to that association that are available in the University of Alberta fails to reveal any reference to fungi. However, it is noted that a woman, identified only as Miss Moodie of Calgary, gave a very interesting lecture, illustrated by mounted specimens, upon edible and poisonous plants of Alberta, in 1913. One is tempted to presume that Miss Moodie, who may have been a noteworthy naturalist at that time, included one or more toadstools in her lecture.

Francis J. Lewis (1875–1955), with a D SC from the University of Liverpool, where he was a lecturer in geographical botany, came to Canada in 1912 to become the first professor of botany in the University of Alberta. Although Lewis's major interest was in the mosses, he became a member of the British Mycological Society in 1924 and collected fungi in Alberta as an incidental hobby associated with his collection and study of mosses. Lewis left the university in 1937 to become a professor of botany in Egypt.[140]

For the period covered by this essay, botanist Ezra H. Moss (1892–1963) did more purely mycological work in Alberta than anyone else.

A native of Ontario, Moss attended the University of Toronto for his three degrees, the theses for two of which dealt with fungi. Developmental studies in the genus Collybia was the research topic for his MA, and "Uredinia and haustoria of the Pucciniastraceae" was the thesis title for the PH D that he was awarded in 1925. Except for periods of leave, Moss was employed in the Department of Biology (later Botany), University of Alberta from 1922 until he retired in 1957. When he published part of his thesis in *The Annals of Botany* in 1926, his title was lecturer in botany. From that position, he rose through the ranks to become an associate professor in 1933, and professor and head of a new Department of Botany in 1938, the year in which he was elected Fellow, in the Royal Society of Canada.

Moss taught "Elementary Mycology" and "Agricultural Botany" for a number of years and, beginning in 1933, a course called "Diseases of Plants." His early mycologically oriented publications dealt mostly with rust fungi, especially the tree rusts.[141] In 1940 he published a small paper on the overwintering of giant puff-balls in Alberta.[142] Moss's personal collection of fungi was deposited in the University Herbarium in Edmonton.

The first woman to be hired on the teaching staff of the University of Alberta, other than those in such traditional positions for females as nursing, household science, etc., was Eleanor Silver Dowding. She became a lecturer in botany in 1924, the year in which she earned an M SC degree there. Dowding went to the University of Manitoba for the PH D that she was awarded in 1931, with a thesis titled "Contributions to our knowledge of: (1) Sex in the Ascomycetes, and (2) the ecology of *Arceuthobium americanum.*" She continued a study of sexuality of fungi during a brief period of employment in Ottawa before returning to Alberta in 1933, when she became a medical mycologist in the Provincial Laboratory of Public Health at the university. After that, her interests were mostly in the realm of medical mycology and bacterial diseases. However, Dowding (later Mrs Keeping) collected fungi and published several papers that were largely mycological in content.[143]

Arthur W. Henry, plant pathologist on the staff of the university, did some pioneering research on soil-borne plant pathogenic fungi and their biological control. He was the first to demonstrate that the biological control of soil fungi was influenced by the temperature of the soil.[144] He was also among the earliest group of plant pathologists-mycologists in the USA to study the fungal-spore content of the upper air, in 1922, by exposing vaseline-smeared glass slides from an aeroplane at altitudes up to eleven thousand feet.[145]

Guthrie B. Sanford (1890–1977), William C. Broadfoot (b.1899), Melville W. Cormack (b.1908), and, to a lesser extent, Lawrence E. Tyner (b.1902) – federal plant pathologists on the university campus

– individually and as joint investigators made contributions to mycology in Alberta. Sanford, who had been a lecturer in the Department of Field Husbandry and an assistant of Professor G.H. Cutler in his barley-breeding program before leaving the university in 1922 for graduate studies in plant pathology, was officer in charge of the Dominion Laboratory of Plant Pathology, established on the campus of the university in 1928. He had MS and PH D degrees from the University of Minnesota, and was primarily interested in the mycology of the soil, especially those fungi that he found to be associated with root diseases. While in Minnesota, he began a long-term study of *Actinomyces scabies,* the cause of potato scab, and suggested that the control of that organism brought about by green manuring was due to the inhibitory action of non-pathogenic organisms in the soil. Thus he was one of the earliest investigators to provide experimental evidence of antibiosis.[146] In Alberta, Sanford continued his study of the potato scab organism, discovered a new root-rot fungus,[147] and published a series of papers on *Rhizoctonia solani,* thus indicating that he was almost as much a mycologist as a plant pathologist.[148]

Broadfoot is also noted for his studies of soil fungi and for research on microbial antagonism, often in collaboration with Sanford.[149] In addition, Broadfoot was an early investigator of snow moulds in Alberta, and as a follow-up on this latter he and Cormack published a paper on a low-temperature basidiomycete.[150]

Of the four plant pathologists named above, Cormack, who had a BSA from Manitoba, an M SC from Alberta, and a PH D from Minnesota, made the most significant contributions to mycology in Alberta. He too was interested in soil fungi and those that produce root rots. In this connection he published on a Cylindrocarpon, on Fusarium species, and on a Phytophthora, an Ascochyta, and a Sclerotinia as parasites of forage legumes, to mention only five of his papers that were significantly mycological in content.[151] For a while, Cormack was also interested in the fleshy fungi, and he collected many species that grow on trees and wood. A dozen or more specimens collected by him are in the mycological section of the herbarium of the University of Alberta.

Tyner, who had B SC and MS degrees from the University of Alberta and a PH D from Minnesota, was primarily interested in the fungi involved in root diseases and in bacteria that cause disease in potatoes, but his publications on these were largely phytopathological in orientation. He, like Cormack, with whom he occasionally went on mycological forays, also collected fleshy fungi, several specimens of which are in the university herbarium. Several others collected fungi in Alberta, and deposited specimens in the university herbarium prior to 1949, but no published record of their collections has been found.

BRITISH COLUMBIA

With its relatively moist environment, a wealth of trees, and great variation in soil types and in the height of land, British Columbia should be a mycologist's paradise. However, one would never know this, judging from the paucity of early mycological literature that may be found in that province.

Several foreign naturalists were early collectors of fungi in British Columbia, among the most noteworthy of whom was Edward Willet Dorland Holway, mentioned above as a collector in Alberta. For several years Holway specialized in the rust fungi and illustrated his publications with his own excellent photomicrographs. He described and named several species, including *Puccinia ornatula*, which he found near a glacier in British Columbia. Holway and John Dearness jointly named a number of rusts, many of which are noted by J.C. Arthur in his *Manual of the rusts of United States and Canada.* Dearness wrote a memorial tribute to his long-time friend, shortly after Holway died, in which it was noted that over nineteen thousand specimens of his collections were donated to the University of Michigan.[152]

A Kansas farmer, Elam Bartholomew (1852–1934), collected fungi in British Columbia and donated nine of them to Ellis and Everhart's Exsiccati.[153]

It was largely due to the inspiring leadership of Dr Josephine Tilden, professor of botany, University of Minnesota, that her department established a marine laboratory on the west coast of Vancouver Island, near Port Renfrew, in 1901. Although Tilden was primarily interested in algae, she collected a few fungi there, as did D.S. Hone and one or two other members of the group who worked out of that laboratory each summer until it closed in 1906. Several of those fungi are listed in Hone's paper on some western Helvellineae.[154]

In 1909, W. Norwood Cheesman, Fellow of the Linnean Society of London and a Justice of the Peace in Selby, England, collected fungi and slime moulds in the Rocky Mountains during a holiday following a meeting of the British Association in Winnipeg. Several of the twenty species of agarics and many of the polypores that he collected were identified for him by G.C. Lloyd of Cincinnati, in the USA.[155]

Albert I. Hill collected *Gymnopilus pallidus* near New Westminster in March 1905 and "new" species of that genus in the same general area during the following month. One of the latter was named *Gymnopilus Hillii* by W.A. Murrill, who also named *Agaricus Hillii*, a fungus that Hill had collected on Mayne Island in the Gulf of Georgia in December 1904. These and at least six additional fungi collected in British

Columbia by Hill are mentioned in Murrill's series of papers on the Agaricaceae of the Pacific Coast.[156]

The great Canadian naturalist John Macoun decided, in February 1912, to semi-retire to Vancouver Island and live with his daughter at Sidney. While getting ready to move there he suffered a paralytic stroke that invalided him for nearly seven weeks. He made a remarkable recovery, but partial paralysis in his right side forced him to walk with the aid of a cane and to write with his left hand. Nevertheless, he collected plants on Vancouver Island, but with more emphasis on cryptogams. In 1914 he collected at least 196 species of fungi that were identified for him by John Dearness, plus an undetermined number of "woody fungi" that were determined by C.G. Lloyd.[157] Macoun is reported to have made the first collection of Pyrenomycetes in British Columbia.[158] Many specimens collected by him between 1912 and 1916 are in the herbarium of the university, on loan from the Provincial Museum in Victoria where they were originally housed, and several hundred are in DAOM.

J.W. Eastham, who joined the British Columbia Department of Agriculture and became provincial plant pathologist in April 1914, had an abiding interest in the collection of plants, including certain fungi. He eventually accumulated more than fifteen thousand specimens, one of the largest private collections in the province. It is now housed in the university herbarium. In 1916 he published a paper on slime moulds, about which he had first become interested four years earlier, while in the Ottawa region.[159] Rust fungi, which he collected with W. Touzeau, and fleshy species, collected with G.E. Woolliams, plus slime moulds, and a few other fungi that Eastham alone collected, are in shoe-boxes, cigar-boxes, and pill-boxes, etc., together with some of his original notes, in the university herbarium, as are several specimens collected by Woolliams alone, and a relatively large collection of slime moulds made by E.M. Halley.

Shortly after Donald C. Buckland (1917–1956) joined the staff of the Forest Pathology Laboratory, Victoria, in 1942, he began to collect fungi that incite diseases in forest and shade trees, a few of which he contributed to the UBC herbarium, but most of which are in the herbarium of the Forest Pathology Laboratory, Victoria (DAVFP), and DAOM.

In 1930 William R. Foster (b.1905), who had B SC and M SC degrees from the University of Alberta, became assistant plant pathologist in the British Columbia Department of Agriculture, stationed in accommodations provided for him in the Dominion Laboratory of Plant Pathology, Saanichton. Although his reputation was built upon his work as a plant pathologist, Foster is mentioned here because he produced Field Crop circular no. 10, *Cereal smuts*, in 1931, and because he isolated a Phytophthora from a strawberry plant and discovered a new species of Phomopsis.[160] Specimens of Puccinia, Phytophthora, Taphrina, Podosphaera, and probably other genera collected by Foster are now in the UBC herbarium, as are a few specimens collected by R.E. Foster.

Although Eastham was the first person to teach plant pathology at the university and must have included at least some mycology in his lectures, it was Frank Dickson (1891–1969), brother of B.T. Dickson at Macdonald College, who really brought plant pathology and mycology to students of the University of British Columbia. Dickson, who had a BA degree from Queen's University, plus some advanced studies and teaching experience at Cornell University, was appointed assistant professor of botany in 1923. He went on study leave from 1 August 1929 until 30 June 1930 to attend courses and continue research at Cornell for the PH D that he was granted later that year. The title of his thesis was "Studies on *Sclerotinia sclerotiorum* (Lib.) de Bary."

At first Dickson taught the one course in mycology and the one in plant pathology that had been Eastham's responsibility, but before long he added courses in both plant pathology and mycology, until there came a time when he was teaching no fewer than five courses, at various levels, in those two subjects. Such a heavy teaching load made it virtually impossible for him to do much research in either plant pathology or mycology, but he guided the studies of at least four of his students who collected fungi or did some mycological work that was published.

A young man who eventually became chairman of the Department of Biology at the University of British Columbia, Thomas M. Taylor, collected specimens of Puccinia and Phytophthora in 1925 and 1926 while he was a student under Dickson. A few specimens collected by him are in the UBC herbarium.

Jean E. Davidson, with advice and encouragement from Dickson, studied and collected fleshy fungi in the vicinity of Vancouver intermittently from 1926 until her results were published in 1930.[161] Davidson's thesis, for the MA degree that she was awarded in 1927, was titled "Agaricaceae of Vancouver district".

Richard Mayers earned an MSA degree in 1932 largely through research on the so-called Indian paint fungus. His thesis was titled "Studies on the biology of *Echinodontium tinctorium* Ellis & Everh."

While Jean Davidson was studying the Agarics, one of Dickson's former undergraduate students, Cecil Yarwood (1907–1981), B SC, began a study of fungi in British Columbia that would lead to his becoming a world authority on the downy mildews and to a distinguished career in the United States, where he earned MS and PH D degrees. Yarwood's first two publications on the mildews were prepared in cooperation with his superior, William Newton (1893–1973), at the Dominion Laboratory of Plant Pathology, Saanichton.[162]

Newton, officer in charge of the laboratory, studied fungi, along with his various other duties. His *Physiology of fungi* was published in 1928, and his *Physiology of Rhizoctonia* in 1931, the year in which he and R.J. Hastings published a paper on *Botrytis tulipae*.[163] After that, most of Newton's publications dealt with plant diseases, especially those incited by nematodes and viruses.

One of Newton's colleagues at the Saanichton laboratory, Walter Jones (1890–1957), was a long-time collector of parasitic fungi in British Columbia. Jones was born in Wales and earned his B SC degree at Aberystwyth before coming to the province in 1924. After a stint of service as a plant disease inspector in the Certified Seed Inspection Service, Jones studied plant pathology at the University of California and was awarded the MS degree in 1929. From that date, and until he retired in 1955, Jones was employed as a plant pathologist at Saanichton. One of his early mycological interests there was *Armillaria mellea* on raspberries, but he soon followed Yarwood's lead and became involved in what was to develop into a long-term interest in the downy mildew fungi.[164] Jones's knowledge of the fungi that incite plant diseases and his familiarity with the names of local plants were combined to produce the bulletin *A Check list of plant diseases in the coastal areas of British Columbia* for the British Columbia Department of Agriculture in 1945. Most of the fungi collected by Jones, Foster, and others associated with the Saanichton laboratory were eventually transferred to DAOM.

In 1923, N.L. Cutler published a list of twenty-four fungi that he considered to be important tree-destroyers in the Vancouver forestry district, and another two dozen that he found associated with timber rots there.[165]

Forest pathologist John E. Bier, together with Irene Mounce and Mildred Nobles, studied *Poria wierii*, the cause of root-rot of Douglas fir, and rots caused by *Armillaria mellea*.[166] Bier, for a while the only forest pathologist in the Forest Pathology Laboratory established in Victoria in 1940, discovered *Lentinus kauffmanii*, a previously undescribed species, during his studies on brown pocket rot of sitka spruce on the Queen Charlotte Islands.[167] Together with his assistants in forest pathology, R.E. Foster and P.J. Salisbury, he summarized three years of those studies in Canada Department of Agriculture technical bulletin no. 56, in 1946. Several fungi collected by Bier, including a specimenof *Poria xantha*, are in the UBC herbarium, but most of Bier's collections, and those of his forest pathology colleagues, are in DAVFP, Victoria, and DAOM, Ottawa.

Irene Mounce, a native of British Columbia, spent the summer of 1938 collecting and culturing forest tree fungi in her home province. Three specimens that she collected bear accession numbers F642, F627, and F637 in the UBC herbarium, but most of her collections and cultures went to the National Herbarium in Ottawa. Shortly after that she became involved in studies of a variety of plant diseases and, with J.E. Bosher, published a paper on a seedling blight in British Columbia that was caused by *Alternaria radicina*.[168] Mounce stopped her mycological research, or at least she stopped publishing anything that was mycologically oriented, when she got married in 1945.[169]

In the provincial capital, the Victoria Natural History Society was such a thriving institution in the early 1940s that its members decided to publish their own journal. Their *Victoria Naturalist* records an annual fungus foray, beginning with volume 1 in 1944. For many years those forays took place in the Hudson's Bay Woods, on Vancouver Island, and were led by George A. Hardy (1888–1966) of the Provincial Museum, Victoria. His accounts of the forays, including names of many of the fungi that were found or collected by members of the foray group, appear in several volumes of the *Victoria Naturalist*. In volume 6:6, 63–4, he commented on "the seventh annual Fungus Foray which was held, as in other years, at the Hudson's Bay Woods ... although the season was somewhat disappointing a total of at least 40 species were noted." He then named many of the species. Hardy wrote several popular articles that were published in the *Victoria Naturalist*. For example, there is one about the fly agaric in volume 14; one about a bird's nest fungus in volume 15; and one dealing with ten coloured russulas in volume 19.

Near the end of the Second World War, Hardy published a semi-popular article on a bracket fungus in *Canadian Nature*,[170] but he is best known as the author of the British Columbia Provincial Museum handbook no. 4, *Some mushrooms and other fungi of British Columbia*, published in 1946.

EPILOGUE

In tracing the misty beginnings of mycology in each of the provinces of Canada, it soon became evident that there has been a similar pattern of development in each. In the beginning fungi were studied and collected purely for the pleasure those activities gave to the individuals, most of whom were professionals or semi-professionals in fields quite unrelated to mycology. Although pleasure is still the motivating factor that induces hundreds of amateur and professional mycologists to go on fungus forays each year, the overall picture of mycology has undergone a profound change.

The transition from the domination of amateurs in the mycology of Canada, so evident in the middle of the nineteenth century, to that of the professional mycologists in the third and fourth decades of the twentieth century was virtually complete by the fifth decade. The amateurs gradually separated from the growing number of professionals when the latter began to deal with the microscopic, the genetic, the physiological, and the cytological aspects of the fungi, and in so doing developed a whole new language within the science.

Amateur mycologists still play an active and important role in the mycology of Canada, but mostly through their own clubs and associations. There are a dozen or more local groups scattered across the country whose major functions are to organize and conduct fungus forays. A good example of such a group is Le Cercle des Mycologues de Québec, which serves a relatively local group around Quebec City. An example of a larger more broadly based group, in fact the largest, is the North American Mycological Association, which caters primarily to amateurs while welcoming professional mycologists and serving the whole continent.

The trend toward more science, or more professionalism, in Canadian mycology coincided with the development of plant pathology as a science. This may be seen in any one of a number of theses, based on modern methods of mycological research, by students working toward advanced degrees in some aspect of botany or plant pathology in the late 1930s and the 1940s. A good example is the doctoral research of Douglas B.O. Savile (b.1909), who exposed a number of misconceptions through his use of advanced staining techniques, including the first use of the DNA-specific Feulgen method to study the cytology of rust fungi.[171] In 1946, Savile, who had BSA and M SC degrees from McGill University, and who was destined to become one of Canada's most highly respected mycologists, was still employed as a plant pathologist, although his work was largely mycological. For example, his study of some foliage parasites of trees, and three additional papers that were almost entirely mycological in content, are listed in the *Proceedings of the Canadian Phytopathological Society* for 1948. Many young scientists like Savile who made significant contributions to mycology were budding plant pathologists, or they were labouring under that appellation because there were very few employment opportunities for mycologists, as such, at that time.

This was a situation more or less peculiar to Canada, because in Britain plant pathologists were frequently designated as "mycologists."

It was not until the agricultural advisory services in Britain were reorganized in 1946 that the provincial mycologists became known as plant pathologists. In Canada, nearly the reverse was true, in that a number of people doing work that was almost entirely mycological for the Dominion Department of Agriculture were designated as plant pathologists. Several of them were not officially recognized as mycologists until 1951. That was the beginning of the decade in which mycology in Canada made its greatest leap forward, both in the number of individuals employed and in the advancement of professionalism within the science. But that decade is beyond the time frame of this essay.

General Summary
and Conclusions

Long before the microbial cause of plant diseases had been discovered, farmers could see that insects were damaging their crops. It is therefore understandable that insect control preceded the control of disease in plants, and that farmers and amateur entomologists would be the first to attempt to control the ravages of the insects they could see, and also the diseased or damaged conditions of undetermined origin. Thus it came about that entomologists were among the first people in Canada to practise plant disease control and to publish their methods and results.

For several decades the most intensive care of plants was provided by horticulturists in their hotbeds, cold frames, and greenhouses. They protected plants from many of the diseases that came to be recognized by plant pathologists as being of a physiological nature – that is, diseases that are due to excesses or deficiencies in such environmental factors as water, light, and heat rather than to living organisms. In a relatively cold climate, the most obvious of the "physiological diseases" in Canada are those resulting from low temperature. For horticulturists, learning how to control physiological problems was but a prelude to the control of other diseases, as their causes were determined. Thus horticulturists were also among the early contributors to the genesis of plant pathology in Canada and elsewhere.

Following the potato famine in Ireland and the large influx of Irish immigrants into Canada, Canadians generally, and biologists in particular, became acutely aware of the devastation that can result from a failure of the potato crop, a major source of carbohydrates. Any failure or threat of a failure in this crop aroused fear in the minds of the whole community. It was largely for this reason that the greatest stimulus to a study of plant diseases and their control came from widely publicized diseases of potatoes and their associated political and economic ramifications.

As sources of carbohydrates for most Canadians, the grain crops ranked in second place after potatoes. Grain was also a major source of farm income, whether directly from off-the-farm sales or indirectly through sales of poultry and livestock that had been fed grain and forage-crop materials. Because smut and rust diseases appeared to pose the greatest threat to the grain crops, they were the first grain diseases to be intensively studied by mycologists and plant pathologists. In those studies it soon became apparent that neither a plant pathologist nor a mycologist, working alone, was likely to learn how to control such perplexing diseases. Consequently, the team approach to plant disease control began to evolve, with plant breeders and other specialists often taking a leading role. This was especially true in the control of the grain diseases, and, to a slightly lesser extent, the potato diseases. Some of the most instructive information about plant diseases during the early decades of the development of plant pathology in Canada were in the publications dealing with diseases of potatoes and of wheat.

By the fourth decade of the twentieth century, mycologists had identified the fungi involved in most forest tree diseases and wood rots; nevertheless, studies on those diseases and their control came at a relatively late stage in the development of plant pathology in Canada. The main reason for this was that the loss of trees to disease was not considered to be a direct threat to the economy or the well-being of the people. For nearly a century, Canadians perceived the forests as a near-inexhaustible source of wood for export and fuel to warm their homes. For these and various other reasons, the study of forest pathology in Canada was delayed to such an extent that no forest pathologists were involved in the study of such devastating tree diseases as the beech bark disease and yellow birch dieback until after most beech and yellow birch trees had died in the Maritime provinces.

The fact that plant parasitic nematodes could induce a plant disease, or adversely affect a disease caused by other factors, was not recognized by plant pathologists in Canada until after their attention was focused on nematodes by entomologists. In fact, entomologists played major roles in plant nematology in Canada prior to the 1950s.

Several devastating or spectacular plant diseases were imported into Canada before anyone in authority was aware of how they got here or how to control them. However, when it became known that disease-inciting organisms were being carried from country to country in infected plants or the soil attached to them, rules and regulations designed to exclude or control the movement of such plants and plant materials into Canada were promulgated. Those laws included restrictions on the movement of plants within the country. As the progressively minded Canadian agriculturists and biologists and a number of educators became aware of the need for plant disease control, they agitated for the teaching of this subject in the schools and colleges. They were so successful in this regard that by 1950 some plant pathology was being taught in schools and colleges in every province of Canada.

During the latter part of the nineteenth century and the early decades of the twentieth, amateur mycologists gave taxonomic mycology a sound beginning in Canada. Their contributions to descriptive mycology, a prerequisite to fungus life history studies and ultimately to the formulation of measures for the control of fungi that incite plant diseases, were invaluable to the genesis of plant pathology and to the maturation of professional mycology. Until about the third decade of the twentieth century, most descriptions of plant pathogenic fungi were provided by mycologists. Then, concurrently with the development of professionalism in plant pathology, more and more plant pathogens were being described by plant pathologists.

In spite of this invasion by plant pathologists into the realm of mycologists, by 1920 mycology had become generally accepted as a science, even though there were very few professional mycologists in the country. The same could not be said, with the same degree of certainty, about plant pathology, even though there were more practising plant pathologists than mycologists in governmental services and in universities. The majority of Canadians who joined the American Phytopathological Society, before the Canadian Phytopathological Society was formed in 1929, were botanists, mycologists, plant physiologists, and others with only a cursory interest in plant diseases. It would be another decade or more before many Canadians began referring to themselves as plant pathologists.

When most of the organisms producing disease in plants, including bacteria and viruses, were thought to have been determined, a few plant pathologists began studies on how the organisms create disease; what enzymes were being produced and how the enzymes, or other disease-inducing entities, could be controlled. Thus another phase in the development of plant pathology began, with a gradual change in emphasis from mycology to biochemistry. Many individuals did not make the change, but more and more of those who entered the field of plant pathology in the late 1940s skipped the mycological stage almost entirely, except for the university course they may have had to take in that subject.

Paralleling this changed emphasis were changes in the methods of disease control. Interest in fungicides had risen steadily as improved sprayers were developed. In the early 1930s emphasis was on copper, lime, and sulphur compounds. This was followed, in the late 1930s and the 1940s, by a period in which organic fungicides were intensively studied. When it was shown that nematode infestations of plant roots could be eliminated by immersing the affected roots in hot water, there was a brief revival of interest in heat therapy. The potential of antibiotics for plant disease control was just beginning to be studied in the late 1940s.

By 1950 most of the early empirical methods of plant disease control had been replaced by more precise scientific techniques which were accompanied by vastly improved methods, machinery, and materials, including chemical compounds. The success of the genetic approach to disease control had been emphatically confirmed. Nevertheless, although mycologists and plant pathologists had learned how to identify most disease-inducing organisms, and plant pathologists had learned a great deal about disease prevention and control, they had not learned how to cure infectious plant diseases. Exceptions to this were diseases in which the pathogen was confined to seeds and those in which the infections were close to the surface of the affected part, and those that could be cured by heat therapy. Partly as a consequence of this, there were no self employed plant pathologists in Canada. Governments at both the federal and provincial levels bore the financial responsibility for providing advice to growers on plant disease control and for the research and research facilities that were necessary to sustain the developing science of plant pathology. There were relatively few amateur plant pathologists, in comparison to the number of amateur mycologists.

By mid-century plant pathologists were still learning how to bring plant physiology, biochemistry, biophysics, organic chemistry, cytology, microbiology, genetics, and related disciplines into full play, in a coordinated manner, so as to gain an understanding of the complex relationships between the host plant and its parasites. They still needed to know much more about how those relationships came about, how they developed, and what could be done to prevent, limit, or otherwise control the destructive aspects of that kind of symbiosis. The acquisition of this knowledge and understanding would occupy much of the attention of the new breed of plant pathologists that came on the scene shortly after the Second World War.

Notes

INTRODUCTION

1 For an authoritative commentary on definitions and terminology in plant pathology, see G.C. Ainsworth, *An introduction to the history of plant pathology* (Cambridge 1981), 4–5.
2 Highlights in the history of the fungal nature of plant diseases have been provided by E.C. Large, *The advance of the fungi* (New York 1962).
3 I.L. Conners, editor, *Plant pathology in Canada* (Winnipeg 1972), 230–42.
4 R.H. Estey, "A History of the Québec Society for the Protection of Plants," *Phytoprotection* 64 (1983), 1–22.

CHAPTER ONE

1 A.F. Mason, *Spraying, dusting and fumigating plants* (New York 1928), 2.
2 H.Y. Hind, *Essay on the Insects and Diseases Injurious to the Wheat Crops* (Toronto 1857). That advertisement is reproduced in the introduction to this small book.
3 Ibid.
4 Leon Provancher, *Essai sur les insectes et les maladies qui affectent le blé* (Montreal 1857).
5 Leon Provancher, *Traité Elémentaire de Botanique à l'usage des maisons d'Education* (Quebec 1858), 94.
6 W.H. Harrington, "The Abbé Provancher," *Canadian Entomologist* 24 (1892), 1–131. See also V.A. Huard, *La Vie et l'oeuvre de l'Abbé Provancher* (Paris 1926).

7 C.J.S. Bethune, "Fungus diseases and insect pests," in *Report of the Ontario Agricultural College* 34 (1908), 24–32.

8 W. Saunders, "The Apple-Tree Blight," in *Report of the Entomological Society of Ontario for 1874* (1875), 372–3.

9 William W. Judd, editor, *Minutes of the Entomological Society of Ontario while headquartered in London, Ontario, 1872–1906* (Toronto 1976).

10 John Dearness, "Some fungus foes of the flower garden. A locality list with notes," *Phytopathology* 16 (1926), 75–7.

11 W.F. Tamblyn, "John Dearness," *Mycologia* 47 (1955), 909–15.

12 R.H. Estey, "James Fletcher (1852–1908) and the genesis of plant pathology in Canada," *Canadian Journal of Plant Pathology* 5 (1983), 120–4. See also C.J.S. Bethune, "James Fletcher," *Canadian Entomologist* 40 (1908), 433–7.

13 A. Mallis, *American entomologists* (New Brunswick, NJ 1971), 113–16.

14 R. Glen, compiler, "Entomology in Canada up to 1956: A review of developments and accomplishments," *Canadian Entomologist* 88 (1956), 298.

15 W. Lochhead, "President's Address," *Annual Report of the Quebec Society for the Protection of Plants* 10 (1918), 15.

16 *Report of the Minister of Agriculture, Province of Quebec* (1913), 265.

17 V.A. Huard, "Les Principales Espèces d'insectes nuisibles et de maladies végétales," *Annual Report of the Quebec Society for the Protection of Plants* 8 (1916), 11.

18 Anon., *The C.S.T.A. who's who 1924* (Ottawa 1924), 108.

19 John G. Coulson, personal communication. For Maheux's papers on history, see *Annual Reports of the Quebec Society for the Protection of Plants*, vols 11, 12, 13, and 16.

20 See *Annual Report of the Ontario Agricultural College* (1900) 15–18, and (1901), 20–5.

21 A.J. Skolko, "A Review of the current Canadian research on insects in relation to plant diseases," *Phytoprotection* 45 (1964), 96.

22 M.E. DuPorte, "Some insect carriers of plant diseases," *Annual Report of the Quebec Society for the Protection of Plants* 11 (1919), 59–65.

23 Personal communication.

24 F. Letourneau, "Fire Blight," *Annual Report of the Quebec Society for the Protection of Plants* 8 (1916), 42–5.

25 Ibid. 6 (1914), 31–3; 7 (1915), 38–42; 9 (1917), 42.

26 *Report of the Minister of Agriculture, Province of Quebec* (1926), 17

27 Anon., *The C.S.T.A. Who's Who 1924* (Ottawa 1924), 102.

28 C.E. Petch, "Some results from spraying and dusting apples in Quebec." *Annual Report of the Quebec Society for the Protection of Plants* 15 (1923), 94–6.

29 Anon., "C.E. (Charlie) Petch," *Macdonald Farm Journal* 12:12 (1951), 23.

30 *Annual Report of the Ontario Agricultural College* (1906), 41 and 44; also ibid.
(1914), 20.

31 For a photo of William Brodie, and his obituary notice, see *Canadian
Entomologist* 41 (1909), 377–8. For examples of his publications on
galls, see ibid., 157–60 and 249–52; also ibid. 24 (1892), 12–13 and
137–9, and ibid. 9 (1877), 11–12.

32 A. Cosens, "A contribution to the morphology and biology of insect galls,"
Transactions of the Canadian Institute 9 (1912), 297–387. This is, essentially, his 1913 PH D thesis. See also *Canadian Entomologist* 45 (1913),
380–4, and ibid. 47 (1915), 354–5.

33 *Annual Report of the Ontario Agricultural College* (1908), 24–5. See also ibid.
(1909), 26.

34 Ibid. (1913), 28.

35 *Report of the Minister of Agriculture* (Ontario) (1941), 14.

36 M.B. Waite, "Results of Recent Investigations in Pear Blight," *Botanical
Gazette* 16 (1891), 259.

37 L. Caesar, "Insects as agents in the dissemination of plant diseases," *Annual
Report of the Entomological Society of Ontario* 49 (1919), 60–6.

38 Anon., "Synopsis of work of the Department of Agriculture and cooperative organizations in B.C., 1873 to 1912," in *Report of the
Department of Agriculture (B.C.) for 1913–14* (1915), 5.

39 That was an unnumbered bulletin of the Department of Agriculture, issued
in 1907.

40 R.C. Treherne, "A review of applied entomology in British Columbia,"
Proceedings of the Entomological Society of British Columbia 4 (1914),
68–9.

41 Brittain's list of plant diseases, in British Columbia Department of
Agriculture bulletin no. 7, includes "Those due to insects and other
animal parasites. See also I.L. Conners, editor, *Plant pathology in Canada,*
(Winnipeg 1972), 217.

42 R.M. Winslow, "Report of the Horticultural Branch," in *Report of the
Department of Agriculture (B.C.) for 1913 and 1914* (1915), R34–5.

43 W.H. Lyne, "Report of the Fruit Inspection Branch," ibid., R90–1.

44 See Department of Agriculture reports for several years.

45 Anon., *Report of the British Columbia Department of Agriculture for 1921* (1922),
U69.

46 J.W. Eastham, "Diseases (and pests) of cultivated plants," British Columbia
Department of Agriculture, bulletin no. 68 (1916).

47 Anon., *Proceedings of the Entomological Society of British Columbia* 60 (1963),
55.

48 Anon., *Regina Leader,* 12 March 1903, 1 and 8.

49 Anon., *Report of the Dominion Department of Agriculture for 1913,* 507. See
also ibid., (1914), 864–5.

50 W.H. Brittain, "Report of the professor of zoology and provincial entomologist," in *Report of the Secretary of Agriculture* (Nova Scotia) (1913),
34.

51 Ibid. (1914), 28–51.

52 A. Gibson, "The development of applied entomology in Canada
1914–1923," *Annual Report of the Quebec Society for the Protection of Plants*
16 (1924) 24–56. See also anon., *The C.S.T.A. Who's Who* (Ottawa 1924)
for lists of their publications.

53 Anon., *Raymond Paddock Gorham, 1885–1946*, catalogue MC211, Provincial
Archives, Fredericton, N.B.

54 R.E. Balch and J.S. Prebble, "The bronze leaf borer and its relation to
the dying of birch in New Brunswick forests," *Forestry Chronicle* 16
(1940), 179–201.

55 L.S. Hawboldt, "Aspects of yellow birch dieback in Nova Scotia," *Journal
of Forestry* 45 (1947), 414–22.

56 Anon., *Proceedings of the Canadian Phytopathological Society* 6 (1936), 5.

57 A.D. Baker, "Some observations on the development of the nematology
section at Ottawa, and on some of the needs of Nematology,"
Phytoprotection 44 (1963), 5–11.

58 Anon., *Consolidated Regulations of Ontario, 1950* (Toronto 1950), regulation
477. This regulation had its inception in the "Sugar Beet Nematode
Regulations" that were approved by an Order in Council dated 12 August
1941.

CHAPTER TWO

1 R. McNeil, *Practical tests on gardening for Manitoba and North-West Territories*
(Winnipeg 1884).

2 S.B. Johnston, "Fires in Orchards," *Nova Scotia Journal of Agriculture* 1
(1869), 435.

3 P.H. Gosse, *The Canadian Naturalist* (London 1840), 327.

4 Anon., *Annual Report of the Minister of Agriculture* (Quebec), (1909), 95.

5 Anon., "Charles Gibb, B.A.," *Canadian Record of Science* 4 (1891), 188.

6 Elsie Pomeroy, *William Saunders and his five sons* (Toronto 1956), 46–7.
See also W. Saunders, *Central Experimental Farm* (Ottawa 1887), 6.

7 W. Saunders, *Progress in the breeding of hardy apples for the Canadian Northwest*,
Dominion Department of Agriculture bulletin 68 (Ottawa 1911).

8 K.J. Kirkland, editor, *Scott Research Station (1910–1985)*, Canada
Department of Agriculture historical series no. 21.

9 Ibid., p. 22.

10 A. MacKay, "Experimental Farm for the North-West Territories," in
Experimental Farm Reports for 1888 (1889), 102.

11 W.E. Johnson and A.E. Smith, *Indian Head Experimental Farm 1886–1986,* Canada Department of Agriculture historical series no. 23, p. 28.

12 M.R. Kilcher, *Swift Current Research Station 1920–1985,* ibid., historical series no. 25, p. 20.

13 Sharon Ramsay, *Brandon Research Station 1886–1986,* ibid., historical series no. 31, p. 39.

14 Anon., "Report of the Historical Committee," *Report of the Fruit Growers' Association of Ontario for 1922* (1923), 9.

15 R.P. Leopold, "Economy in spraying," *Annual Report of the Quebec Society for the Protection of Plants* 12 (1920), 48–50. See also ibid. 9 (1917), 42–4.

16 W.S. Blair, *Report of the Horticulturist,* in Experimental Farms Reports (1896), 313.

17 W.S. Blair, "Dusting vs liquid spraying," *Agricultural Gazette of Canada* 5 (1918), 226.

18 D.S. Blair, "Winter injury to apple trees in eastern Canada," *Scientific Agriculture* 16 (1935), 8–15.

19 J.F. Snell, *Macdonald College of McGill University: A history from 1909–1955* (Montreal 1963), 104–5.

20 F.M. Clement, "Winter injury in orchards," *Annual Report of the Quebec Society for the Protection of Plants* 5 (1913), 24–6.

21 W. Saunders, "Essay on the Cultivation of the Plum," *Annual Report of the Commissioner of Agriculture and Public Works, Ontario, for 1873* (1874), 233–8.

22 Anon., "Report of the Committee Appointed to Proceed to London to Visit Mr. Saunders's Fruit Farm," ibid., 255.

23 W. Saunders, "Proposed Remedies for Scab of Apple," *Canadian Horticulturist* 7 (1884), 127–8.

24 P.C. Dempsey, "Experiments in Hybridizing," *Annual Report of the Commissioner of Agriculture and Public Works, Ontario, for 1873* (1874), 287–9.

25 Ibid. (1884), 165.

26 Ibid. (1885), 23.

27 P.M.A. Millardet, "The Discovery of the Bordeaux Mixture." (1885), Phytopathological Classic no. 3 (1933).

28 E.F. Palmer, *The first fifty years, 1906 to 1956, Horticultural Experiment Station and Products Laboratory,* Department of Agriculture (Toronto 1956).

29 P.E. Bucke, "Hybridizing," *Annual Report of the Commissioner of Agriculture and Arts, Ontario, for 1875* (Toronto 1876), 268–9.

30 G. Lawson, "Some Points in Vegetable Physiology Bearing Upon the Methods Available for Systematic Improvement of Fruits," *Transactions and Reports of the Fruit Growers' Association and International Show Society of Nova Scotia* (1884), 24–7.

31 K. Cox, *A history of the Nova Scotia Agricultural College* (Truro 1965), 6. See also Ronald S. Longley, *Acadia University, 1838–1938* (Wolfville 1939), 102.

32 E.E. Faville, "Course of Study, Nova Scotia School of Horticulture," *Transactions and Reports of the Fruit Growers' Association, Nova Scotia* (1894), 126.

33 F.C. Sears, "Report of the Nova Scotia School of Horticulture," *Report of the Secretary of Agriculture, Nova Scotia, for 1900* (1901), 46.

34 P.J. Shaw, "Report of the provincial horticulturist for 1921–22," *Journal and Proceedings of the House of Assembly, Nova Scotia*, session 1923, part 2, appendix 8 (Halifax 1923), 62.

35 W. Ferguson, "Letter from a fruit grower in Nova Scotia," *Annual Report of the Commissioner of Agriculture and Public Works for 1873* (Halifax 1874), 342

36 P.M.A. Millardet, "Concerning the History of Treatment of Mildew with Copper Sulphate" (1885), Phytopathological Classic no. 3 (1933), 23.

37 A.G. Turney, "Report of the horticulturist," *Annual Report of the Department of Agriculture, New Brunswick, for 1919* (1920), 67–76. See also ibid. (1913), 46.

38 F.P. Sharp, "The Plum," *Report on Agriculture for the Province of New Brunswick* (1890), 85–7.

39 See *Report of the Commissioner of Agriculture and Arts for the Province of Ontario for 1881* (1882), 568.

40 Ibid. for 1882 (1883), 128–32.

41 J.P. McMurrich, "Report of the Professor of Biology," in *Annual Report of the Commissioner of Agriculture and Arts of the Province of Ontario for 1882* (1883), 130.

42 J.H.L. Truscott, "A storage rot of celery," *Report of the Ontario Vegetable Growers' Association for 1935–36* (1936), 25–32. See also *Report of the Ontario Agricultural College* (1936), 67.

43 V.W. Jackson, "Rusts and smuts of grain crops," Manitoba Department of Agriculture and Immigration extension bulletin no. 44 (1919), 11.

44 J.H. Ellis, *The ministry of agriculture in Manitoba 1870–1970* (Winnipeg 1970), 364. See also *Development of horticulture on the Canadian prairies: An historical review*, compiled by the Western Canada Society of Horticulture (Winnipeg 1956), 25

45 J.W. Eastham, "Some recollections of plant diseases in British Columbia," *C.P.S. NEWS* 2 (2) (1950), 1–6.

46 Anon., *Statutes of the Province of British Columbia passed in the session held in the fifty-fifth year of the reign of Her Majesty, Queen Victoria* (Victoria 1892), chapter 23.

47 Anon., *Annual Reports of the British Columbia Department of Agriculture for 1913–14* (1915), 34–7.

48 Ibid. for the relevant years.

49 Personal Communications and taped interviews.

50 W.T. Macoun, "Report of the horticulturist," *Experimental Farms Reports* (Ottawa 1910), 127.

51 Anon., *Fifty years of progress, Dominion Experimental Farms* (Ottawa 1939), 35.

52 John Craig, "Report of the Horticulturist for the year 1895," *Experimental Farms Reports* (Ottawa 1896), 121.

53 Ibid. for 1891 (1892), 141–8.

54 W.T. Macoun, "Spraying potatoes for the prevention of blight and rot," *Report of the Central Experimental Farm* (Ottawa 1905), 131–2.

55 For a synopsis of Macoun's life, see *Scientific Agriculture* 13 (1933), 790.

56 M.B. Davis, "Fire pots as protection against frost," *Agricultural Gazette of Canada* 2:1 (1915), 9–13.

57 Ibid. 2:7 (1915), 627–9.

58 Anon., *Annual Report of the Fruit Growers' Association of Ontario.* (1914), 78.

59 Anon., *Report of the Horticultural Division, for the year ended March 31, 1921* (Ottawa 1921), 8–10.

60 H. Hill, "Minor elements affecting horticultural crops," *Scientific Agriculture* 17 (1936), 148–53. See also ibid. 29 (1949), 376–84.

61 T.H. Anstey, *One hundred harvests* (Ottawa 1986), 78.

62 W.R. Phillips and P.A. Poapst, "Low temperature research," In *Division of Horticulture Progress Report 1934–48* (Ottawa c1949), 163–193.

63 Ibid., 164.

64 Mary MacArthur, "Freezing of commercially packaged asparagus, strawberries and corn," *Fruit Products Journal* 24 (1945), 238–40.

65 D.V. Fisher, J.E. Britton and H.J. O'Reilly, "Peach harvesting and storage investigations," *Scientific Agriculture* 24 (1943), 1–15.

66 R.S. Willison, "Brown rot and other fungal wastage in harvested peaches," *Scientific Agriculture* 24 (1944), 221–33. See also ibid. 21 (1941), 624–45.

67 For an example of Willison's work on virus diseases, see *Phytopathology* 34 (1944), 1037–49, and for the peach canker work, see *Canadian Journal of Research* "c" 15 (1937), 324–39.

68 C.A. Eaves, "The present status of gas storage research with particular reference to studies conducted in Great Britain and preliminary trials at the Central Experimental Farm, Canada," *Scientific Agriculture* 15 (1935), 542–56.

69 Anstey, *One hundred harvests,* 328.

70 G.H. Berkeley, G.O. Madden, and R.S. Willison, "Verticillium wilts in Ontario," *Scientific Agriculture* 11 (1931), 739–59. For work of Chamberlain, see *Canadian Journal of Research* "c" 16 (1938), 118–24, and *Phytopathology* 38 (1928), 776–92.

71 G.H. Harris, "Causes of raspberry failure in the coastal area of British Columbia," *Scientific Agriculture* 16 (1936), 353–7. See also ibid. 20 (1940), 344–54.

72 A.A. Hildebrand and L.W. Koch, "A microscopic study of infection of the roots of strawberry and tobacco seedlings by microorganisms of the soil," *Canadian Journal of Research* "c" 14 (1936), 11–26. Also ibid., 19 (1941), 225–32.

73 L.W. Koch, "Spur blight of raspberries in Ontario caused by Didmella applanata," *Phytopathology* 21 (1931), 247–87.

74 R.E. Fitzpatrick, Frances C. Mellor, and Maurice F. Welsh, "Crown rot of apple trees in British Columbia – rootstock and scion resistance trials," *Scientific Agriculture* 24 (1944), 533–41.

75 H. Katznelson and L. Richardson, "The microflora of the rhizosphere of tomato plants in relation to soil sterilization," *Canadian Journal of Research* "c" 21 (1943), 249–55.

76 H. Katznelson, "Inhibition of micro-organisms by a toxic substance produced by an aerobic spore-forming bacillus," *Canadian Journal of Research* "c" 20 (1942), 169–73.

CHAPTER THREE

1 E.C. Large, *The advance of the fungi* (New York 1962) has an excellent commentary on potato diseases and their effect on the human population.

2 William Evans, *A Treatise on the Theory and Practice of Agriculture* (Montreal 1835), 80.

3 Anon., *The New Brunswick Almanac for the Year of Our Lord 1840* (St John 1840), 75.

4 Anon., *Report of the Visitor of Schools for Kings County (P.E.I.) for the year ending January 28, 1848* (Charlottetown 1848).

5 Anon., "Notes on Potato Disease," *Agricultural Journal and Transactions of the Lower Canada Agricultural Society* (1848), 276.

6 R.J. Morgan, "Poverty, wretchedness and misery: The great famine in Cape Breton, 1845–1851," *Nova Scotia Historical Review* 6 (1986), 88.

7 G. Buckland and W. McDougall, *The Canadian Agriculturist* (Toronto 1849), 176.

8 H.Y. Hind, *Two Lectures on Agricultural Chemistry* (Toronto 1850), 68.

9 Amasa Walker, *A synopsis of the several communications on the cause and cure of the Potato Rot, received by the Executive of Massachusetts* (Boston 1852), 1.

10 Anon., *Journal and Transactions of the Board of Agriculture of Upper Canada* 1 (1856), 220.

11 Ibid., 39.

12 H.Y. Hind, *Lectures on Agricultural Chemistry*, second edition (Toronto 1851), 140.

13 P.C. Dempsey, "The Dempsey Potato," in *Report of the Fruit Growers' Association of Ontario, for 1882* (Toronto 1883), 38.

14 W. Bustin, "The Potato Disease," *Journal of Agriculture for Nova Scotia* 1 (1868), 314.

15 A. Macfarlane and G. Lawson, "First Annual Report of the Board of Agriculture," *Journal of Agriculture* (Nova Scotia) 1 (1865), 19 and 81.

16 Anon., *Report of the Department of Agriculture, Ontario, for 1896*, volume 1 (Toronto 1897), 149.

17 J. Fletcher, *Papers by the Entomologist* (Ottawa 1891), 11.

18 Ibid. (Ottawa 1892), 17.

19 G. Maheux, "How the protection of plants has progressed in Quebec," *Annual Report of the Quebec Society for the Protection of Plants* 11 (1919), 56

20 D.W. Beadle, *Canadian Fruit, Flower and Kitchen Gardener* (Toronto 1872), 242.

21 That allegation was in an advertisement about the "'Eldorado Potato'" in the *Canadian Horticulturist* 29 (1906), 75.

22 H.T. Güssow, "The problems of plant disease," in *Evidence of H.T. Güssow before the Standing Committee on Agriculture and Colonization* (Ottawa 1910), 74.

23 H.T. Güssow, in *Experimental Farms Reports for the year ending March 31, 1910* (Ottawa 1910), 262.

24 G.C. Morgan and E.H. Peters, "The potato wart disease problem in Newfoundland." *Annual Report of the Quebec Society for the Protection of Plants* 38 (1957), 62–8. See also M.C. Hampson and K.G. Proudfoot, "Potato wart disease, its introduction to North America, distribution and control problems in Newfoundland," *FAO Plant Protection Bulletin* 22 (1974), 53–64.

25 Smith addressed a number of meetings on this topic. See *Annual Report of the Nova Scotia Fruit Growers' Association* (Halifax 1910), 16.

26 Anon., *Acts of the Parliament of the Dominion of Canada*, Edward VII, chapter 31 (Ottawa 1910), 247.

27 H.T. Güssow, *Report of the Dominion Botanist for 1912* (Ottawa 1913), 195.

28 Ibid., 196.

29 William Lochhead, "President's address," *Annual Report of the Quebec Society for the Protection of Plants* 5 (1913), 10–11.

30 This was mentioned by J.F. Hockey in the *Twenty-first Annual Report, Canadian Plant Disease Survey* (Ottawa 1941), 48.

31 H.T. Güssow, "Powdery scab of potatoes," *Phytopathology* 3 (1913), 18–19.

32 H.T. Güssow, "Potato Inspection in Practice," in Dominion of Canada Department of Agriculture pamphlet no. 2, (Ottawa 1915), 7.

33 Ibid. See also *C.P.S. NEWS* 2 (1950), 1–6.

34 H.T. Güssow, *Report of the Division of Botany for 1915* (Ottawa 1915), 41.

35 Anon., *Report of the Secretary for Agriculture, Nova Scotia, for 1914* (Halifax 1914), 47.

36 The comment about the "'worst place for blight in Canada'" is in anon., *Potato blight epidemics throughout the world*, USA Department of Agriculture handbook 174 (Washington 1960), 147.

37 R.P. Gorham, "Powdery scab of the potato," *New Brunswick Department of Agriculture bulletin no. 3* (1914).

38 H.T. Güssow, *Report of the Division of Botany for 1914* (Ottawa 1914), 83.

39 See the news item written by H.T. Güssow in the *Agricultural Gazette of Canada* 2 (1915), 629.

40 T.H. Anstey, *One hundred harvests* (Ottawa 1986), 286.

41 G.C. Cunningham, "Report of the Dominion Field Laboratory, Fredericton, N.B.," in *Interim Report of the Division of Botany to March 31, 1921* (Ottawa 1921), 57.

42 S.F. Clarkson and Grace A. Olts, *History of potato industry legislation in New Brunswick* (Fredericton 1963), 3.

43 H.T. Güssow, *Annual Report of the Division of Botany* (Ottawa 1916), 1149.

44 P.A. Murphy, "The field laboratory for Prince Edward Island and its work," *Annual Report of the Quebec Society for the Protection of Plants* 9 (1917), 87.

45 Ibid., 92.

46 M. Cumming, *Report of the Secretary for Agriculture, Nova Scotia, for 1915* (Halifax 1916), 43.

47 Ibid. for 1922 (1923), 62.

48 G.H. Duff and Catherine G. Welch, "Sulphur as a control agent for common scab of potato," *Phytopathology* 17 (1927), 297–314.

49 W.A. DeLong, "Sulphur and soil acidity," *Scientific Agriculture* 2 (1923), 354–6.

50 J.F. Hockey, "Agricultural research in the Annapolis Valley area 1909/1961," Agriculture Canada historical series no. 2 (Ottawa 1967), 11. Also, *American Potato Journal* 5 (1928), 294.

51 Anon., *Report of the Department of Agriculture, Nova Scotia, for 1930* (Halifax 1931), 40, and 90.

52 Personal communications with Dr Güssow's daughter and I.L. Conners.

53 B.T. Dickson, "Report of the delegate to the Canadian Branch of the American Phytopathological Society," *Annual Report of the Quebec Society for the Protection of Plants* 12 (1920), 25.

54 J.H. Craigie, "Hans Theodor Güssow 1879–1961," *Transactions of the Royal Society of Canada*, series 3, 56 (1962), 191–5.

55 Anstey, *One hundred harvests, 286.*

56 Anon., *Report of the Dominion Horticulturist for 1922* (Ottawa 1922), 28.

57 Anon., *Interim Report, Division of Botany for 1921* (1921), 6.

58 H.T. Güssow, *Report of the Division of Botany for 1921* (Ottawa 1921), 6.

59 For a history of this transitional period, see Mary B. Bourdon, *Charlottetown Research Station 1909–1984*, Agriculture Canada historical series no. 19 (Ottawa 1984), 32.

60 Much of Hurst's work at this time was outlined by Güssow in his *Progress Report of the Dominion Botanist for the years 1935 to 1937 inclusive* (Ottawa 1938), 76, 77, 82, 87, and 88.

61 G.W. Ayers, "Studies on the life history of the club root organism, *Plasmodiophora brassicae*," *Canadian Journal of Research* "c" 22 (1944), 143–9. Also, *Scientific Agriculture* 19 (1939), 722–35.

62 Personal communication. Also, obituary notices in *C.P.S. NEWS* 34 (1980), 2, and *Canadian Journal of Plant Pathology* 2 (1980), 102.

63 A.D. Baker, "The potato-rot nematode, *Ditylenchus destructor* Thorne, 1945, attacking potatoes in Prince Edward Island," *Scientific Agriculture* 26 (1946), 138–8.

64 H.T. Güssow, *Report of the Dominion Botanist for 1924* (Ottawa 1925), 27.

65 For a tribute to Richardson see, G.G. Dustan and R.S. Willison, *A history of the entomology and plant pathology laboratories on the Niagara Peninsula, 1911/1960* Canada Agriculture historical series no. 4 (Ottawa 1968), 32–3.

66 J.L. Howatt and W.A. Hodgson, "Testing potatoes to late blight *Phytophthora infestans* (Mont.) de Bary," *Proceedings of the Canadian Phytopathological Society* 18 (1950), 11.

67 D.J. MacLeod, "The mosaic type of viruses affecting potatoes in Canada and Great Britain," PH D thesis, Cambridge University, England, 1944.

68 Personal conversation with Frank Dunlop. See also *Scientific Agriculture* 15 (1935), 435–6.

69 R.R. Hurst and D.J. MacLeod, "Turnip brown-heart," *Scientific Agriculture* 17 (1937), 209–14.

70 J.F. Monroe, "Some field experiments of a potato rot," *Annual Report of the Quebec Society for the Protection of Plants* 3 (1911), 41–2.

71 W. Lochhead, "A stem rot of potatoes (*Corticum vagum*, B & C var *Solani* Burt)," ibid. 2 (1910), 32–5.

72 B.T. Dickson, "The black dot disease of potato," *Phytopathology* 16 (1926), 23–40.

73 J.G. Coulson, "Some effects of gypsum upon the growth and common scab of the potato," *Proceedings of the Canadian Phytopathological Society* 1 (1930), 55.

74 B. Baribeau, "Bacterial ring rot of potatoes," *American Potato Journal* 25 (1948), 71–82.

75 P.A. Murphy, "New or little known diseases of potato which cause 'running out,'" *Phytopathology* 10 (1920), 316–17.

76 Savile's work was nominally supervised by Racicot, but the research results were produced by him alone. Personal communication.

77 D.B.O. Savile and H.N. Racicot, "Bacterial wilt and rot of potatoes," *Scientific Agriculture* 17 (1937), 518–22.

78 This is from a conversation with J.G. Coulson, and confirmed in a tape recorded interview with I.L. Conners.

79 C. Perrault, "The uneven distribution of viruses in potato tubers," *Proceedings of the Canadian Phytopathological Society* 17 (1949), 16.

80 To arrive at this total, see G.R. Johnson, "History of spud varieties," *Grower* (April 1979), 16, and his paper in the March issue, p. 22.

81 H.J. Panton, "Report of the Professor of Natural History and Geology," in *Report of the Ontario Agricultural College and Experimental Farm for 1886* (1887), 73–4.

82 F.C. Harrison, "A bacterial rot of potato caused by *Bacillus solanisaprus*," *Centralblatt fuer Bakteriologie* 17 (1906), 34–9, 120–8. See also the *American Potato Journal* 47 (1970), 337–43.

83 C.A. Zavitz, "Report of the professor of field husbandry and director of field experiments," in *Report of the Ontario Agricultural College and Experimental Farm for 1910* (1911), 226–7.

84 J.E. Howitt, "Some observations made in inspecting for leaf roll and mosaic," *Phytopathology* 10 (1920), 316. Also ibid., 317.

85 R.E. Stone and J.E. Howitt, "Plant diseases and fungi comparatively new in Ontario," *Phytopathology* 10 (1920), 317 (abstract).

86 J.E. Howitt and W.G. Evans, "Corrosive sublimate and time of treatment in relation to yield and control of Rhizoctonia," ibid 16 (1926), 755.

87 H.T. Güssow, "Report of the Dominion botanist for 1912," in *Experimental Farms Reports* (Ottawa 1912), 201–2.

88 G.H. Genereux, "Soil treatments with mercurials for the control of common scab of potatoes," M SC thesis, McGill University 1940.

89 F.L. Drayton, "The Rhizoctonia lesions on potato tubers," *Phytopathology* 5 (1915), 59–63.

90 Anon., *Who's Who in the Agricultural Institute of Canada* (Ottawa 1948), 53.

91 See memorial note, *Proceedings of the Canadian Phytopathological Society* 37 (1971), 27.

92 V.W. Jackson, "Potato Diseases in Manitoba," Manitoba extension bulletin no. 14, (1917).

93 G.R. Bisby and I.L. Conners, "Plant Diseases new to Manitoba," *Scientific Agriculture* 8 (1928), 456–8.

94 G.R. Bisby, J.F. Higham, and H. Groh, "Potato seed treatment in Manitoba," ibid. 3 (1923), 219–21.

95 G.R. Bisby, "Manitoba potato diseases and their control," Manitoba extension bulletin no. 66 (1923).

96 John Bracken, "Saskatchewan," in Dominion Department of Agriculture pamphlet no. 2 (1915), 24.

97 J.K. Finlayson, "Potato crop inspection," *Report of the Saskatchewan Department of Agriculture* (1927), 228.

98 Howard Harding, editor, "Saskatoon Research Station 1917–1985," Agriculture Canada historical series no. 20 (1986), 20.

99 Anon., *Report of the Saskatchewan Department of Agriculture* (1934), 48.

100 E.A. Howes, *Annual Report of the Alberta College of Agriculture* (1920), 54.

101 Sanford's PH D thesis, University of Minnesota, 1925, was "A study of some factors relative to the pathogenicity of *Actinomyces scabies* (Thaxter) Güssow." He was still publishing on that general theme in 1947. See *Scientific Agriculture* 27 (1947), 533–4.

102 G.B. Sanford, "Some factors affecting the pathogenicity of *Actinomyces scabies*," *Phytopathology* 16 (1926), 525–47.

103 G.B. Sanford and S.B. Clay, "Purple dwarf, an undescribed potato disease in Alberta," *Canadian Journal of Research* "C"19 (1941), 68–74. See also *Scientific Agriculture* 22 (1942), 772–4, and the memorial tribute to Sanford in *Proceedings of the Canadian Phytopathological Society* 45 (1978), 43.

104 G.B. Sanford and J.G. Grimble, "Observations on phloem necrosis in potato tubers," *Canadian Journal of Research* "C" 22 (1944), 162–70.

105 D.B.O. Savile, "Alteration of potato starch grain structure under influence of disease," *American Journal of Botany* 29 (1942), 286–7.

106 I.L. Conners and D.B.O. Savile, *Twenty-second annual report of the Canadian plant disease survey, 1942* (1943), mimeographed.

107 J.L. Eaglesham, "Bacterial ring rot," *Canadian Plant Disease Survey* 25 (1946), 63–64. See also anon., *Annual Report of the Alberta Department of Agriculture for 1943* (Edmonton 1944), 8, and ibid. for 1945 (1945), 22.

108 J.W. Marritt, "Treating, packaging and storing potato eye sets," *Scientific Agriculture* 24 (1944), 526–32.

109 T.A. Sharp, "Report of the Superintendent of the Experimental Farm for British Columbia," in *Experimental Farms Reports for 1891* (Ottawa 1891).

110 J.W. Eastham, M.H. Ruhmann, and B. Hoy, "Diseases and Pests of Cultivated Plants," British Columbia Department of Agriculture bulletin no. 68 (1916), 36.

111 Anon., *Report of the British Columbia Department of Agriculture* (1921), U76. See also J.W. Eastham, "Some recollections of plant disease in British Columbia," *C.P.S. NEWS* (March 1950), 1–6.

112 C. Tice, "Potato certification in B.C.," British Columbia Department of Agriculture circular no. 32 (1921). See also anon., *Report of the British Columbia Department of Agriculture for 1924* (1924), K69.

113 W. Newton, "The physiology of Rhizoctonia," *Scientific Agriculture* 12 (1931), 178–82.

114 W. Newton and H.I. Edwards, "Virus studies i. The production of antisera in chickens by inoculation with potato 'x,'" *Canadian Journal of Research* "c" 14 (1936), 412–14.

115 J.W. Eastham, "The storage rots of potatoes," in *Report of the Dominion botanist* (1913), 480–3. See also ibid. (1913), 485–7.

116 Anon., *Report of the British Columbia Department of Agriculture* (1926), N36, and (1928), Q34.

117 W. Jones and H.S. MacLeod, "Armillaria dry rot of potato tubers in British Columbia," *American Potato Journal* 14 (1937), 215–17. Jones's paper on pink rot is in *Scientific Agriculture* 25 (1945), 557–60.

118 H.R. McLarty, "Witches' broom of potatoes," *Scientific Agriculture* 6 (1926), 394–5.

119 G.E. Woolliams, "An experiment to develop disease-free strains of potatoes at Summerland, B.C.," in *Report of the Dominion Botanist for 1927* (1929), 224–5.

120 W.R. Foster and H.S. MacLeod, "A new stem-end rot of potato," *Canadian Journal of Research* 7 (1932), 520–3.

121 H.T. Güssow and W.R. Foster, "A new species of Phomopsis," ibid. 6 (1932), 253–4.

122 N.S. Wright, "A stemphylium leaf spot of potatoes in British Columbia," *Scientific Agriculture* 27 (1947), 130–5. Wright's paper on witches' broom was abstracted in *Proceedings of the Canadian Phytopathological Society* 17 (1949), 11.

123 H.Y. Hind, *Narrative of the Canadian Red River exploring expedition of 1857, and of the Assiniboine and Saskatchewan exploring expedition of 1858*, vol. 1 (London 1860), 149.

124 V.O. Buyniak, "Doukhobor immigration: The potato dilemma," *Saskatchewan History* 38 (1985), 72–6.

125 Anon., *Report of the Special Committee to whom was referred the Report of the Lower Canada Agricultural Society* (Quebec 1852), 40.

126 H.T. Güssow, *Report of the Division of Botany* (Ottawa 1914), 831–49. See also *Annual Report of the Quebec Society for the Protection of Plants* 6 (1914), 43–9.

127 W.P. Fraser, "Storage rots of potatoes and other vegetables," *Annual Report of the Quebec Society for the Protection of Plants* 6 (1914), 50–1.

128 E. Smith, "Farm storage of fruits and vegetables," British Columbia Department of Agriculture bulletin no. 58 (1914).

129 H.N. Racicot, "A spotting and shrinking disease of potatoes in storage," *Annual Report of the Quebec Society for the Protection of Plants* 18 (1926), 55–6.

130 L.T. Richardson and W.R. Phillips, "Low temperature breakdown of potatoes in storage," *Scientific Agriculture* 29 (1949), 149–66.

131 Anon., *Report of the Ontario Agricultural College for 1944* (1945), 76–7.

132 E.H. Garrard, "A storage rot of potatoes caused by a fluorescent organism resembling *Pseudomonas fluorescens* (Flugg) *Migula*," *Canadian Journal of Research* "C" 23 (1945), 79–84. See also anon., "Bacterial diseases of plants," Ontario Department of Agriculture bulletin no. 278 (Toronto 1951).

133 D.F. Forward, "The response of potato tubers to a period of anaerobiosis," *Canadian Journal of Botany* 31 (1953), 33–62.

134 G.W. Ayers, "Fusarium storage rot of potatoes," *Proceedings of the Canadian Phytopathological Society* 17 (1949), 11.

135 L.T. Richardson and W.R. Phillips, "Low temperature breakdown of potatoes in storage," ibid. 15 (1947), 18.

CHAPTER FOUR

1 S.N. Kramer, "Sumerian 'Farmers Almanac,'" *Scientific American* 185 (1951), 54–5.

2 R.P. Gorham, "Birth of agriculture in Canada," *Canadian Geographical Journal* 3 (1932), 3–17.

3 Joyce Marshall, *Word from New France: The selected letters of Marie de l'Incarnation* (Oxford 1967).

4 D.C. Harvey, *The French regime in Prince Edward Island* (New Haven 1926), 107.

5 Ibid., 172.

6 Jared Eliot, *Essays on field husbandry in New England, and other papers, 1748–1762* (New York 1934), 236.

7 H.M. Woolman and H.B. Humphrey, *Summary of literature on bunt, or stinking smut, of wheat*, United States Department of Agriculture bulletin no. 1210 (Washington 1924).

8 Samuel Deane, *The New England Farmer; or, Georgical Dictionary* (Worcester, Mass. 1791), 173.

9 Vernon C. Fowke, *Canadian Agricultural Policy* (Toronto 1946), 304.

10 R.L. Jones, "French-Canadian agriculture in the St. Lawrence Valley, 1815–1850," *Agricultural History* 16 (1942), 137–48.

11 Anon., *The Acts of the General Assembly of His Majesty's Province of New Brunswick* (Fredericton 1824), 604.

12 William Evans, *A Treatise on the Theory and Practice of Agriculture* (Montreal 1835), 58.

13 Richard Remnant, *A Discourse or Historie of Bees – Whereunto is added the causes, and cure of blasted wheat.* (London 1637), 47.

14 John Young, *The Letters of Agricola* (Halifax 1822), 319.

15 H.Y. Hind, *Two Lectures on Agricultural Chemistry* (Toronto 1850), 68.

16 H.G.L. Strange, *A Short History of Prairie Agriculture* (Winnipeg 1954), 60.

17 *Nor'-West Farmer and Miller* 11 (1892), 69.

18 Ibid. 10 (1891), 129.

19 Frank Shutt, "Report of the Chemist," in *Experimental Farms Reports for 1906* (Ottawa 1907), 183.

20 Ibid.

21 Anon., *Statutes of the Province of Ontario* (Toronto 1891), chapter 48.

22 Leonard S. Klinck, "The susceptibility of certain cereals to smut," *Annual Report of the Quebec Society for the Protection of Plants* 2 (1910), 14.

23 E. Campagna, "New white smut of wheat collected at Ste-Anne-de-la-Pocatière," ibid. 18 (1926), 71–2.

24 E. Campagna and René-O Lachance, "Annotated list of the Ustilaginales observed in Quebec," ibid. 27 (1935), 50–2.

25 W.P. Fraser, "A smut of western rye grass," *Phytopathology* 10 (1920), 316.

26 W.P. Fraser and P.M. Simmonds, "Cooperative experiments with copper-carbonate dust and other substances for smut control in 1923," *Scientific Agriculture* 4 (1924), 257–63.

27 I.L. Conners, "Organic mercury compounds for the control of loose smuts of wheat and barley and barley stripe," *Phytopathology* 16 (1926), 63–4.

28 H.M. Tory, *Report of the President and Financial Statement, N.R.C., 1928–29* (Ottawa 1930), 31.

29 W.F. Hanna and W. Popp, "Relationship of the oat smuts," *Nature* 126 (1930), 843–4.

30 W.F. Hanna and W. Popp, "Experiments in the control of cereal smuts by seed treatment," *Scientific Agriculture* 15 (1935), 424–34. See also ibid. 745–52.

31 W.F. Hanna, "The association of bunt with loose smut and ergot," *Phytopathology* 28 (1938), 142–6.

32 W.F. Hanna, "Effect of vernalization on the incidence of loose smut in wheat," *Scientific Agriculture* 16 (1936), 404–7.

33 O.S. Aamodt, "Varietal trials, physiologic specialization, and breeding spring wheats for resistance to *Tilletia tritici* and *T. levis*," *Canadian Journal of Research* 5 (1931), 501–28.

34 A.W. Platt, "The effect of freezing temperatures and of defoliation on the subsequent growth of wheat plants," *Scientific Agriculture* 17 (1937), 420–30. See also ibid., 616–25.

35 A.W. Platt and J.C. Darroch, "The seedling resistance of wheat varieties to artificial drought in relation to yield," *Scientific Agriculture* 22 (1941), 521–7.

36 *Report of the Division of Botany for the years 1935–37* (Ottawa 1938), 23.

37 Anon., "Rust and Mildew of Wheat," *Gleaner*, 4 October 1842.

38 Anon., "Rust in Wheat," *Agricultural Journal and Transactions of the Lower Canada Agricultural Society* 2 (1849), 236.

39 See *Journal and Transactions of the Board of Agriculture, Upper Canada* 1 (1855), 39

40 William Evans, "The Devastating Growth of Puccinia," *Agricultural Journal and Transactions of the Lower Canada Agricultural Society* 2 (1849), 40.

41 James H. Peters, *Hints to the Farmers of Prince Edward Island* (Charlottetown 1851).

42 P.H. Gosse, *The Canadian Naturalist* (London 1840), 292.

43 James F.W. Johnson, *Report on the Agricultural Capabilities of the Province of New Brunswick* (Fredericton 1850), 163.

44 Ibid.

45 A.F. Scott, "Prize Essay," *Journal and Transactions of the Board of Agriculture, Upper Canada* 1 (1856), 359.

46 H.Y. Hind, *Essay on the Insects and Diseases Injurious to the Wheat Crop* (Toronto 1857), chapter 5.

47 Ibid., 59.

48 Ibid., 67.

49 That information is in the Preface of the above-mentioned book by Hind.

50 J.W. Dawson, *Contributions toward the Improvement of Agriculture in Nova Scotia* (Halifax 1856), 28.

51 Ibid., 38.

52 William Marshall, *Planting and Ornamental Gardening* (London 1785), 79.

53 Panton's comment is in the *Report of the Ontario Agricultural College and Experimental Farm for 1886* (1886), 76.

54 J. Hoyes Panton, "Rust (Puccinia graminis)," Ontario Agricultural College bulletin no. 36 (Guelph 1888).

55 Evidence of Mr James Fletcher, entomologist and botanist before the Select Standing Committee on Agriculture and Colonization, Ottawa 1887.

56 Saunders's letter is in the National Archives as RG 17, vol. 2100, 10/3/99–21/4/99.

57 Anon., *Report of the Department of Agriculture, Saskatchewan, for 1919* (Regina 1919), 101.

58 Anon., *Report of the Ontario Agricultural Commission* (Toronto 1818), 339.

59 Alexander Ross, *The Red River Settlement: Its Rise, Progress and Present State, with some account of the native races and its general history to the present day* (London 1856 [1957]), 133–40.

60 A.H.R. Buller, *Essays on wheat* (New York 1919), 1.

61 Ross, *The Red River Settlement*, 23.

62 H.Y. Hind, *Narrative of the Canadian Red River Exploring Expedition of 1857, and of the Assiniboine and Saskatchewan Exploring Expedition of 1858* (London 1860), 1:227.

63 *Nor' West Farmer and Miller* 12 (1893), 214.

64 Anon., *Report of the Department of Agriculture, Saskatchewan, for 1919* (Regina 1919), 101.

65 That list was not published until 1906, when it appeared in the *Transactions of the Nova Scotia Institute of Science* 11 (1906), 122–43.

66 T.C. Haliburton, *An Historical and Statistical Account of Nova Scotia* (Halifax 1829), 405.

67 W.P. Fraser, "The rusts of Nova Scotia," *Proceedings and Transactions of the Nova Scotia Institute of Science* 12 (1910), 313–445.

68 For a story of her accomplishments, see R.H. Estey, "Margaret Newton: Distinguished Canadian scientist and first woman member of the Quebec Society for the Protection of Plants," *Phytoprotection* 68 (1987), 79–85.

69 J.H. Craigie, "William Pollock Fraser 1867–1943," *Canadian Field Naturalist* 58 (1944), 1–3.

70 T. Johnson, *Rust research in Canada and related plant-disease investigations,* Canada Department of Agriculture publication no. 1098 (Ottawa 1961), 17.

71 T.C. Vanterpool, "William Pollock Fraser, 1867–1943," *Mycologia* 36 (1944), 313–17.

72 See reports of the Dominion botanist for 1921, 1922, and 1923.

73 W.P. Thompson, *Report of the Biological Department in relation to the Agricultural College 1917–18.* This report to Dean Rutherford is in the archives of the University of Saskatchewan.

74 Margaret Newton, "Studies in wheat rust (*Puccinia graminis tritici*)," *Transactions of the Royal Society of Canada* series 3, section 5, 16 (1922), 153–210.

75 W.P. Fraser, "Culture experiments with heteroecious rusts in 1922, 1923, and 1924," *Mycologia* 17 (1925), 78.

76 W.P. Thompson, "Cytological conditions in wheat in relation to the rust problem," *Scientific Agriculture* 5 (1925), 237–39. See his review in *Transactions of the Royal Society of Canada* section 5, 43 (1949), 25–44. His obituary notice is in ibid. (1971), 109–14.

77 E.S. Archibald, *Report of the Director, Dominion Experimental Farms for the year ending March 31, 1925* (Ottawa 1925), 10.

78 Letter in National Archives, RG 17, vol. 2822.

79 Johnson, *Rust research,* 18.

80 Ibid. See also I.L. Conners, editor, *Plant pathology in Canada* (Winnipeg 1972), 128

81 Z.A. Patrick and B.H. MacNeill, "Dixon Lloyd Bailey, 1896–1984," *Canadian Journal of Plant Pathology* 6 (1984), 182.

82 C.H. Goulden and K.W. Neatby, "A study of disease resistance and other varietal characters of wheat – application of the analysis of variance, and correlation," *Scientific Agriculture* 9 (1929), 575–86.

83 C.H. Goulden, *Modern methods for testing a large number of varieties,* Canada Department of Agriculture technical bulletin no. 9 (Ottawa 1937).

84 Johnson, *Rust research,* 18–19.

85 Ibid., 32–3, and F.J. Greaney, "Field experiments on the prevention of cereal rusts by sulphur dusting (1930–1932)," *Scientific Agriculture* 14

(1934), 496. See also "In memoriam, Frank James Greaney (1897–1976)," *Proceedings of the Canadian Phytopathological Society* 43 (1976), 41.

86 Johnson, *Rust research*, 36–7.

87 W.A. Hagborg, "The longevity of three bacterial pathogens of cereals under certain storage conditions," *Proceedings of the Canadian Phytopathological Society* 13 (1945), 12. See also the obituary notice in *Canadian Journal of Plant Pathology* 4 (1982), 392.

88 J.H. Craigie, "Discovery of the function of the pycnia of the rust fungus," *Nature* 120 (1927), 765–7.

89 J.H. Craigie, "Epidemiology of stem rust in western Canada," *Scientific Agriculture* 25 (1945), 285–402.

90 R.H. Estey, "A.H.R. Buller: Pioneer leader in plant pathology," *Annual Review of Phytopathology* 24 (1986), 17–25.

91 This is from an unpublished typescript by T. Johnson, in the Library of the Rust Research Laboratory, Winnipeg.

92 K.W. Neatby, "Factor relations in wheat for resistance to *Puccinia graminis tritici, Puccinia glumarum* and *Erysiphe graminis*," *Phytopathology* 26 (1936), 360–74.

93 S.B. Helgason, "Plant science," in anon., *The University of Manitoba Faculty of Agriculture seventy-fifth anniversary 1906–1981* (Winnipeg c1982), 34.

94 J.B. Harrington, "The effect of temperature on the expression of factors governing rust reaction in a cross between two varieties of *Triticum vulgare*," *Canadian Journal of Research* 5 (1931), 200–7. See also pages 208–18, and *Scientific Agriculture* 16 (1936), 538–48. The obituary notice in *Agro News* for February 1980 credits Harrington with "agrologist."

95 Winifred Kincade, *Saskatchewan monuments* (Regina c1963), 267–73.

96 See an interesting story on the origin of Red Fife in A.H.R. Buller, *Essays on wheat* (New York 1919), 206–12

97 H.G.L. Strange, *A short history of prairie agriculture* (Winnipeg 1954), 89.

98 Johnson, *Rust research*, 41. Also, P.M. Simmonds, "A review of the investigations conducted in western Canada on root rots of cereals," *Scientific Agriculture* 26 (1939), 569.99

99 W.P. Fraser, "Take-all of wheat in western Canada," *Phytopathology* 14 (1924), 347.

100 Anon., *Who's Who in the Agricultural Institute of Canada* (Ottawa 1948), 180.

101 R.C. Russell, "Take-all – a destructive disease of wheat," Dominion of Canada Department of Agriculture pamphlet no. 58 (Ottawa 1927).

102 W.P. Fraser, P.M. Simmonds and R.C. Russell, "The take-all disease in Canada," *Phytopathology* 16 (1926), 80 (abstract).

103 F.J. Greaney and J.E. Machacek, "Production of a white fertile saltant of *Helminthosporium sativum* by means of ultra-violet radiation,"

Phytopathology 23 (1933), 379. See obituary notice, anon., "John Emil Machacek (1902–1970)," *Canadian Journal of Plant Pathology* 36 (1970), 34.

104 For an example of Wallace's work with Greaney, see *Scientific Agriculture* 24 (1943), 126–34. For work with Machacek, see ibid. 26 (1946), 59–78. His obituary notice is in the *Canadian Journal of Plant Pathology* 4 (1982), 312.

105 P.M. Simmonds, "A review of the investigations conducted in western Canada on root rots of cereals," *Scientific Agriculture* 26 (1939), 51–8.

106 R.C. Russell, "A nematode discovered on wheat in Saskatchewan," *Scientific Agriculture* 7 (1927), 385–6.

107 P.M. Simmonds, R.C. Russell, and B.J. Sallans, "A comparison of different types of root rot of wheat by means of root excavation studies," ibid. 15 (1935), 680–700.

108 H.W. Mead, "A biological method of detecting the presence of fungicides in seeds," *Scientific Agriculture* 25 (1945), 458–60. See also Howard Harding, editor, *Saskatoon Research Station 1707–1986*, Agriculture Canada historical series no. 20 (1986), 19.

109 R.J. Ledingham, "The effect of seed treatment and date of seeding on the emergence and yield of peas," *Scientific Agriculture* 26 (1946), 248–57.

110 T.C. Vanterpool, "*Asterocystis radicis* in the roots of cereals," *Phytopathology* 20 (1930), 677–80.

111 T.C. Vanterpool, "Toxin formation by species of Pythium parasitic on wheat," *Proceedings of the World Grain Show and Conference* 2 (1933), 294–8.

112 T.C. Vanterpool, "Studies on browning root rot of cereals, vi," *Canadian Journal of Research* "c" 18 (1940), 240–57.

113 G.B. Sanford, "Foot and root rots of wheat," in *Report of the Dominion Botanist for 1927* (Ottawa 1927), 114. Also, *Scientific Agriculture* 7 (1927), 292–4.

114 One example of this is G.B. Sanford and W.C. Broadfoot, "On the prevalence of pathogenic forms of *Helminthosporium sativum* and *Fusarium culmorum* in the soil of wheat fields and its relation to the root-rot problem," *Canadian Journal of Research* 10 (1934), 264–74.

115 G.B. Sanford, "Some soil microbiological aspects of plant pathology," *Scientific Agriculture* 13 (1933), 638–41.

116 W.C. Broadfoot, "Studies on foot and root rot of wheat," *Canadian Journal of Research* 8 (1933), 545–52.

117 Ibid. 10 (1934), 115–24.

118 H.T. Robertson, "Histological study of the root rots of wheat during the post seedling stage," in *Report of the Dominion Botanist for 1930* (1931), 93–4.

119 L.E. Tyner, "The effect of crop debris on the pathogenicity of cereal root-rotting fungi," *Canadian Journal of Research* "c" 18 (1940), 289–306. See also ibid. 21 (1943), 18–25.

120 For an example of Cormack's work with Sanford see ibid. 18 (1940), 562–6. For a reference to his work on forage crops that will lead to others, see *Scientific Agriculture* 26 (1946), 448–59

121 See comment on this work in G.C. Ainsworth, *Introduction to the history of plant pathology* (Cambridge 1981), 154. Also, *Phytopathology* 33 (1943), 332–3; ibid. 53 (1963), 124, and *Proceedings of the Canadian Phytopathological Society* 10 (1940), 10–11.

122 A.W. Henry and W.R. Foster, "Leptosphaeria foot rot of wheat in Alberta," *Phytopathology* 19 (1929), 689.

123 This is from a *curriculum vitae* prepared by Dr Henry.

124 Personal communication.

125 Robert Newton and his students and colleagues have three papers in the *Canadian Journal of Research* 5 (1931), 327, 333, and 337.

126 G.R. Bisby, N. James, and M.I. Timonin, "Fungi isolated from Manitoba soil by the plate method," *Canadian Journal of Research* 8 (1933), 253–7.

127 A.G. Lochhead, "Rhizosphere microorganisms in relation to root-disease fungi," in *Plant Pathology Problems and Progress 1908–1958*, C.S. Holton, G.W. Fisher, R.W. Fulton, Helen Hart, and S.E.A. McCallan, editors (Madison, Wisconsin 1959), 327–38.

128 Anon., *Royal Gazette* (Charlottetown 1840), 24 September.

129 James Fletcher, "Red leaf or blight of Oats," in *Central Experimental Farm Report* (Ottawa 1890), 178.

130 T.H. Anstey, *One hundred harvests* (Ottawa 1986), 217–18.

131 B.T. Dickson, R. Summerby, and J.G. Coulson, "Control of oat smut," *Phytopathology* 13 (1923), 292.

132 W.L. Gordon, "Studies concerning injury to seed coats after smut disinfection," *Report of the Quebec Society for the Protection of Plants* 16 (1924), 79–94.

133 J.G. Coulson and E.A. Lods, "Oat smut infection in relation to size of grain," *Phytopathology* 15 (1925), 302–3.

134 B.T. Dickson, "Two Epimycetous fungi," *Transactions of the Royal Society of Canada* section 5, 20 (1926), 113–15.

135 William Popp, "Crown rust of oats in Eastern Canada," *Report of the Quebec Society for the Protection of Plants* 18 (1926), 38–54.

136 William Popp and W.F. Hanna, "Studies on the physiology of oat smuts," *Scientific Agriculture* 15 (1935), 424–34.

137 A. Kelsall, "Experiments on the dust method of smut control," *Scientific Agriculture* 3 (1923), 303–5.

138 J.E. Howitt, "Experiments in sprinkling seed oats with formalin for the prevention of smut," *Ontario Agricultural College and Experimental Farm Report* (1917), 29

139 J.E. Howitt and R.E. Stone, "Experiments in oat smut control in 1923," *Phytopathology* 14 (1924), 346.

140 J.D. MacLachlan, "Manganese deficiency in soils and crops. ii The use of various compounds to control manganese deficiency in oats," *Scientific Agriculture* 24 (1943), 86–94.

141 S.A. Bedford, "Report of the Experimental Farm for Manitoba," in *Experimental Farms Reports for 1885* (Ottawa 1886).

142 D.L. Bailey, "Physiological specialization in *Puccinia graminis avenae* Erikss. and Henn.", PH D thesis, University of Minnesota 1924.

143 D.L. Bailey, "Report of the Dominion Field Laboratory of Plant Pathology, Winnipeg," in *Report of the Dominion Botanist for 1923* (Ottawa 1924), 25.

144 W.L. Gordon and J.N. Welsh, "Oat stem rust investigations in Canada," *Scientific Agriculture* 13 (1932), 228–35.

145 J.N. Welsh, "The effect of smut on rust development and plant vigour in oats," *Scientific Agriculture* 13 (1932), 154–64. For an example of work with Gordon, see ibid., 228–35.

146 H.M. Tory, *Report of the President, National Research Council, 1925–26* (Ottawa 1927), 25.

147 B. Peturson, "Physiologic specialization in *Puccinia coronata avenae,*" *Scientific Agriculture* 15 (1935), 806–10.

148 Mabel Ruttle, "A cytological study of *Puccinia coronata* Cda., on Banner and Cowra oats," unpublished, undated manuscript, presumably of a master's thesis, at the University of Saskatchewan.

149 W.P. Fraser and G.A. Ledingham, "Studies on the crown rust *Puccinia coronata,*" *Scientific Agriculture* 13 (1933), 313–23.

150 T.C. Vanterpool, "*Asterocystis radicis* in the roots of cereals," *Phytopathology* 20 (1930), 677–80.

151 P.M. Simmonds, "A seedling blight disease of oats caused by *Fusarium culmorum,*" *Phytopathology* 18 (1928), 480. He also wrote the text for Dominion Department of Agriculture bulletin no. 105 (1928).

152 G.B. Sanford, "A preliminary note on an unreported root rot of oats," *Scientific Agriculture* 14 (1933), 50. See also ibid. 15 (1935), 370–6.

153 C.L. Huskins, "Blindness or blast in oats," *Scientific Agriculture* 12 (1931), 191–9.

154 R.A. Derick and J.L. Forsyth, "A study of the causes of 'blast' in oats," *Scientific Agriculture* 15 (1935), 814–24. See also ibid. 20 (1939), 157–65.

155 R.A. Derick and R.M. Love, "Artificially induced fatuoids in a dwarf mutant oat," *Scientific Agriculture* 17 (1937), 703–6.

156 T. Johnson and A.M. Brown, "Studies on oat blast," *Scientific Agriculture* 20 (1940), 532–50.

157 D.F. Putnam and L.J. Chapman, "Oat seedling diseases in Ontario. 1 The oat nematode *Heterodera schachtii* Schm," *Scientific Agriculture* 15 (1935), 633–51.

158 Johnson, *Rust research,* 48.

159 James Fletcher, "Blight on oats and barley," *Central Experimental Farm Report* (Ottawa 1894), 186.

160 J.A. Clark, "Report of the Experimental Station, Charlottetown, P.E.I.," in *Experimental Stations Reports for 1928* (Ottawa 1929), 35.

161 J.E. Howitt and R.E. Stone, "Results of experiments on the control of barley stripe," *Phytopathology* 18 (1928), 477.

162 E.C. Beck, "The application of serological methods to the differentiation of closely related smut fungi," *Canadian Journal of Research* "C" 16 (1938), 391–404.

163 Sharon Ramsay, *Brandon Research Station 1886–1986,* Agriculture Canada historical series no. 31 (Ottawa 1986), 32–3.

164 O.S. Aamodt and W. Johnston, "Reaction of barley varieties to infection with covered smut (*Ustilago hordei* Pers. K. & S.)," *Canadian Journal of Research* 12 (1935), 590–613.

165 A.M. Brown and M. Newton, "The dwarf leaf rust of barley in Western Canada (*Puccinia anomala* Rostr.)," *Phytopathology* 18 (1928), 481.

166 Margaret Newton and W.J. Cherewick, "Erysiphe Graminis in Canada," *Canadian Journal of Research* "C" 25 (1947), 73–93.

167 P.M. Simminds, "Detection of loose smut fungi in embryos of barley and wheat," *Scientific Agriculture* 26 (1946), 51–8.

168 R.C. Russell, "The whole embryo method of testing barley for loose smut as a routine test," ibid. 30 (1950), 361–6.

169 R.R. Hurst, "Report of the Dominion Field Laboratory of Plant Pathology, Charlottetown, P.E.I," in *Report of the Dominion Botanist for 1925* (1925), 21.

170 A.W. Henry, "Reaction of Linum species of various chromosome numbers to rust and powdery mildew," *Phytopathology* 18 (1928), 478.

171 A.W. Henry, "Inheritance of immunity from flax rust," *Phytopathology* 20 (1930), 707–21.

172 Johnson, *Rust research,* 63.

173 Ibid.

174 W.C. Broadfoot, "Studies on the parasitism of *Fusarium lini* Bolley," ibid. 16 (1926), 951–78.

175 T.C. Vanterpool, "Flax diseases in Saskatchewan," *Annual Report of the Saskatchewan Department of Agriculture* (1947), 92.

176 J.E. Machacek and F.J. Greaney, "Summary of co-operative experiments on treatment of flax seed," *Phytopathology* 38 (1948), 803–7.

177 W.E. Sackston, "Pasmo in western Canada," *Flax Facts* 12 (1948), 1–2. Sackston had a note on flax diseases in the *Canadian Plant Disease Survey* each year for several years, beginning in 1945.

178 For examples of Bisby's work on sunflower diseases, plus references to the work of others, see *Scientific Agriculture* 2 (1920), 58–61, and ibid. 4 (1924), 381.

179 J.A. Godbout, "Agriculture in Quebec, past – present – future," *Canadian Geographical Journal* (April 1944), 157–81.

180 R.–O Lachance and A. Payette, "Quelques observations sur l'anthracnose du lin," *Annual Report of the Quebec Society for the Protection of Plants* 28 (1943), 53–5. See also ibid. 29 (1944), 101–4, and 105–9.

181 John Young, *The Letters of Agricola* (Halifax 1822 [1922]), 40.

182 Robert Russell, *North America, Its Agriculture and Climate* (Edinburgh 1857), 42.

183 Catharine Parr Traill, *The Backwoods of Canada* (Toronto 1883), 194.

184 W.F. Hanna, "Studies in the physiology and cytology of *Ustilago zeae* and *Sorosporium reilianum*," *Phytopathology* 19 (1929), 415–43.

185 G.P. McRositie, "The thermal death point of corn from low temperature," *Scientific Agriculture* 19 (1939), 699.

186 That advertisement, signed by Peter McGowan, secretary, Central Agricultural Society, appeared in the *Royal Gazette*, Charlottetown, 5 July 1836.

187 J.W. Dawson, *Contributions toward the Improvement of Agriculture in Nova Scotia*, second edition (Halifax 1856), 71.

188 This comment was made by R.L. Keating in a personal letter to superintendent of fairs and institutes, Regina, Saskatchewan, 30 September 1909.

189 J.C. Ready, *Clover dodder*, British Columbia Department of Agriculture circular no. 5 (Victoria 1913), 2.

190 E.G. Anderson, "The life history of *Cuscuta campestris* Yuncker," *Annual Report of the Quebec Society for the Protection of Plants* 29 (1944), 40–3.

191 R. Newton and W.R. Brown, "Is the apparent winter-killing of sweet clover and red clover a result of disease injury?" *Scientific Agriculture* 5 (1924), 93–6.

192 Bisby's first paper on forage crop diseases is in the *Western Canada Society of Agronomy Proceedings* 5 (1925), 62. He and Conners published their paper "Plant diseases new to Manitoba" in *Scientific Agriculture* 8 (1928), 456–8.

193 R.R. Hurst, "The resistance of varieties and strains of clovers to mildew," in *Report of Dominion Botanist for 1930* (Ottawa 1931), 183–4.

194 W.J. Cherewick, "Rhizoctonia root rot of sweet clover," *Phytopathology* 31 (1941), 673–4.

195 G.B. Sanford, "A root rot of sweet clover and related crops caused by *Plenodomus meliloti* Dearness and Sanford," *Canadian Journal of Research*, 8 (1933), 337.

196 W.C. Broadfoot, "Experiments in the chemical control of snow-mould of turf in Alberta," *Scientific Agriculture* 16 (1936), 615–18.

197 W.C. Broadfoot and M.W. Cormack, "A low-temperature basidiomycete causing early spring killing of grasses and legumes in Alberta," *Phytopathology* 31 (1941), 1058.

198 For examples of Cormack's papers published before 1948, see *Canadian Journal of Research* "c" 15 (1937), 403–24 and 493–510; *Scientific Agriculture* 22 (1942), 775–86; ibid. 26 (1946), 448–59; and *Phytopathology* 35 (1945), 838–55. Also, *AIC Review* 12 (1957), 44.

199 H.R. McLarty, J.C. Wilcox, and C.G. Woodbridge, "A yellowing of alfalfa due to boron deficiency," *Scientific Agriculture* 17 (1937), 515–17.

200 From a taped interview with Dr McLarty.

201 See *Ontario Agricultural College Report* (1945), 64.

202 R.A. Boothroyd, "Studies concerning mosaic diseases of plants," master's thesis, McGill University 1932.

203 J.N. Bird, "Red clover improvement in Quebec," *Agriculture* 3 (1946), 2–8.

204 W.P. Fraser, "The rusts of Nova Scotia," *Proceedings and Transactions of the Nova Scotia Institute of Science* 12 (1906–10), 356–60.

205 W.P. Fraser, "A smut of western rye grass," *Phytopathology* 10 (1920), 316.

206 T.C. Vanterpool, "Pythium root rot of grasses," *Scientific Agriculture* 22 (1942), 674–87.

207 J.T. Slykhuis, "Studies of *Fusarium culmorum* blight of crested wheat and brome grass seedlings," *Canadian Journal Research* "c" 25 (1947), 155–80.

208 See *Report of the Division of Botany, Canada* Department of Agriculture (Ottawa 1929), 165–86, also 125–6.

209 J.N. Bird. "Influence of rust injury on the vigour and yield of timothy," *Scientific Agriculture* 14 (1934), 550–9.

210 W. Jones, "Downy mildew of orchard grass," *Proceedings of the Canadian Phytopathological Society* 17 (1949), 13.

211 J.D. Gilpatrick and A.W. Henry, "The effect of nutritional factors on the development of Ophiobolus graminis Sacc." *Proceedings of the Canadian Phytopathological Society* 17 (1949), 14.

212 L.E. Lopatecki, "The nutrition of Phytophthora species," ibid., 13–14.

CHAPTER FIVE

1 K.W. Woodward, "Forestry in Nova Scotia," *Forestry Quarterly* 4 (1906), 10–13.

2 G. Wynn, *Timber Colony* (Toronto 1981).

3 J. Macoun, "The Forests of Canada and their distribution, with notes on the more interesting species," *Transactions of the Royal Society of Canada,* section 4, 12 (1894), 3. For another commentary on the loss of Canadian forests, see A.R.M. Lower, *The North American assault on the Canadian forest: A history of the lumber trade between Canada and the United States* (New York 1938).

4 E.J. Zavitz, "The development of forestry in Ontario," *Forestry Chronicle* 15 (1939), 38.

5 See the *Annual Report of the Fruit Growers' Association of Ontario for 1882* (1883), 165–7.

6 R. Hartig, "Important Diseases of Forest Trees" (1874), Phytopathological Classic no. 12 (1974).

7 J. Eliot, *Essays upon field husbandry in New England, and other papers 1748–1762* (New York 1934), 236.

8 G.E. Bowes, *Peace River chronicles* (Vancouver 1963), 53.

9 R. Campbell, "The fleshy fungi of 1909," *Report of the Quebec Society for the Protection of Plants* 2 (1910), 74.

10 Anon., *The organization, achievements and present work of the Dominion Experimental Farms* (Ottawa 1926), 7. See also anon., "An Act respecting Experimental Farm Stations," *Revised Statutes of Canada* (Ottawa 1887), chapter 57.

11 A.D. Rodgers, III, *Burnhard Eduard Fernow, a story of North American forestry* (Princeton 1951), 381 and 394.

12 Anon., *Fifty years of progress on the Dominion Experimental Farm 1886–1936* (Ottawa 1939), 130–2. See also *Canadian Journal of Forestry* 1 (1905), 155.

13 C. Leavitt, *Forest protection in Canada, 1912* (Toronto 1913), 131.

14 R.W. Lyons, "Nursery practice discussion," *Forestry Chronicle* 3 (1927), 27–8.

15 R. Pomerleau, "Fungi responsible for damping-off of conifer seedlings at the Berthierville forest nursery," *Report of the Quebec Society for the Protection of Plants* 26 (1934), 59–62. See also *La Forêt Québécoise* 3 (1941), 13–22.

16 L.P.V. Johnson and G.M. Linton, "Experiments on chemical control of damping-off in *Pinus resinosa* Ait," *Canadian Journal of Research* "c" 20 (1942), 559–71.

17 M. Gibson, "Education in forestry," *Forestry Chronicle* 22 (1946), 238–45. See also pp. 246–9, 253–60.

18 *Report of the Ontario Agricultural College* (Guelph 1900), 20.

19 W.P. Fraser, "Diseases of forest and shade trees," *Annual Report of the Quebec Society for the Protection of Plants* 4 (1913), 76–84.

20 An outline of the life and work of J.H. Faull has been provided by Anna F. Faull, "Joseph Horace Faull, 1870–1961," *Journal of the Arnold*

Arboretum 43 (1962), 223–33. His comment about the paucity of information about many tree diseases is in the *Journal of Forestry* 20 (1922), 67–70.

21 R.E. Stone, "Witches' broom of the Canada balsam and the alternate hosts of the causal organism," *Phytopathology* 10 (1920), 315. Stone's work on white pine is in *Phytopathology* 8 (1918), 438–40. See also pp. 27–9.

22 W.H. Rankin, "Butt rots of the balsam fir in Quebec province," ibid., 314.

23 N.L. Cutler, "A contribution to the knowledge of the tree-destroying fungi of the Vancouver forestry district," *Phytopathology* 13 (1923), 294.

24 T.S. Buchanan, *Forest disease research in the United States and British Columbia* (Washington 1976), 213.

25 H.W. Eades, *British Columbia softwoods, their decays and natural defects.* Canadian Forestry Service bulletin no. 80 (Ottawa 1932).

26 A.W. McCallum, "*Napicladium tremulae.* A new disease of the poplar," *Phytopathology* 10 (1920), 318.

27 A.W. McCallum, "Armillaria root rot," in *Report of the Division of Botany, Dominion Department of Agriculture* (Ottawa 1922), 8–9.

28 A.W. McCallum, "A study of decay in the balsam fir," *Phytopathology* 15 (1925), 302.

29 A.W. McCallum, "Studies in forest pathology. 1 Decay in balsam fir (*Abies balsamea* Mill)." Dominion Department of Agriculture bulletin no. 104 (Ottawa 1928).

30 C. Heimburger and A.W. McCallum, "Balsam fir butt rot in relation to some site factors," *Pulp and Paper Magazine, Canada* 41 (1940), 301–3.

31 I.L. Conners, editor, *Plant pathology in Canada* (Winnipeg 1972), 115. See also N.T. Gridgeman, *Biological studies of the National Research Council of Canada: The early years to 1952* (Ottawa 1979), 27.

32 J.H. Faull, "Notes on forest diseases in Nova Scotia," *Forestry Chronicle* 6 (1930), 107. See also *Blister Rust News* 14 (1930), 63.

33 R.B. Thomson and H.B. Sifton, "Resin canals in the Canadian spruce (*Picea canadensis*)," *Philosophical Transactions, Royal Society* (B) 214 (1925), 63–111.

34 J.E. Bier and Mildred Nobles, "Brown pocket rot of Sitka spruce," *Canadian Journal of Research* "C" 24 (1945), 115–20.

35 J.H. Faull, "Notes on forest diseases in Nova Scotia," *Forestry Chronicle* 6 (1930), 108.

36 J.H. Faull, "Some problems of forest pathology in Ontario. Needle blight of white pine," *Journal of Forestry* 20 (1922), 67–70. See also René Pomerleau, "Joseph Horace Faull 1870–1961," *Transactions of the Royal Society of Canada* 1 (1965), 83–7.

37 J.H. Faull, "Taxonomy and geographical distribution of the genus Uredinopsis," *Contributions of the Arnold Arboretum* 11 (1938), 1–120.

38 Anon., "In memoriam, David Richmond Sands (1883–1966)," *Proceedings of the Canadian Phytopathological Society* 33 (1966), 3.

39 Anon., *Report of the Minister of Agriculture, Ontario, for 1914* (Toronto 1915), 18 and 22.

40 Ibid. for 1915 (1916), 18 and 20.

41 H.T. Güssow, *Report of the Dominion Botanist for 1910* (19??), 278–9.

42 W.A. McCubbin, "The white pine blister rust in Canada," *Report of the Fruit Growers' Association of Ontario for 1916* (Toronto 1917), 81.

43 W. Lochhead, "Three pests threatening Quebec," *Annual Report of the Quebec Society for the Protection of Plants* 4 (1912), 33.

44 J.H. Faull and G.H. Graham, "Bark disease of chestnut in British Columbia," *Forestry Quarterly* 12 (1914), 201.

45 McCubbin, "The white pine blister rust," 81.

46 J.H. Grisdale, "Report of the Acting Dominion Botanist," in *Report of the Dominion Experimental Farms* (Ottawa 1918), 38.

47 Ibid., 39.

48 E.S. Archibald, "Report of the Acting Dominion Botanist," in *Report of the Dominion Experimental Farms* (Ottawa 1920), 58.

49 R. Pomerleau, "Present status of the white pine blister rust in the province of Quebec," *Annual Report of the Quebec Society for the Protection of Plants* 23 (1932), 176–98.

50 Henri Roy, "The white pine blister rust situation in Quebec," ibid. 10 (1918), 25–30. See also p. 49.

51 R.H. Campbell, *Report of the Director of Forestry for the year 1918* (Ottawa 1919), 11.

52 H.T. Güssow, "Report of the Dominion Botanist," in *Report of the Dominion Experimental Farms for 1922* (Ottawa 1923), 54.

53 T.S. Buchanan, *Forest disease research in the western United States and British Columbia, Canada* (Washington 1976), 163.

54 J.W. Eastham, "Some recollections of plant disease in British Columbia," *C.P.S. NEWS* 2 (March 1950) 3.

55 J.W. Eastham, "Report of provincial plant pathologist, Vancouver," in *Report of the British Columbia Department of Agriculture for 1921* (1922), U66.

56 H.T. Güssow, "Report of the Dominion Botanist," in *Report of the Dominion Experimental Farms for 1922* (Ottawa 1923), 7.

57 J.W. Eastham, "Some recollections of plant disease in British Columbia," *C.P.S. NEWS* 2 (March 1950), 4.

58 Buchanan, *Forest disease research*, 168.

59 J.W. Eastham, "White pine blister rust in B.C.," *Agricultural Journal* 7 (1922–23), 57.

60 A.W. McCallum, "The present status of white pine blister rust in Canada," *Phytopathology* 13 (1923), 291.

61 W.A. McCubbin, "The white pine blister rust in Canada," *Annual Report of the Quebec Society for the Protection of Plants* 8 (1916), 64–72

62 W.A. McCubbin, "Dispersal distance of urediniospores of *Cronartium ribicola* as indicated by their rate of fall in still air," *Phytopathology* 8 (1918), 35–6. His work was also outlined in the *Experimental Farms Reports for 1917* (Ottawa 1917), 1098–1105.

63 J.F. Hockey, "Currant rust control," *Scientific Agriculture* 9 (1929), 455–7.

64 A.W.S. Hunter and M.B. Davis, "Breeding rust resistant black currants," *Proceedings of the American Society of Horticultural Science* 42 (1943), 467.

65 J.L. Farrar, "Some historical notes on forest tree breeding in Canada," *Forestry Chronicle* 45 (1969), 393. See also *Report of the Director of Forestry for 1940* (Ottawa 1941), 117.

66 G.H. Duff, "Some factors affecting the viability of urediniospores of *Cronartium ribicola,*" *Phytopathology* 8 (1918), 288–9.

67 I.L. Conners, editor, *Plant pathology in Canada* (Winnipeg 1972), 100. Also personal communications.

68 A.W. McCallum, "White pine blister rust," in *Report of the Division of Botany, Dominion Department of Agriculture* (Ottawa 1921), 13–15.

69 Ibid., 14.

70 C.F. Coons, *The John R. Booth story*, Ontario Ministry of Natural Resources (Toronto 1978).

71 A.W. McCallum, "Woodgate rust in Canada," *Phytopathology* 19 (1929), 414.

72 E.H. Finlayson, *Report of the Director of Forestry* (Ottawa 1940), 117.

73 E. Silver Dowding, "*Wallrothiella arceuthobii,* a parasite of the jack pine mistletoe," *Canadian Journal of Research* 5 (1931), 219–30.

74 J.H. Faull, "Needle blight of white pine," *Phytopathology* 10 (1920), 315. Also, *Report of the Division of Botany* (Ottawa 1922), 9–10.

75 W.R. Haddow and M.A. Adamson, "Note on the occurrence of needle blight and late fall browning in red pine (*Pinus resinosa* Ait.)," *Forestry Chronicle* 25 (1939), 107–10. See also *Transactions of the Royal Canadian Institute* 22 (1938), 21–81, and ibid. 23 (1941), 161.

76 J.R. Dickson, "White pine blister rust control in Ontario as post-war employment," *Forestry Chronicle* 19 (1943), 40–3.

77 That discovery is mentioned by G.H. Lachmund, "*Cronartium comptoniae* Arth. in western North America," *Phytopathology* 19 (1929), 453–66.

78 G.H. Lachmund, "Damage to *Pinus monticola* by *Cronartium ribicola* at Garibaldi, British Columbia," *Journal of Agricultural Research* 49 (1934), 239–49.

79 C.G. Hewitt, "Note on the occurrence of the fetted beech coccus (*Cryptococcus fagi* Baerens, Dougl.) in Nova Scotia," *Canadian Entomologist* 46 (1914), 15–16.

80 J. Ehrlich, "The beech bark disease. A Nectria disease of *Fagus* following *Cryptococcus fagi* (Baer)," *Canadian Journal of Research* 10 (1934), 593–692. See also *Phytopathology* 23 (1933), 10.

81 C.B. Hutchings, "The shade tree insects of eastern Canada for the year 1925, with remarks on their activities and prevalence," *Annual Report of the Quebec Society for the Protection of Plants* 18 (1926), 113.

82 G.W.I. Creighton, "Forestry in Nova Scotia," *Journal of Forestry* 35 (1937), 671.

83 D.P. Penhallow, "Birch rope; an account of a remarkable tumor growing upon the white birch," *Proceedings and Transactions of the Royal Society of Canada*, section 4 (1906), 239–55.

84 Anon., *Report of forest insect investigations, Dominion Entomological Laboratory* (Fredericton 1940), 273, and 305. See also R.E. Balch and J.S. Prebble, "The bronze birch borer and its relation to the dying of birch in New Brunswick forests," *Forestry Chronicle* 16 (1940), 179–201.

85 R.F. Morris, "Tree injection experiments on the study of birch dieback," *Forestry Chronicle* 27 (1951), 313–29.

86 R. Pomerleau, "History of hardwood species dying in Quebec," in *Report of symposium on birch dieback*, Canada Department of Agriculture (Ottawa 1953), 10–11. See also *Proceedings of the Royal Society of Canada* section 26, 43 (1949).

87 J. Fletcher, "The poplar rust (*Melampsora populina*, Jacq. Lev.)," in *Report of the Central Experimental Farm* (Ottawa 1901), 259.

88 A.W. McCallum, "*Napicladium tremulae*, a new disease of the poplar," *Phytopathology* 10 (1920), 318.

89 E.H. Moss, "Observations on two poplar cankers in Ontario," ibid. 12 (1922), 425–7.

90 Thompson's 1930 master's thesis was titled "Leaf spot of *Populus balsamifera.*"

91 G.E. Thompson, "A leaf blight of *Populus tacamahaca* Mill. caused by an undescribed species of *Linospora,*" *Canadian Journal of Research* "C" 17 (1939), 232–8.

92 Bier's PH D thesis, in 1938, was titled "Hypoxylon canker of poplar and preliminary investigations on hypoxylon cankers of maple and oak." It was published as Dominion Department of Agriculture publication 691, technical bulletin no. 27, in 1940.

93 J.E. Bier, "Septoria canker of introduced and native hybrid poplars," *Canadian Journal of Research* "C" 17 (1939), 195–204. His paper on canker of maple is in *Forestry Chronicle* 15 (1939), 122–3.

94 C.G. Riley and J.E. Bier, "Extent of decay in poplar as indicated by the presence of sporophores of the fungus *Fomes igniarius* Linn," *Forestry Chronicle* 12 (1936), 249–53.

95 W.A. Munro, *Report of the Superintendent, Dominion Experimental Station, Rosthern, Saskatchewan, for the year 1928* (Ottawa 1929), 25.

96 The work of Van Camp is mentioned by E.H. Finlayson, *Report of the Director of Forestry for the year 1929–30* (Ottawa 1930), 67. It is the only disease research noted by J.H. White, "Forest research in Canada during 1929," *Forestry Chronicle* 6 (1930), 104.

97 G.B. Sanford, "The Cytospora disease of Russian poplar," in *Report of the Dominion Botanist for 1930* (Ottawa 1931), 36–7.

98 J.R. Dickson, *The Riding Mountain Forest Reserve*, Forestry Branch bulletin no. 6 (Ottawa 1909), 24.

99 C. Heimburger, "Report on poplar hybridization, 1936," *Forestry Chronicle* 12 (1936), 285–90. See also J.L. Farrar, "Some historical notes on forest tree breeding in Canada," ibid. 45 (1969), 392–4, and 413.

100 R. Pomerleau, "Studies on the ink-spot disease of poplar," *Canadian Journal of Research* "c" 18 (1940), 199–214. For Pomerleau's early notes on tree diseases in Quebec, see *Le Naturaliste Canadien* 58 (1931), 73–82; ibid. 59 (1932), 49–58; ibid. 61 (1934), 305–8; ibid. 64 (1937), 261–89; and ibid. 65 (1938), 24–31, 57–70, 167–88.

101 H. Boulton, editor, *A century of wood preserving* (London 1930), 150 pp.

102 F.W. Cumberland, "Preservation of Timber," *Canadian Journal of Industry, Science and Art* n.s. 1 (1856), 559–61.

103 J. Robinson, "On preserving timber from decay," *Canadian Journal of Industry, Science and Art* n.s. 2 (1857), 8–11

104 A.P. Reid, "The economy of timber and preservation of structures from fire and decay," *Proceedings and Transactions of the Nova Scotia Institute of Natural Science* 3 (1874), 326–33.

105 W.S. Finch, *The Common Sense Wood Preservative* (Toronto 1889), not paginated.

106 D.P. Penhallow, "The mycelium of dry rot," *Canadian Record of Science* 9 (1905), 318.

107 A.H.R. Buller, "The destruction of wood by fungi," *Science Progress* 11 (1909), 1–18. Most of Buller's earlier research on wood destroying fungi was done in Birmingham, England.

108 R.J. Blair, "Decay in the timber of pulp and paper mill roofs," *Phytopathology* 10 (1920), 314. See also *Paper Trade Journal* 19 (1921), 51.

109 H.W. Eades, *British Columbia softwoods, their decays and natural defects*. Canadian Forestry Service bulletin no. 80 (Ottawa 1932). An early history of the Forest Products Laboratories of Canada is included in the *Report of the Director of Forestry for 1926* (Ottawa 1927), 27, and 31–2. See also *Forestry Chronicle* 6 (1930), 7–12.

110 J.H. Faull, "*Fomes officinalis*, a timber destroying fungus," *Transactions of the Royal Canadian Institute* 11 (1916), 185–209.

111 J.H. White, "The biology of *Fomes applanatus* (Pers.) Wallr," ibid. 12 (1919), 133–74. See also J.W.B. Sisam, *Forestry and forestry education in a developing country: A Canadian dilemma* (Toronto 1982).

112 Clara W. Fritz, "Cultural criteria for the distinction of wood-destroying fungi," *Transactions of the Royal Society of Canada*, section 5, 17 (1923), 191–288.

113 E.A. Atwell, "The occurrence of *Cadophora fastigiata* in Canada," *Phytopathology* 21 (1931), 761.

114 Clara Fritz and E.A. Atwell, *Decay in red stained Jack pine ties under service conditions*, Department of Natural Resources circular no. 58 (Ottawa 1941), 19. See also circular no. 37 (1933).

115 Clara Fritz, "Rate of deterioration of pulpwood in storage," *Pulp and Paper Magazine of Canada* 34 (1936), 217–18. See also ibid. 33 (1933), 191–210.

116 Anon., "Clara W. Fritz (1889–1974)," *Proceedings of the Canadian Phytopathological Society* 41 (1974), 38. See also *Canadian Mining Journal* 63 (1942), 719–20.

117 J.H. Faull and Irene Mounce, "*Stereum sanguinolentum* as the cause of 'Sapin Rouge' or red heart of balsam," *Phytopathology* 14 (1924), 349–50.

118 Irene Mounce, "Microscopic characters of sporophores produced in culture as an aid in identifying wood-destroying fungi," *Transactions of the Royal Society of Canada*, Section 5, 26 (1932), 177–81.

119 J. Ginns, "Irene Mounce, 1894–1987," *Mycologia* 80 (1988), 607–8.

120 J.A. Roy, "Methods of prolonging the durability of fence posts," *Annual Report of the Quebec Society for the Protection of Plants* 13 (1921), 57–65.

121 C. Edwards and J. Walker, "Preservatives for farm fence posts," *Scientific Agriculture* 24 (1944), 366–72.

122 Irene Mounce, J.E. Bier, and Mildred Nobles, "A root-rot of Douglas fir caused by *Poria weirii*," *Canadian Journal of Research* "c" 18 (1940), 522–33.

123 C.G. Riley, "A preliminary investigation into the rate of deterioration of pulpwood in log jams," *Annual Report, Woodlands Section, Canadian Pulp and Paper Association* 15 (1933), 4–5.

124 C.G. Riley, "Investigation into rate of deterioration in insect-killed spruce on the Gaspé peninsula," *Pulp and Paper Magazine of Canada* 36 (1935), 130–2.

125 C.G. Riley, "Deterioration in piled pulpwood," ibid. 41 (1940), 450.

126 T.A. McElhanny, "Forest products research in Canada," *Forestry Chronicle* 6 (1930), 7–12.

127 C.A. Sankey, *Paprican: The first fifty years* (Pointe Claire, Quebec 1976), 214 pp.

128 A.R.M. Lower, *The North American Assault of the Canadian Forest: A history of the lumber trade between Canada and the United States* (New York 1968), 162.

129 L.S. Hawboldt and A.J. Skolko, "Investigation of yellow birch dieback in Nova Scotia in 1947," *Journal of Forestry* 46 (1948), 659–71.

130 R. Pomerleau, "History of the Dutch elm disease in the Province of Quebec," *Forest Chronicle* 37 (1961), 356–67.

131 J.E. Bier, "The development of forest pathology in British Columbia," *Proceedings of the Canadian Phytopathological Society* 17 (1949), 12.

CHAPTER SIX

1 N.A. Cobb, *Contributions to a science of nematology: Part 1, Antarctic free-living nematodes of the Shackleton expedition* (Baltimore 1914). See also *Journal of the American Medical Association* 98 (1932), 75.

2 The original minute books are in the library of the University of Guelph. The microscope slides prepared by Dearness are in the possession of Regional History, University of Western Ontario.

3 J. Dearness, "Annual Address of the President," *Annual Report of the Entomological Society of Ontario* 27 (1897), 24.

4 J.C. Chapais, "Anguillulae or eel-worms," *Annual Report of the Quebec Society for the Protection of Plants* 1 (1909), 37–9.

5 G.C. Hewitt, "Report of the Division of Entomology," in *Experimental Farms Reports for 1913* (Ottawa 1914), 499–518.

6 Ibid. for 1914, 864–5.

7 Personal communication from Ottawa nematologist R.V. Anderson, 30 July 1980. See also A. Johnston, *To serve agriculture: The Lethbridge Research Station 1906–1976*, Canada Department of Agriculture historical series no. 9 (1977), 32–3.

8 J.W. Noble, "Greenhouse Insects," *Annual Report of the Entomological Society of Ontario for 1917* (1918), 29.

9 W.A. Ross and L. Caesar, "Insects of the season in Ontario," ibid. for 1918 (1919), 23–7.

10 H.T. Güssow, "Common tulip diseases," *Report of the Horticultural Societies (Ontario)* 2 (1911), 115.

11 H.T. Güssow, "'Take all,' flag smut and ear 'cockle' of wheat," *Agricultural Gazette of Canada* 6 (1919), 1–4.

12 H.T. Güssow, "Eelworm on carrot, *Heterodera radicicola*," in *Report of the Division of Botany for 1922* (1923), 12.

13 J.W. Crow, "Report of the professor of pomology," in *Annual Report of the Department of Agriculture for 1912* (Ontario), vol. 1 (1913), 102.

14 W.A. McCubbin, *The diseases of tulips*, Dominion of Canada Experimental Farm bulletin no. 35 (Ottawa 1918).

15 A.A. Hildebrand, "Recent observations on strawberry root rot in the Niagara peninsula," *Canadian Journal of Research* "c" 11 (1934), 18–31.

16 A.A. Hildebrand and L.W. Koch, "A microscopic study of infection of the roots of strawberry and tobacco seedlings by micro-organisms," *Canadian Journal of Research* "c" 14 (1936), 11–26.

17 R.C. Russell, "A nematode discovered on wheat in Saskatchewan," *Scientific Agriculture* 6 (1927), 385–6.

18 G. Thorne, "*Heterodera punctata* n.sp. A nematode parasite on wheat from Saskatchewan," *Scientific Agriculture* 8 (1928), 707–11.

19 T.C. Vanterpool, "*Ditylenchus radicicola* (Greff) Filipjev, a root-gall nematode new to Canada, found on wheat and other gramineae," *Scientific Agriculture* 28 (1948), 200–5.

20 R.R. Hurst, "Report of the Dominion Field Laboratory of Plant Pathology, Charlottetown, P.E.I.," in *Report of the Dominion Botanist for 1926* (1927), 24–8.

21 A.D. Baker, "Records of plant parasitic nematodes in the Dominion of Canada," *Canadian Insect Pest Review* 23, supplement no. 1 (1945), 143.

22 I.H. Crowell, *Report of the Canadian Plant Disease Survey* 19 (1939), 66.

23 H.N. Racicot, "Nematode diseases of potatoes in Canada," *Proceedings of the Canadian Phytopathological Society* 14 (1946), 17 (abstract).

24 D.F. Putnam and L.J. Chapman, "Oat seedling diseases in Canada. 1 The oat nematode *Heterodera schachtii* Schm." *Scientific Agriculture* 15 (1935), 633–51. See also *Ontario Agricultural College Report* (1937), 63.

25 L.J. Chapman, "Oat nematode on winter wheat," *Scientific Agriculture* 18 (1938), 527–8.

26 J.E. Howitt, "Nematode (*Heterodera schachtii*)," *Canadian Plant Disease Survey* 17 (1937), 11.

27 O.J. Robb, "Control of nematodes," *Annual Report of the Vegetable Growers' Association* 21 (1926), 41–3.

28 L. Daviault, untitled note, *Canadian Insect Pest Review* 12 (1934), 108.

29 T.W.M. Cameron, "Nematodes and Plants," *Annual Report of the Quebec Society for the Protection of Plants* 26 (1934), 13–22.

30 E.H.J. Marchant, "An estimated number of nemas in the soils of Manitoba," *Canadian Journal of Research* 11 (1934), 394.

31 H.T. Güssow, *Studies on diseases of ornamental plants. The stem eelworm (Tylenchus dipsaci Kühn, Bastian) of narcissi.* Dominion Department of Agriculture pamphlet no. 114, new series (1931).

32 R.J. Hastings, J.E. Bosher, and W. Newton, "The nematode disease of narcissi and crop sequence," *Canadian Journal of Research* 8 (1933), 101.

33 W. Newton, R.J. Hastings, and J.E. Bosher, "Nematode infestation symptoms on barley as a means of determining the efficiency of chemicals as lethal agents against *Tylenchus dipsaci* Kühn," *Canadian Journal of Research* 9 (1933), 37–42. See also 31–6.

34 R.J. Hastings and W. Newton, "The effect of temperature upon the pre-adult bulb nematode *Anguillulina dipsaci* (Kühn, 1858) Gern. & v. Ben., 1859, in relation to time and moisture," *Canadian Journal of Research* 10 (1934), 793–7.

35 R.J. Hastings and W. Newton, "The influence of a number of factors upon the activation of dormant or quiescent bulb nematodes, *Anguillulina dipsaci* (Kühn 1858) Gern. & v. Ben., 1859," *Proceedings of the Helminthological Society of Washington* 1 (1934), 52–4.

36 R.J. Hastings, "Transfer of the bulb nematode – *Ditylenchus dipsaci* from *Tropaeolum polphyllum*, a new host, to potatoes," *Scientific Agriculture* 21 (1940), 115–16.

37 A.D. Baker, "Records of plant parasitic nematodes in the Dominion of Canada," *Canadian Insect Pest Review* 23 (1945), supplement 1, 153.

38 R.J. Hastings and J.E. Bosher, "A study of the pathogenicity of the meadow nematode and associated fungus *Cylindrocarpon radicicola* Wr.," *Canadian Journal of Research* "c" 16 (1938), 225–9.

39 R.J. Hastings, "The biology of the meadow nematode *Pratylenchus pratensis* (De Man) Filipjev," *Canadian Journal of Research* "d" 17 (1939), 39–44. See also ibid. 16 (1938), 225–9.

40 J.E. Bosher and W. Newton, "Host preference of the root knot nematode in British Columbia," *Plant Disease Reporter* 17 (1933), 2. See also *Scientific Agriculture* 13 (1933), 594–5.

41 W. Newton, J.E. Bosher, and R.J. Hastings, "The treatment of glasshouse soils with chloropicrin for the control of *Heterodera marioni* (Cornu) Goodey, and other soil pathogens," *Canadian Journal of Research* "c" 15 (1937), 182–6.

42 H.D. Brown, "The sugar beet nematode *Heterodera schachtii*. A new parasite in Canada," *Scientific Agriculture* 12 (1932), 544–52.

43 Anon., *Ontario Department of Agriculture Report for 1939* (1939), 118. See also ibid. (1942), 12.

44 A.D. Baker, "Records of Plant-Parasitic Nematodes in the Dominion of Canada," *Canadian Insect Pest Review* 23 (1945), supplement 1

45 Personal communication with the late Dr Baker.

46 A.D. Baker, *The sugar-beet nematode in Ontario*, Division of Entomology processed publication no. 30 (Ottawa 1945), 1–4.

47 A.D. Baker, "The potato-rot nematode, *Ditylenchus destructor* Thorne, 1945, attacking potatoes in Prince Edward Island," *Scientific Agriculture* 26 (1946), 138.

48 V.E. Henderson, "Some host relationships of the potato-rot nematode, *Ditylenchus destructor* Thorne, 1945," *Nature* 167 (1951), 952.

49 R.R. Hurst, "Potato rot nematode, *Ditylenchus destructor*, in p.e.i.," *Proceedings of the Canadian Phytopathological Society* 15 (1947), 17 (abstract).

50 H.N. Racicot, "Nematode diseases of potatoes in Canada," *Proceedings of the Canadian Phytopathological Society* 14 (1946), 17 (abstract).

51 A.D. Baker, "Some observations on the development of the nematology section at Ottawa, and on some of the needs of Nematology," *Phytoprotection* 44 (1963), 5–11.

52 A.D. Baker, "Records of plant-parasitic nematodes in the Dominion of Canada," *Canadian Insect Pest Review* 23 (1945), 140–53.

CHAPTER SEVEN

1 W.O. Raymond, editor, *Winslow Papers A.D. 1776–1826* (Saint John 1901), 7:372.

2 T.B. Atkins, editor, *Acadia and Nova Scotia* (Cottonport, Louisiana 1754), 219.

3 John Spencer, governor, *Proclamation* (Fort Deer, 8 January 1814). From the original in the Hudson's Bay Company Archives, Winnipeg, Manitoba.

4 See the essay, in this book, "Potato Diseases and the Early Development of Plant Pathology in Canada."

5 Anon., *The Provincial Statutes of Lower Canada* (Quebec 1805), 4:122.

6 A.W.H. Eaton, *The History of Kings County Nova Scotia* (Salem, Mass. 1873), 197.

7 Anon., *Statutes of the Province of Ontario* (Toronto 1879), chapter 33.

8 Ibid. (Toronto 1880), chapter 53.

9 See the *Annual Report of the Fruit Growers' Association of Ontario for 1887*, 122.

10 Anon., *Acts of the General Assembly of Prince Edward Island* (Charlottetown 1895).

11 Anon., *The Statutes of the Province of Nova Scotia, 1896* (Halifax 1896), chapter 8.

12 A.M. Smith, "A Word of Warning to Peach Growers of Ontario," *Canadian Horticulturist* 1 (1878), 15–16.

13 Anon., *Statutes of the Province of Ontario* (Toronto 1881), chapter 81. Also, ibid. (1887), chapter 202.

14 Ibid. (Toronto 1891), chapter 48, section 5.

15 Ibid. (Toronto 1893), chapter 42.

16 Ibid. (Toronto 1896), chapter 61.

17 D. Nicol, "Hedges," *Annual Report of the Ontario Fruit Growers' Association for 1892* (1893), 131.

18 E.C. Large, *The advance of the fungi* (London 1940), 123.

19 Anon., *Mr Fletcher's Evidence, Ottawa, 15th June, 1887, before the Select Committee appointed by the House of Commons to obtain information as to the Agricultural Interests of Canada* (Ottawa 1887), 48.

20 Anon., *Statutes of the Province of Ontario* (Toronto 1900), chapter 48.

21 Anon., *The Revised Statutes of Ontario* (Toronto 1927), volume 3, chapter 311.

22 T. Johnson, *Rust research in Canada and related plant-disease investigations*, Canada Department of Agriculture publication no. 1098 (Ottawa 1961), 15.

23 Anon., *Acts of the Legislature of the Province of Manitoba* (Winnipeg 1917), chapter 65.

24 This was done by an Order in Council dated 19 April 1919. See *Canada Gazette* 52 (1919), 3367.

25 J.W. Eastham, "Some Recollections of Plant Diseases in British Columbia," *C.P.S. NEWS* 2:2 (1950), 2.

26 Anon., *Statutes of the Province of British Columbia* (Victoria 1892), chapter 23.

27 Ibid. (Victoria 1894), chapter 20.

28 Ibid. (Victoria 1909), chapter 38.

29 See the *Annual Report of the Department of Agriculture for British Columbia for 1921* (1922), U3–U4.

30 Anon., *Acts of the Parliament of the Dominion of Canada* (Ottawa 1939), chapter 21.

31 T. Cunningham, "The work of the inspector of fruit pests," in *Annual Report of the British Columbia Department of Agriculture* (Victoria 1913), 25.

32 Ibid. (1915), 11 and 77.

33 See *Statutes of the Dominion of Canada 1917*, volume 1, (Ottawa 1918), lxxx–lxxxii.

34 See *Canada Gazette* 52 (1919), 3367. Also, *Experimental Farms Reports* (Ottawa 1920), 58.

35 J.W. Eastham, "Some Recollections," 4. See also the essay on forest pathology in this book.

36 See *Annual Report of the British Columbia Department of Agriculture* (Victoria 1925), 11.

37 Anon., *The Revised Statutes of British Columbia* (Victoria 1924), volume 1, chapter 8, 91.

38 See *Annual Report British Columbia Department of Agriculture for 1930* (1931), 36.

39 Ibid. (1935), Y36.

40 *Statutes of the Province of British Columbia* (Victoria 1936), volume 3, chapter 306, 4549.

41 D.A. Roberts and C.W. Boothroyd, *Fundamentals of plant pathology* (San Francisco 1972), 145.

42 T. Dano, "A Hundred years of plant quarantine in Indonesia, 1877–1977," FAO information letter no. 117 (1977), 8.

43 Anon., *Acts of the Parliament of the Dominion of Canada* (Ottawa 1910), chapter 31.

44 See *Canada Gazette* 45 (Ottawa 1912), 92.

45 Ibid. 44 (1911), 2945.

46 Anon., *Acts of the General Assembly of Newfoundland* (St John's 1911), chapter 14.

47 See *Report of the Department of Agriculture and Mines, Dominion of Newfoundland for the year 1920* (St John's 1921), 12.

48 See *Canada Gazette* 49 (Ottawa 1915), 1142.

49 W.H. Brittain, "Legislation in force in Nova Scotia to prevent the importation and spread of insects and diseases injurious to plants," *Annual Report of the Secretary for Agriculture, Nova Scotia, for 1916* (1917), 202.

50 See *Report of the Minister of Agriculture, Dominion of Canada for 1934–35* (Ottawa 1935), 5.

51 Anon., *The Revised Statutes of Ontario* (Toronto 1937), volume 2, chapter 346.

52 Anon., *Acts of the General Assembly of Prince Edward Island* (1940), chapter 9. Also ibid. (1946), chapter 3.

53 See *Report of the Minister of Agriculture, Dominion of Canada for 1919–20* (Ottawa 1920), 3 and 4.

54 Anon., *Revised Statutes of Canada* (Ottawa 1927), chapter 181.

55 Anon., *Acts of the General Assembly of Newfoundland* (St John's 1931), chapter 19.

56 For a commentary on these early congresses and conventions, see G.C. Ainsworth, *Introduction to the history of plant pathology* (Cambridge 1981), 183–5.

57 H.T. Güssow, "International plant disease legislation as viewed by a scientific officer of an importing country," *Report of the International Congress of Phytopathology and Economic Entomology*, Holland 1923 (1924), 96–107.

58 H.T. Güssow, "International plant disease legislation – is it practical?" *Proceedings of the International Congress of Plant Science*, Ithaca, New York 1926 (1929), 1334–42.

59 This note is in the Public Archives of Canada, filed as RG17 vol. 2869, 9–4 part 2.

60 Ibid.

61 Ibid.

62 H.T. Güssow, "Plant quarantine legislation – a review and a reform," *Phytopathology* 26 (1936), 465–82.

CHAPTER EIGHT

1 *Provincial Statutes of Canada*, vol. 2 (Ottawa 1847), chapters 60 and 61.

2 Dorothy M. Duke, *Agricultural periodicals printed in Canada 1836–1961*, Information Division, Canada Department of Agriculture (Ottawa 1962), 101.

3 R.H. Estey, "A History of the Quebec Society for the Protection of Plants," *Phytoprotection* 64 (1983), 1–22.

4 R.H. Estey, "James Fletcher (1852–1908) and the genesis of plant pathology in Canada," *Canadian Journal of Plant Pathology* 5 (1983), 120–4.

5 H.T. Güssow, "The rise of plant pathology in the Dominion of Canada," *Annual Report of the Quebec Society for the Protection of Plants* 9 (1917), 78–82.

6 Correspondence between federal and provincial officials regarding the potential threat to the potato industry is recorded in the *Evening Telegram*, St John's, Newfoundland, 20 November 1909, 4.

7 J.W. Dawson, *Scientific Contributions Toward the Improvement of Agriculture in Nova Scotia* (Pictou 1853), 264; *Contributions Toward the Improvement of Agriculture in Nova Scotia* (Halifax 1856), 280; and *First Lessons in Scientific Agriculture for Schools and Private Institutions* (Montreal 1864).

8 *Annual Report of the Secretary for Agriculture, Nova Scotia* (1900), 57.

9 J.H. Ellis, *The Ministry of Agriculture in Manitoba 1870–1970* (Winnipeg 1970), 192.

10 A.J. Madill, *History of agricultural education in Ontario* (Toronto 1930), 42

11 H.Y. Hind, *Essay on the Insects and Diseases Injurious to the Wheat Crops* (Toronto 1857)

12 W.L. Morton, *Henry Youle Hind 1823–1908* (Toronto 1980), 19 and 123–4.

13 From typed pages, A73–0026/44(06), archives, University of Toronto.

14 See University of Toronto calendars for 1876 and 1877.

15 W.R. Shaw, "Peach Yellows," *Transactions of the Canadian Institute* series 4, 2 (1892), 209–20.

16 R.H. Estey, "James Fletcher (1852–1908) and the genesis of plant pathology in Canada," *Canadian Journal of Plant Pathology* 5 (1983), 120–4.

17 Dorothy Forward, *The history of botany in the University of Toronto* (Toronto 1977), 5–6.

18 J.H. Faull, "Presidential address, Plant pathology: Its status and outlook," *Transactions of the Royal Society of Canada,* Section 4, 14 (1920), 1–16. See also Anna A. Faull, "Joseph Horace Faull 1870–1961," *Journal of the Arnold Arboretum* 43 (1962), 222–33.

19 H.T. Güssow, "Report of the Division of Botany," in *Dominion Department of Agriculture Report of 1914* (Ottawa 1915), 833.

20 University of Toronto calendar (1925–26), 236.

21 Forward, *The History of Botany,* 55.

22 J.H. Grisdale, "Report of the Acting Botanist," in *Report of the Dominion Experimental Farms for 1920* (Ottawa 1921), 57.

23 University of Toronto, School of Graduate Studies calendar (1929–30), 57.

24 Anon., "Report of the Horticultural Experimental Station," in *Report of the Ontario Department of Agriculture for 1942* (Toronto 1943), 11.

25 D.L. Bailey, "Herbert Spencer Jackson (1883–1931)," *Proceedings of the Canadian Phytopathological Society* 19 (1951), 4–5. See also *Transactions of the Royal Society of Canada* 46 (1952), 87–9.

26 See University of Toronto calendars for those years.

27 Ibid. for 1936. Also, Forward, *The History of Botany*, 92.

28 *Report of the Ontario School of Agriculture and Experimental Farm, for the year ending 31st October, 1876*, 28.

29 *Annual Report of the Commissioner of Agriculture and Arts, Ontario, for 1880*, 316.

30 Ibid., 440.

31 Ibid., 376.

32 Ibid. for 1881, 445.

33 Ibid. for 1881, 4.

34 Ibid. for 1882, 130.

35 *Annual Report of the Department of Agriculture, Ontario, for 1898*, x.

36 Ibid. 10.

37 Ibid. for 1901, 10.

38 *Annual Report of the Ontario Agricultural College and Experimental Farm for 1904*, 173.

39 Ibid., 123.

40 Ibid.

41 Ibid., 117.

42 Anon., *Report of the Ontario Fruit Growers' Association for 1910* (1910), 61–6; *Phytopathology* 1 (1910), 155–8; and ibid. 19 (1929), 413.

43 A.H. Richardson, "E.J. Zavitz honoured by McMaster University," *Forestry Chronicle* 28 (1952), 90–1.

44 *Annual Report of the Ontario Agricultural College and Experimental Farm for 1906* (1907), 47 and 54.

45 Ibid. for 1908, 24.

46 Ibid. for 1910, 28.

47 *Annual Report of the Minister of Agriculture, Ontario, for 1941* (1942), 14.

48 J. Mills and T. Shaw, *The First Principles of Agriculture* (Toronto 1890).

49 *Annual Report of the Ontario Agricultural College and Experimental Farm for 1927* (1928), 6–7. See also ibid. for 1928, 7.

50 Ibid. for 1913, 34.

51 Anon., "In memoriam, David Richmond Sands," *Proceedings of the Canadian Phytopathological Society* 33 (1966), 3.

52 *Annual Report of the Ontario Agricultural College and Experimental Farm for 1931* (1932), 27.

53 W.D. McHoull, "The founding and early history of Queen's University," Master's thesis, Queen's University 1935.

54 A.H. MacKay, "Memoir of the late Professor Lawson," *Proceedings of the Royal Society of Canada* 13 (1896), appendix B. For a list of Lawson's publications, see ibid. 12 (1895), 49–52.

55 University of Queen's College, Kingston, calendar (1879–80), 120.

56 Ibid. (1881–82), 159.

57 Queen's College and University, Kingston, calendar (1885–86), 253.

58 Ibid. (1895–96), 65.

59 L.E. Wehmeyer, *The Fungi of New Brunswick, Nova Scotia and Prince Edward Island* (Ottawa 1950), 2.

60 Queen's University calendar (1897–98), 64.

61 Anon., "Retiring Professors," *Queen's University Journal* 34 (1907), 443–4.

62 Anon., *Notes in McIlwraith Field Naturalists Club*, London, Ontario, Public Library, Accession No. 218.

63 Personal communication from Professor MacClement's son, Dr D. MacClement.

64 The "Minute Books" of the early sections of the Entomological Society of Canada are in the archives of Queen's University.

65 Queen's University calendar (1910–11), 107.

66 Queen's University calendar (1912–13), 112.

67 Ibid. (1916–17), 163 and 165.

68 Ibid. (1918–19), 161.

69 Ibid. (1932–33), 136.

70 Ibid., 146.

71 Anon., *Annual Report of the Quebec Society for the Protection of Plants* 2 (1911), 6.

72 D.L. Bailey, "Canadian Plant Pathology in Retrospect and Prospect," *Agricultural Institute Review* (September 1945), 6.

73 *Annual Report of the Ontario School of Agriculture and Experimental Farm, 1876* (1877), 25.

74 McMaster University calendar (1896).

75 John Craig, "Black knot of the plum and cherry," *Central Experimental Farm Bulletin* 23 (1895), 28–34.

76 C.M. Johnston, *McMaster University*, vol. 1 (Toronto 1976), 83.

77 McMaster University calendar (1912) and subsequent relevant years.

78 McMaster University, Arts-Theology calendar (1914–15), 37.

79 Personal communication with H.R. McLarty.

80 See more in R.J. Moore and W.F. Grant, "Lulu Odell Gaiser 1896–1965," *Canadian Journal of Genetics* 7 (1965), 361–2

81 B.O. Dodge and L.O. Gaiser, "The question of nuclear fusion in the blackberry rust *Caeoma nitens*," *Journal of Agricultural Research* 32 (1926), 1003–24.

82 C.M. Johnston, *McMaster University*, vol. 2. (Toronto 1981), 25, 63, and 106.

83 For an account of the early days at the University of Western Ontario, see James J. Talman and Ruth D. Talman, *"Western" – 1878–1953* (London, Ontario 1953), and J.R.W. Gwynne-Timothy, *Western's first century* (London, Ontario 1978)

84 W.F. Tamblyn, *These sixty years* (London, Ontario 1938), 5.

85 W.F. Tamblyn, "John Dearness," *Mycologia* 47 (1955), 909–15.

86 J.J. Talman, "John Dearness 1852–1954," *Proceedings and Transactions of the Royal Society of Canada* 51, series 3 (1957), 80.

87 Gwynne-Timothy, *Western's first century*, 267.

88 University of Western Ontario calendar (1922–23), 64.

89 Gwynne-Timothy, *Western's first century*, 267.

90 University of Western Ontario calendar (1923–24), title page.

91 Anon., *Prospectus of McGill College, Session of 1856–57* (Montreal 1856), 11–12.

92 A. Walker, *Synopsis of the Several Communications on the Cause and Cure of the Potato Rot* (Boston 1852), 27–36.

93 J.W. Dawson, *First Lessons in Scientific Agriculture for Schools and Private Institutions* (Montreal 1864).

94 Penhallow's early publications are listed in *Proceedings and Transactions of the Royal Society of Canada for 1894*, 12 (1895), 63–4.

95 D.P. Penhallow, "The Spot Disease of the Fameuse," *Transactions and Reports of the Fruit Growers' Association of Nova Scotia for 1887*, 63–8.

96 D.P. Penhallow, "Distribution of the reserve material of plants in relation to disease," *Canadian Record of Science* 1 (1888), 193.

97 D.P. Penhallow, "A Review of Canadian Botany from 1800 to 1895. Part 2," *Proceedings and Transactions of the Royal Society of Canada*, section 4, 3 (1897), 26.

98 For a biographical sketch of Penhallow, see ibid. for 1911, vii–x, and also Carrie M. Derick, "Dr. David Pearce Penhallow," *Canadian Record of Science* 9 (1916), 387–90.

99 Anon., *Annual Report of the Principal and Fellows of McGill University for 1911–12* (1913), 8, 34. See also Margaret Gillett, "Carrie Derick (1862–1941) and the Chair of Botany at McGill," in Marianne G. Ainley, editor, *Despite the odds* (Montreal 1990), 74–87.

100 *Annual Report of the Principal and Fellows of McGill University for 1919–20* (1921), 56.

101 Macdonald College announcement (1907–8) (Ste-Anne-de-Bellevue 1907), 18–19.

102 J.F. Snell, *Macdonald College of McGill University* (Montreal 1963), 171.

103 Macdonald College announcement (1942–43), 941 (as bound with the McGill calendar).

104 R.H. Estey, "A history of the Quebec Society for the Protection of Plants," *Phytoprotection* 64 (1983), 1–22.

105 The five short articles are all on one page of the *Canadian Horticulturist* 29 (1906), 154.

106 Macdonald College announcement (1907–8), 19–20

107 Ibid., 31.

108 D.W. Hamilton, *Lesson topics in nature study and elementary agriculture for rural schools* (Macdonald College 1915), 80 and 82.

109 J.C. Arthur, "Cultures of Uredineae in 1910," *Mycologia* 4 (1912), 7–33.

110 R.H. Estey, "Margaret Newton: Distinguished Canadian scientist and first woman member of the Quebec Society for the Protection of Plants," *Phytoprotection* 68 (1987), 79–85.

111 See Macdonald College announcements for those years.

112 Ibid. (1920–21), 56.

113 Ibid.

114 Ibid. (1922–23), 85.

115 J.G. Coulson, personal communication.

116 Snell, *Macdonald College of McGill University*, 63.

117 B.T. Dickson, "Plant Pathology in Canada," *Scientific Agriculture* 5 (1925), 211–17.

118 See news item in *Macdonald College Magazine* 18 (1927), 43.

119 Abstracts of their joint publications are in *Phytopathology* 13 (1913), 291; ibid. 14 (1924), 198; and ibid. 15 (1915), 302.

120 L.S. Klinck, "The susceptibility of certain weeds to smut," *Annual Report of the Quebec Society for the Protection of Plants* 2 (1910), 14–15.

121 B.T. Dickson and G.P. McRostie, "Further studies on mosaic," *Phytopathology* 12 (1922), 42.

122 J.G. Coulson and L.C. Raymond, "Progress report on the investigation of brown heart of swede turnips at Macdonald College," *Scientific Agriculture* 17 (1937), 299–301

123 *Annual Report of the Governors, Principal and Fellows of McGill University for 1920* (1920), 61.

124 Macdonald College announcement (1923–24), 68.

125 Ibid. (1927–28), 50.

126 T.C. Vanterpool, "Streak or winter blight of tomato in Quebec," *Phytopathology* 16 (1926), 311–31.

127 J.E. Machacek, "Studies on the association of certain phytopathogens," PH D thesis, McGill University, 1928.

128 Suit published in the 24th and 25th *Annual Report of the Quebec Society for the Protection of Plants*, and in *Scientific Agriculture* 15 (1935), 345–57.

129 Letter from J.G. Coulson to T.F. Ward, bursar, Macdonald College, dated 31 August 1936, in the college archives.

130 In a personal communication, Dr Brodie implied that having a higher degree and better qualifications than the head of the department created such real or imaginary problems for him that he had to leave.

131 I.H. Crowell, "The Geographic distribution of the genus Gymnosporangium," *Canadian Journal of Research* "C" 18 (1940), 469–88.

132 I.H. Crowell, "Report of the Common Names Committee," *Proceedings of the Canadian Phytopathological Society* 11 (1944), 7.

133 J.G. Coulson, personal communication.

134 Snell, *Macdonald College of McGill University*, 131.

135 Much of Thatcher's thesis was published in the *American Journal of Botany* 26 (1939), 449–58. See also *Canadian Journal of Research* "C" 20 (1942), 283–311.

136 Anon., "Report of the Agricultural School of Ste-Anne-de-la-Pocatière for the year 1899–1900," in *Report of the Commissioner of Agriculture, Province of Quebec, for 1900* (1901), 4.

137 J.B. Roy, *Histoire de la pomologie au Québec*, Agriculture Québec publication no. A-309 (1978).

138 Anon., "Report of the Agricultural School of Ste-Anne-de-la-Pocatière for the year 1913–14," in *Report of the Commissioner of Agriculture, Province of Quebec, for 1914* (1914), 3–4.

139 Anon., *Who's Who in the Agricultural Institute of Canada* (Ottawa, 1948), 18. See also *Report of the Minister of Agriculture, Province of Quebec, for 1919*, 44.

140 Anon., "Report of the Agricultural School of Ste-Anne-de-la-Pocatière for the year 1923–24." in *Report of the Commissioner of Agriculture, Province of Quebec, for 1924*. 3.

141 Personal communication. See also *Report of the Minister of Agriculture, Province of Quebec, for 1927*, 50, and tributes to Campagna in *Agro Nouvelles* 23:8 (1988), 2 and 13–14.

142 See *Who's Who in the Agricultural Institute of Canada*. (Ottawa 1948), 28. Also, *Annual Report of the Quebec Society for the Protection of Plants for 1938–39*, 15.

143 R. Pomerleau, "Georges Maheux, 8 décembre 1889–11 octobre 1977," *Le Naturaliste Canadien* 104 (1977), 573–5.

144 Personal Communication.

145 G. Boron, "Agricultural School at Oka," in *Report of the Commissioner of Agriculture, Province of Quebec, for 1897*, 35–6. See also Father Louis-Marie, *L'Institut D'Oka. Cinquantenaire 1893–1933* (Montreal 1944), 498.

146 *Report of the Minister of Agriculture, Province of Quebec, for 1906* (1907), 10.

147 Ibid. for 1907 (1908), 14.

148 Ibid. for 1912 (1912), 14.

149 Father Leopold, *La Culture Fruitière dans la Province de Québec* (Montreal 1914).

150 *Report of the Minister of Agriculture, Province of Quebec, for 1916* (1917), 17. Also, *Agricultural Gazette of Canada* 3 (1916), 745.

151 *Report of the Minister of Agriculture, Province of Quebec, for 1920–21*, 20.

152 Ibid. for 1921–22, 4.

153 Ibid. (1923–24), 25. (See more about Father Leopold's work in the history of the school at Oka by Father Louis-Marie.)

154 Ibid. (1934), 154.

155 Ibid. (1935), 142.

156 Ibid. (1942), 109.

157 John White, "Speed the plough: Agricultural societies in pre-Confederation New Brunswick," MA thesis, University of New Brunswick 1977, 12. See also Howard Trueman, *Early agriculture in the Atlantic provinces* (Moncton 1907), 25 R.P. Gorham, "Early agriculture in New Brunswick," *C.S.T.A. Review* 9 (1936), 3, gives 1789 as the date.

158 John White, above. See also Trueman, *Early Agriculture in the Atlantic Provinces*, 25.

159 Anon., "An act to amend chapter 58 Revised Statutes, 1900," in *The Statutes of Nova Scotia Passed in the First Year of the Reign of His Majesty King Edward VII* (Halifax 1901), chapter 18.

160 See "Respecting aid to establish an agricultural school for the Maritime provinces," in *Consolidated Statutes of New Brunswick, 1903* (Fredericton 1904), chapter 47.

161 For an account of Macdonald's many benefactions, see Stanley B. Frost, *McGill University*, vol. 2 (Montreal 1984), 3, 35, 62–6, etc.

162 *Annual Report of the New Brunswick Department of Agriculture* (1918), 16.

163 Ibid. (1915), 20–2.

164 Ibid. (1916), 110.

165 R.P. Gorham, *Insecticides and fungicides for orchards and garden crops*, New Brunswick Department of Agriculture bulletin no. 2 (1913).

166 *Annual Report of the New Brunswick Department of Agriculture* (1925), 18.

167 Ibid., 46.

168 Ibid. (1915), 102.

169 Ibid. (1918), 130–1.

170 H.T. Güssow, *Report of the Dominion Botanist for the year 1924* (Ottawa 1925), 19.

171 Personal communication with Dr MacLeod.

172 *Annual Report of the New Brunswick Department of Agriculture for 1915* (1916), 110.

173 University of New Brunswick calendar (1909), 29.

174 Ibid., 70.

175 Ibid. (1920), 59.

176 A.G. Bailey, *The University of New Brunswick Memorial*, vol. 1 (Fredericton 1950), 88–93.

177 M.L. Prebble, "Forest entomology and pathology at UNB," in David G. Bryant, editor, *The fiftieth anniversary of the Faculty of Forestry at the University of New Brunswick* (Fredericton 1958), 41.

178 John Dearness, *The nature study course, with suggestions for teaching it based on notes of lectures to teachers-in-training* (Toronto 1905), 48.

179 Harvey W. MacPhee, *The School of Agriculture, 1885–1905* (Truro 1985).

180 Kenneth Cox, *A history of the Nova Scotia Agricultural College* (Truro 1965), 2.

181 Ibid.

182 H.W. Smith, "Report of the professor of biology," in *Annual Report of the Secretary for Agriculture, Nova Scotia, for the year 1909* (Halifax 1910), 33.

183 Cox, *A history of the Nova Scotia Agricultural College*, 76.

184 *Annual Report of the Secretary for Agriculture, Nova Scotia, for the year 1912* (Halifax 1913), 17.

185 Ibid. for 1916, 103–119. Smith's 1910 paper is in the *Annual Report of the Nova Scotia Fruit Growers' Association, for 1910*, 69–88.

186 H.W. Smith, "Annual Report of the Principal of the Provincial School of Agriculture," in *Annual Report of the Secretary of Agriculture, Nova Scotia, for the year 1896* (Halifax 1897), 36.

187 H.W. Smith, "Report of the professor of biology for 1906," in *Journal and Proceedings of the House of Assembly of the Province of Nova Scotia for 1907*, appendix 8 (Halifax 1907), 25.

188 *Annual Report of the Secretary of Agriculture, Nova Scotia, for the year 1912* (Halifax 1913), 21–2.

189 Ibid. for 1917, 50.

190 Ibid. for 1919, 63.

191 Ibid. for 1908, 21.

192 Ibid. for 1924, 26.

193 Anon., *Report of the Department of Natural Resources, Province of Nova Scotia, for 1926* (Halifax 1927), 31.

194 Cox, *A history of the Nova Scotia Agricultural College*, 6. For a slightly different version, see Ronald S. Longley, *Acadia University, 1838–1938* (Wolfville 1939), 102.

195 Anon., *Transactions and Reports of the Fruit Growers' Association and International Show of Nova Scotia, for 1894* (Halifax 1894), 126.

196 *Annual Report of the Secretary of Agriculture, Nova Scotia, for the year 1901* (Halifax 1902), 65–6.

197 *Journal and Proceedings of the House of Assembly of the Province of Nova Scotia, Session 1903–04* (Halifax 1904), 57.

198 *Annual Report of the Secretary of Agriculture, Nova Scotia, for the year 1904* (Halifax 1905), 51.

199 "First annual report of the Nova Scotia Agricultural College," in *Journal and Transactions of the House of Assembly of the Province of Nova Scotia, for 1906*, appendix 8 (Halifax 1906), 46–7.

200 Ibid. for 1908, appendix 8, 2.

201 *Annual Report of the Secretary of Agriculture, Nova Scotia, for the year 1908* (Halifax 1909), 31.

202 "Report of the provincial horticulturist, for 1921–22," in *Journal and Proceedings of the House of Assembly of the Province of Nova Scotia, Session 1923*, part 2, appendix 8 (Halifax 1923), 62.

203 "Report of the provincial horticulturist," in *Annual Report of the Secretary of Agriculture, Nova Scotia, for the year 1924* (Halifax 1924), 60–4.

204 "The semi-annual report of the superintendent of education for Nova Scotia," *Journal of Education* (October 1918), 50.

205 Hugh P. Bell, "The origin and histology of Bordeaux spray russeting of the apple," *Canadian Journal of Research* "c", 19 (1941), 493–9.

206 *Journal of the House of Assembly of Prince Edward Island, Second Session of the Seventeenth General Assembly* (Charlottetown 1848), 17, 24.

207 *Islander*, Charlottetown (21 August 1846).

208 Anon., "Minutes of the Royal Agricultural Society of Prince Edward Island," meeting of 20 June 1849.

209 Ibid., meeting of 2 April 1851.

210 James H. Peters, *Hints to the Farmers of Prince Edward Island*, second edition (Charlottetown 1851), 20–1.

211 Lorne Callbeck, "P.E.I.'s university: A story of progress," *Atlantic Advocate* 66:10 (1976), 56–9.

212 Anon., *Annual Report of the Public Schools of Prince Edward Island, for 1888* (Charlottetown 1889), xxi.

213 J.A. Clark, "Apple growing in Prince Edward Island," *Scientific Agriculture* 23 (1943), 251.

214 S.N. Robertson, "Report of the principal of Prince of Wales College and Normal School," in *Annual Report of Public Schools of Prince Edward Island for the year 1900* (Charlottetown 1902), 119.

215 Anon., *Annual Report of the Public Schools of Prince Edward Island, for 1909* (Charlottetown, 1909), appendix F, 39.

216 Ibid. for 1910, appendix c, 37.

217 Prince of Wales College and Normal School calendar, 1939–40 (Charlottetown 1939), 16

218 Mary B. Bourdon, "Charlottetown Research Station 1909–1984," Agriculture Canada historical series no. 19, (1984), 10.

219 Ibid., 21.

220 Ibid., 32.

221 Paul A. Murphy, *Late blight and rot of potatoes caused by the fungus Phytophthora infestans*, de Bary, Dominion of Canada Department of Agriculture circular no. 10 (1916).

222 Bourdon, "Charlottetown Research Station," 32.

223 I.L. Conners, editor, *Plant pathology in Canada* (Winnipeg 1972), 43.

224 H.T. Güssow, *Report of the Dominion Botanist for 1924* (Ottawa 1925), 18.

225 See obituary notice for Hurst in *Proceedings of the Canadian Phytopathological Society* 28 (1961), 4.

226 See obituary notice for Howatt, ibid. 27 (1960), 4.

227 K.L. Hatch and J.A. Haselwood, *Elementary agriculture with practical arithmetic* (Toronto 1910), chapter 11.

228 W.J. Spence, *University of Manitoba historical notes 1877–1917* (Winnipeg 1918), 22.

229 Minute book of the Council of the University of Manitoba, for 14 April 1885, special meeting.

230 Manitoba University calendar and syllabus (1888), examiners, May 1888 (not paginated).

231 Ibid. (1885), structural botany examination, May 1885, question 4 (not paginated).

232 Ibid. (1886), question 2 (not paginated).

233 Ibid. (1895), question 5 (not paginated).

234 Anon., *Annual Report of the Manitoba Historical and Scientific Society for 1886*, meeting of 9 February.

235 R.B. Thomson, "A.H. Reginald Buller (1874–1944)," *Transactions of the Royal Society of Canada* 39 (1945), 79–81.

236 Manitoba University calendar and syllabus (1905), botany examination, April 1905 (not paginated).

237 Ibid. for April 1906.

238 T. Johnson, *Rust research in Canada and related plant-disease investigations* Canada Department of Agriculture publication no. 1098 (Ottawa 1961), 15.

239 R.H. Estey, "A.H.R. Buller: Pioneer leader in plant pathology," *Annual Review of Phytopathology* 24 (1986), 17–25.

240 E.H. Lange, "Seventy-five years of agricultural education in Manitoba," In anon., *The University of Manitoba Faculty of Agriculture seventy-fifth anniversary, 1906–1981* (Winnipeg c1980), 5.

241 Ibid., 73.

242 Manitoba Agricultural College calendar (1917–18) (Winnipeg 1917), 33, 37.

243 Ibid. (1908–9), 163.

244 J.H. Ellis, *The Ministry of Agriculture in Manitoba 1870–1970* (Winnipeg 1970), 189.

245 Ibid., 184.

246 Manitoba Agricultural College calendar (1909–10), 46.

247 Ibid. (1912–13), 53.

248 Ibid. (1910–11), 17.

249 Norman James, "Bacteriology," in anon., *A record of the years 1906–Golden Jubilee–1956.* (Winnipeg 1956), 27.

250 Manitoba Agricultural College calendar (1914–15), 71.

251 Ibid.

252 Bryce J. Sallans, personal communication.

253 Manitoba Agricultural College calendar (1920–21), 9.

254 Ibid. (1933–34), 44.

255 Ibid. (1935–36), 50.

256 University of Manitoba calendar (1933–34), 12.

257 A.H.R. Buller, "Botany," in *Departmental Reports, University of Manitoba, for 1919–30*, 12–13.

258 G.R. Bisby, A.H.R. Buller, and J. Dearness, *Fungi of Manitoba* (London 1929), 194.

259 Ellis, *The Ministry of Agriculture in Manitoba*, 250.

260 Charles W. Lowe, "Department of Botany," in *University of Manitoba, president's report, for year ending 30 April 1937*.

261 Manitoba Agricultural College calendar (1941–42), 48.

262 Jean E. Murray, "The early history of Emmanuel College," *Saskatchewan History* 9 (1956), 81.

263 Michael Hayden, *Seeking a balance. The University of Saskatchewan 1907–1982* (Vancouver 1983), 8.

264 Murray, "The early history of Emmanuel College," 101.

265 Ibid., 96.

266 Lucy H. Murray, "St. John's College of Qu'Appelle," *Saskatchewan History* 11 (1958), 20.

267 William H. Coard, *North-Western Agricultural College and Experiment Station, Regina, North-West Territories, Canada*, bulletin no. 1 (1903), not paginated. See also the *Saskatoon Phoenix*, 13 March, 4, and 8 May, 4.

268 This one-page, undated, typewritten statement, and bulletin no. 1, to which it was attached, were seen in the University of Saskatchewan Archives, as were the University of Saskatchewan catalogues.

269 W.P. Thompson, *The University of Saskatchewan: A personal history* (Toronto 1970), 83.

270 Arthur S. Morton, *Saskatchewan: The making of a university*, revised and edited by Carlyle King (Toronto 1959), 68.

271 L.E. Kirk, "Early years in the College of Agriculture," *Saskatchewan History* 12 (1959), 23.

272 University of Saskatchewan calendar (1914–15), 93.

273 A.W. Henry, personal communication.

274 For a brief biographical sketch of Willing, see J.H. Morgan, *The Canadian men and women of the time*, second edition (Toronto 1912).

275 The president's report, University of Saskatchewan, for 1920–21, has comments on Willing and his work.

276 Willing's report to the president of the university, dated 11 June 1914.

277 Thompson, *The University*, 54.

278 University of Saskatchewan calendar (1915–16), 81.

279 Thompson, *The University*, 140.

280 University of Saskatchewan calendar (1916–17), 48, 96.

281 Ibid. (1915–16), 48.

282 Thompson's report to the president of the university, dated 13 April 1917.

283 Thompson, *The University*, 168.

284 A.W. Henry, personal communication.

285 C. King, *The first fifty: Teaching, research and public service at the University of Saskatchewan 1909–1959* (Toronto 1959), 85.

286 University of Saskatchewan calendar (1919–20), 44.

287 Ibid. (1920–21), 42. See also Thompson's report to the president, dated 1 May 1919.

288 For a review of Margaret Newton's life and work, see *Phytoprotection* 68 (1987), 79–85.

289 See university calendars for the various years.

290 T.C. Vanterpool, "William Pollock Fraser," *Mycologia* 36 (1944), 313–17.

291 For a brief review of Vanterpool's life, see R.A.A. Morrall and M. Shaw, "Thomas Clifford Vanterpool," *Canadian Journal of Plant Pathology* 6 (1984), 336–7.

292 Thomas C. Vanterpool, personal communication.

293 Ibid.

294 Ibid.

295 University of Saskatchewan calendar (1939–40), staff list.

296 Anon., *Annual Report of the Department of Education, Alberta, for 1907*, 82.

297 Ibid. for 1914, 55.

298 University of Alberta calendar (1918–19); supplement, Summer School for Teachers, 17, 18.

299 J. McCaig, *Elementary agriculture for Alberta schools* (Toronto 1915), chapter 26.

300 J.H. Hutchinson, *Public school agriculture* (Toronto c1927, revised 1930, 1935), chapter 12.

301 Anon., *Annual Report of the Department of Education, Alberta, for 1912*, 169.

302 J.E. Birdsall, *The sixth decade at the Alberta Agricultural Colleges, 1964–74* (1975), 1, 2.

303 E.B. Swindlehurst, *Alberta's Schools of Agriculture* (an undated publication of the Alberta Department of Agriculture), 63, 64.

304 W.H. Johns, *A history of the University of Alberta 1908–1969* (Edmonton 1981), 30.

305 E.B. Swindlehurst, *Alberta agriculture* (Edmonton c1976), 99.

306 Anon., *The Canadian Society of Technical Agriculturists, Who's Who* (Ottawa 1924), 177.

307 Annual Report of the Board of Governors, University of Alberta, for 1934–35, 8.

308 University of Alberta calendar (1920–21), 86.

309 Ibid., 85.

310 Ibid. (1921–22), 125.

311 G.B. Sanford, "Edmonton laboratory." This is in a fourteen-page, unbound, typewritten document that was kindly made available by Mrs Sanford.

312 Much of the story of Robert Newton's life is in typescript form, titled "I passed this way 1889–1964," one copy of which is in the University of Alberta archives, accession no. 71–87. A brief résumé of his life to that date is in the *Macdonald Farm Journal* (June 1942), 27–8.

313 See *Canadian Journal of Research* 10 (1936), 414, and ibid. 5 (1931), 87–110, for examples of his research publications.

314 University of Alberta calendar (1923–24), 159.

315 Cutler's letter to Dr Tory was seen in the university archives.

316 This is largely from the curriculum vitae that Dr Henry kindly prepared of himself, and correspondence with him, plus faculty minutes as noted.

317 Ibid.

318 University of Alberta calendar (1927–28), 171.

319 Evidence of this is seen in such publications as O.S. Aamodt and J.G. Malloch, "Smutty wheat caused by *Ustilago utriculosa* on dock-leaved Persicary," *Canadian Journal of Research* 7 (1932), 578–82.

320 University of Alberta calendar (1935–36).

321 E.H. Moss, "The haustorium of *Cuscuta gronvii*," *Phytopathology* 18 (1928), 478. See also the *Annual Report of the Board of Governors, University of Alberta, for 1934–35*, 8.

322 University of Alberta calendar (1936–37), 121–66.

323 Ibid. (1940–41), 146.

324 J.W. Gibson, "Agriculture in high schools," *Agricultural Gazette of Canada* 3 (1916), 1105.

325 J. Brittain, *Elementary agriculture and nature study* (Toronto 1909), 245–80.

326 J.W. Eastham, "Annual report of the provincial plant pathologist," in *Report of the British Columbia Department of Agriculture for 1915* (1916), R 76.

327 Ibid. for 1919 (1920), 42.

328 Ibid. for 1920 (1921), Q43, and subsequent years.

329 Anon., "John William Eastham," *Proceedings of the Canadian Phytopathological Society* 37 (1970), 32.

330 *Annual Report of the British Columbia Department of Agriculture for 1924* (1925), K34.

331 From personal interview with Dr R.J. Bandoni, and Dickson's former student Dr James D. Menzies. See also the obituary notice in *Forestry Chronicle* 32 (1956), 103–5.

CHAPTER NINE

1 Thomas C. Haliburton, *An Historical and Statistical Account of Nova Scotia*, vol. 2 (Halifax 1829), 405.

2 Titus Smith, "A List of the Principal Indigenous Plants of Nova Scotia," *Halifax Monthly Magazine* 1 (1831), 342–5.

3 This is from the *Dictionary of Canadian Biography* (Toronto 1988), 814–16.

4 John Somers, "Nova Scotian Fungi," *Proceedings and Transactions of the Nova Scotia Institute of Natural Science* 5 (1881), 188–91, 247–53, 332–3. See also ibid. 6 (1886), 286–8; 7 (1890), 18–19, and 464–6.

5 See obituary notices in *Transactions of the Royal Society of Canada* 24 (1930), iii–vii, and *Proceedings and Transactions of the Nova Scotian Institute of Science* 17 (1930), xlvii–liii.

6 Alexander H. MacKay, "Fungi of Nova Scotia," *Transactions of the Nova Scotian Institute of Science* 11 (1906), 122–3.

7 Reginald R. Gates, "Middleton fungi," ibid. 11 (1906), 115–21.

8 Alexander H. MacKay, "Fungi of Nova Scotia, first supplemental list," ibid. 12 (1908), 124–6.

9 Clarence L. Moore, "The Myxomycetes of Pictou County," *Bulletin of the Pictou Academy of Science Association* 1 (1909), 11–16. For his account of the rusts, see ibid., 20–2. For his aquatic fungi, see *Transactions of the Nova Scotian Institute of Science* 12 (1909), 217–38, and for more Myxomycetes, see ibid., 165–206.

10 William P. Fraser and botany class, "The Erysiphaceae of Pictou County," *Bulletin of the Pictou Academy of Science Association* 1 (1909), 51–8.

11 W.P. Fraser, "Collection of the aecial stage of *Calytospora columnaris* (Alb. & Schw.) Kühn," *Science* 30 (1909), 814–15.

12 W.P. Fraser, "The rusts of Nova Scotia," *Proceedings and Transactions of the Nova Scotia Institute of Science* 12 (1910), 313–445.

13 For notes on Fraser's life, and more of his publications, see J.H. Craigie, "William Pollock Fraser 1867–1943," *Canadian Field Naturalist* 58 (1944), 1–3. Also, *Mycologia* 36 (1944), 313–17.

14 Lewis E. Wehmeyer, *The fungi of New Brunswick, Nova Scotia, and Prince Edward Island* (Ottawa 1950).

15 K.A. Harrison, "Fleshy fungi of Kentville, N.S.," in *Report of the Dominion Botanist for 1927* (Ottawa 1928), 32–5.

16 James Fowler, "List of New Brunswick Plants," in *Annual Report of the Secretary of Agriculture, New Brunswick, for 1878* (Saint John 1879), appendix B.

17 For a brief story of Fowler's life, with photo, see *Queen's University Journal* 34 (1907), 443–4.

18 G.U. Hay, "Preliminary list of New Brunswick fungi," *Bulletin of the Natural History Society of New Brunswick* 4 (1901), 341–4. See also Hay's "Study

of Canadian fungi: A review," *Transactions of the Royal Society of Canada,* section 4 (1904) 139–45.

19 See ibid. 7 (1913), xix–xxi, for a sketch of Hay's life. Also, *Bulletin of the Natural History Society of New Brunswick* 7 (1914), 70–1.

20 G.U. Hay, "New Brunswick fungi," *Bulletin of the Natural History Society of New Brunswick* 5 (1907), 112. See also ibid. 6 (1908), 40–3.

21 Both of the van Horne papers are in *Canadian Record of Science* 9 (1916), 157–75, and 328–38.

22 John MacSwain, "Some of Our Fungi." That talk, and one on "Rust of Wheat" and the reference to Miss Pippy's collection, are all in the *Bulletin of the Natural History Society of New Brunswick* 4 (1899).

23 The handwritten minutes of the meetings from 27 November 1899 through February 1901, in the Public Archives of Prince Edward Island, do not give Miss Pippy's full name, nor do they include a membership list.

24 D.A. Watt, "A Provisional Catalogue of Canadian Cryptogams," *Canadian Naturalist and Geologist* 2 (1865), 390–404.

25 Ibid.

26 Robert Campbell, "Canadian fungi," *Canadian Record of Science* 9 (1903), 89–99. See also *Annual Report of the Quebec Society for the Protection of Plants* 2 (1910), 73–5.

27 J. Adams, "Some fungi from Anticosti Island and Gaspé Peninsula," *Canadian Field Naturalist* 49 (1935), 107–8.

28 I.L. Conners, "Additions to the fungus flora of Anticosti Island and Gaspé Peninsula," *Canadian Field Naturalist* 51 (1937), 6–7.

29 Anon., "Nos Champignons," *Le Naturaliste Canadien* 10 (1878), 6–11.

30 Henri Prat, "Les mycorrhizes de l'If du Canada (*Taxus canadensis* Marsh)," *Le Naturaliste Canadien* 61 (1934), 47–56.

31 René Pomerleau, "Some Pyrenomycetes of Quebec," *Annual Report of the Quebec Society for the Protection of Plants* 19 (1927), 52–77. This was his master's thesis.

32 Irene Mounce and H.A.C. Jackson, "Two Canadian collections of *Cantharellus multiplex,"* *Mycologia* 29 (1937), 286. See also J. Walton Groves, "Henry A.C. Jackson," *Mycologia* 54 (1962), 1–4.

33 David P. Penhallow, "The mycelium of dry rot," *Canadian Review of Science* 9 (1903), 318.

34 See Macdonald College announcement for 1908–9 and succeeding years.

35 R.H. Estey, "Margaret Newton: Distinguished Canadian scientist and first woman member of the Quebec Society for the Protection of Plants," *Phytoprotection* 68 (1987), 79–85.

36 Macdonald College announcement for 1920–21. See also B.T. Dickson, "*Onygena equina* (Willd.) Pers." *Mycologia* 12 (1920), 289–91, and *Transactions of the Royal Society of Canada* 19 (1925), 275–7.

37 I.H. Crowell, "A new species of *Gymnosporangium*," *Canadian Journal of Research* "c" 18 (1940), 10–12. See also his "Geographical distribution of the genus Gymnosporangium," ibid. 469–88, and "Two new Canadian smuts," ibid. 20 (1942), 327–8.

38 N. Polunin, editor, *Botany of the Canadian Eastern Arctic*, National Museum of Canada bulletin no. 97, part 2 (Ottawa 1947), 240–1, 244–5, etc. See also *Journal of Botany* 72 (1934), 197–204.

39 W.J. Cody, D.B.O. Savile, and M.J. Sarazin, *Systematics in Agriculture Canada at Ottawa*, historical series no. 28 (Ottawa 1986), 23.

40 C.H. Peck, "New Species of Fungi," *Bulletin of the Torrey Botanical Club* 22 (1895), 198–211. See also G.R. Brassard, "Rev. Arthur C. Waghorne (1851–1900)," *Canadian Botanical Association Bulletin* (supplement) 13 (1980), 17–18.

41 Donald P. Rogers, *A brief history of mycology in North America* (University of Illinois 1981), 20. See also H.B. Humphrey, *Makers of North American botany* (New York 1961), 160–2.

42 John Dearness, "Report on fleshy fungi collected in August 1926," in J.D. Soper, *A faunal investigation of southern Baffin Island*, Canada Department of Mines, Biological Series, Bulletin 53 (1928), 120–3.

43 Polunin, editor, *Botany of the Canadian Eastern Arctic*, 246, 252, and 254.

44 James Hubbard, "On the Fungi," *Canadian Journal of Industry, Science and Art* 8 (1863), 471. See also *Canadian Naturalist and Geologist* 8 (1863), 76–80.

45 Cephas Guillet, "Fungi from the Kawartha Lakes (and a few from Toronto) including several new species," *Ottawa Naturalist* 21 (1907), 57–60.

46 J.B. Ellis and John Dearness, "New Species of Canadian Fungi," *Canadian Record of Science* 5 (1893), 266–72.

47 J. Dearness, "New Species of Tennessee fungi," *Mycologia* 33 (1941), 360–6.

48 J. Dearness, "An annotated list of Anthracnoses," in F.L. Drayton, *A summary of the prevalence of plant diseases in the Dominion of Canada 1920–1924*, Department of Agriculture bulletin no. 71 (1926), 62–76.

49 J.A. Stevenson, *An account of fungus exsiccati containing material from the Americas* (1971), 9.

50 W.F. Tamblyn, "John Dearness," *Mycologia* 47 (1955), 909–15. See also J.J. Talmen, "John Dearness 1852–1954," *Proceedings and Transactions of the Royal Society of Canada* 51 (1957), 79–83.

51 J.A. Parmelee, "The Dearness mycological collection," *Mycologia* 70 (1978), 509–26. See also J. Ginns, "The Canadians John Macoun and John Dearness, and their contributions to North American mycology," *Mycotaxon* 26 (1986), 47–53.

52 Adolph Lehmann, "Parasitic Fungi," *Ottawa Naturalist* 6 (1892), 38–40.

53 John Macoun, "Edible and Poisonous Fungi," *Transactions of the Ottawa Field Naturalists Club*, 2 (1884), 62–7. For a story of his work with the

Geological Survey, see W.A. Waiser, *The field naturalist: John Macoun, the Geological Survey, and natural science* (Toronto 1989). For Macoun's 1884 collection, see J.B. Ellis and B.M. Everhart, "Canadian Fungi," *Journal of Mycology* 1 (1885), 85–7. See also J. Ginns, "The Canadians John Macoun and John Dearness, and their contributions to North American mycology," *Mycotaxon* 26 (1986), 47–53.

54 D.A. Campbell, "Report of the Botanical Section," *Ottawa Naturalist* 12 (1898), 187–9.

55 R.H. Estey, "James Fletcher (1852–1908) and the genesis of plant pathology in Canada," *Canadian Journal of Plant Pathology* 5 (1983), 120–4.

56 A.W. McCallum, "The occurrence of *Bulgaria platydiscus* in Canada," *Mycologia* 11 (1919), 293.

57 H.T. Güssow, "The nature of parasitic fungi and their influence on the host plant," *Ottawa Naturalist* 25 (1911), 130–7. See also *Annual Report of the Quebec Society for the Protection of Plants* 3 (1911), 15–23, and *Phytopathology* 4 (1914), 386. For notes on Güssow's life, see *Transactions of the Royal Society of Canada* 56 (1962), 191, and *Phytopathology* 51 (1961), 739.

58 W.S. Odell, "List of mushrooms and other fleshy fungi of the Ottawa district," *Victoria Memorial Museum Bulletin* 43 (1926). Additions were made by J.W. Groves in *Canadian Field Naturalist* 52 (1938), 57–60.

59 F.T. Shutt, "Note on the food value of certain mushrooms," *Ottawa Naturalist* 18 (1904), 87–8, and ibid. 19 (1905), 87–8. For obituary notice and photo, see ibid. 55 (1941), 130–2.

60 See *Report of the Ontario Agricultural College and Experimental Farm for 1908* (1909), 36 and 41.

61 Ibid. for 1912, 55–7.

62 J.W. Eastham, "The Myxomycetes of the Ottawa district – A preliminary list," *Ottawa Naturalist* 25 (1912), 157–63. See also *Annual Report of the Quebec Society for the Protection of Plants* 5 (1916), 66–71.

63 McCubbin's two publications are Ontario Natural Science bulletin no. 7 (1912), and no. 8 (1913). There may be an error regarding bulletin 7, where the author's name is given as B.W.A. McCubbins. See *Annual Report of the Quebec Society for the Protection of Plants* 5 (1913), 59–60, for his comments on toadstools in nature, and the *Report of the Dominion Botanist* (1913), 497, for a reference to his mushroom display.

64 Thomas Langton, "Partial list of Canadian fungi," *Transactions of the Canadian Institute* 9 (1913), 69–81. See also his "Mushrooms and other fungi," in J.H. Faull, editor, *Natural history of the Toronto region* (Toronto 1913).

65 T.H. Bissonnette, "List of Georgian Bay fleshy fungi and myxomycetes," in *Forty-seventh Annual Report of the Department of Marine and Fisheries* (Ottawa 1915).

66 R.E. Stone, "Studies on the life histories of some species of Septoria occurring on Ribes," *Phytopathology* 6 (1916), 419–27. His final publication on fungi was Ontario Agricultural College *Bulletin* 397 (1938).

67 A. Booker Klugh, "*Morchella bispora* in Canada," *Canadian Field Naturalist* 34 (1920), 119.

68 F.A. Wolf and F.T. Wolf, *The fungi,* vol. 1 (New York 1947), 11. See also *Mycologia* 29 (1937), 305–18, and ibid. 35 (1943), 517–28.

69 D.B.O. Savile, "Frank Lisle Drayton 1892–1970," *Transactions of the Royal Society of Canada* 9 (1971), 48.

70 G.S. Bell, "Lists of the larger fungi. Toronto region," *Transactions of the Royal Canadian Institute* 19 (1933), 275–99.

71 Mary E. Currie, "A critical study of the slime-molds of Ontario," *Transactions of the Royal Canadian Institute* 12 (1920), 247–307.

72 W.N. Cheesman, "A contribution to the mycological flora and the Mycetozoa of the Rocky Mountains," *Transactions of the British Mycological Society* 3 (1911), 267–76.

73 G.R. Bisby, A.H.R. Buller, and John Dearness, *The fungi of Manitoba* (Toronto 1929), 18, 42.

74 I.L. Conners, editor, *Plant pathology in Canada* (Winnipeg 1972), 16. For a biographical sketch of Conners, see J. Ginns, "Ibra Lockwood Conners," *Mycologia* 82 (1990), 158–9.

75 Eleanor S. Dowding, "Gelasinospora, a new genus of Pyrenomycetes with pitted spores," *Canadian Journal of Research* 9 (1933), 294–305.

76 Clara W. Fritz, "Cultural criteria for the distinction of wood destroying fungi," PH D thesis, University of Toronto 1924. See also anon., "In memoriam," *Proceedings of the Canadian Phytopathological Society* 41 (1974), 38.

77 For an example of her work, see Irene Mounce, "Microscopic characters of sporophores produced in culture as an aid in identifying wood destroying fungi," *Transactions of the Royal Society of Canada* 26 (1932), 177–81. See also J. Ginns, "Irene Mounce, 1894–1987," *Mycologia* 80 (1988), 607–8. For more of Mounce's work, see references 32 and 82.

78 For a sketch of Jackson's life, see D.L. Bailey, "Herbert Spencer Jackson (1883–1951)," *Transactions of the Royal Society of Canada* 46 (1952), 87–9, and *Proceedings of the Canadian Phytopathological Society* 19 (1951), 4–5.

79 H.J. Brodie, "The occurrence in nature of mutual aversion between mycelia of Hymenomycetous fungi," *Canadian Journal of Research* "C" 13 (1925), 187–9. For a biographical sketch of Brodie, see D.B.O. Savile, "Harold Johnston Brodie, 1907–1989," *Mycologia* 81 (1989), 832–6.

80 Mildred K. Nobles, "Conidial cycles in the Thelephoraceae," PH D thesis, University of Toronto 1935. See also *Mycologia* 27 (1935), 286, and 29 (1937), 387 and 557.

81 Mildred K. Nobles, "Studies in forest pathology vi. Identification of cultures of wood-rotting fungi," *Canadian Journal of Research* "c" 26 (1948), 281–431. For additional early papers, see ibid. 21 (1943), 211, and ibid. 20 (1942), 237.

82 For examples of Macrae's research, see *Canadian Journal of Research* "c" 14 (1936), 215; 15 (1937), 154; 16 (1938), 354; and 20 (1942), 411.

83 For some of Cain's early papers, see *Mycologia* 27 (1935), 227; 40 (1948), 158; *Canadian Journal of Research* "c" 26 (1948), 486; 28 (1950), 566; and *Canadian Journal of Botany* 30 (1952), 338.

84 The six papers, in addition to part of his PH D thesis, which is in the *Canadian Journal of Research* 17 (1939), 125, are in *Mycologia* 28 (1936), 451–62; ibid. 29 (1937), 66–80; ibid. 30 (1938), 46–53, and 416–30; ibid. 32 (1940), 112–13; and ibid. 33 (1941), 510–22. See also J.W. Groves, "North American species of Derma," ibid. 38 (1946), 351–431. For obituary information, see D.B.O. Savile, "James Walton Groves 1906–1970," *Proceedings of the Royal Society of Canada* 8 (1970), 73–7.

85 J.W. Groves and A.J. Skolko, "Notes on seed-borne fungi. i Stemphylium," *Mycologia* 22 (1944), 190–9. See also ibid. 217–34; ibid. 23 (1945), 94–104; ibid. 24 (1946), 74–80. Part of Skolko's PH D thesis is A.J. Skolko, "A critical study of a two-spored Basidiomycete, *Aleurodiscus canadensis* n.sp.," *Canadian Journal of Research* "c" 22 (1944), 251–71.

86 The two papers by Haddow are in *Transactions of the British Mycological Society* 22 (1938), 182–93, and 25 (1941), 179–90.

87 J.G. Rempel, "George Aleck Ledingham 1903–1962," *Transactions of the Royal Society of Canada* 1 (1963), 93. See also *Mycologia* 55 (1963), 365–70, which has a list of Ledingham's publications.

88 J.J. Miller, "Further notes on the maintenance of fungus cultures in soil," *Proceedings of the Canadian Phytopathological Society* 16 (1948), 17. See also *Canadian Journal of Research* "c" 24 (1946), 188 and 213.

89 Some of the information about Koch was provided through personal communications. For examples of his largely mycological publications, see *Phytopathology* 21 (1931), 247–87; *Canadian Journal of Research* 11 (1934), 190–206; *Scientific Agriculture* 15 (1935), 729–44, and 25 (1945), 680–706.

90 For Halliday's contribution, see *Canadian Nature* vols. 2, 5, 6, 10, and 11. For Güssow's, see vol. 3, and for Scoggan's see vol. 12.

91 M.J. Berkeley, "Descriptions of Exotic Fungi in the collection of Sir W.J. Hooker, from Memoirs and Notes of J.F. Klotzch, with additions and corrections," *Annals of Natural History* 3 (1838), 375–401. See also ibid., 322, and Ibid. 7 (1841), 451–4.

92 John Dearness, *Report of the Canadian Arctic Expedition 1913–1918* (Ottawa 1923), vol. 4, part c (Fungi).

93 H.S. Spence, "Sub-Arctic Mushrooms," *Canadian Field Naturalist* 46 (1932), 53–4.

94 N. Polunin, editor, "Botany of the Canadian eastern Arctic. Part ii. Thallophyta and Bryophyta," *National Museum of Canada bulletin* no. 97 (Ottawa 1947). See also *Journal of Botany* 72 (1934), 197–204 *Journal of Ecology* 23 (1934), 161–209.

95 S.M. Pady, "Fungi isolated from arctic air in 1947," *Canadian Journal of Botany* 29 (1951), 46–56.

96 C.N. Bell, "President's Inaugural Address," *Transactions of the Manitoba Historical and Scientific Society* 1 (1889), 3–6

97 G.R. Bisby, A.H.R. Buller, and J. Dearness, *The fungi of Manitoba* (London 1929), 38–9.

98 N. Criddle, "The fly agaric and how it affects cattle," *Ottawa Naturalist* 11 (1906), 203. For an obituary notice, and a list of Criddle's publications, see *Canadian Entomologist* 65 (1947), 193–200.

99 A.H.R. Buller, "The red squirrel of North America as a mycophagist," *Transactions of the British Mycological Society* 6 (1920), 355–62.

100 G.R. Bisby, A.H.R. Buller, J. Dearness, W.P. Fraser, and R.C. Russell, *The fungi of Manitoba and Saskatchewan* (Ottawa 1938), 63 and 83. Biographies of the Criddle family of naturalists have been written by Alma Criddle, in *Criddle-de-diddle-ensis* (Winnipeg 1973).

101 R.H. Estey, "A.H.R. Buller: Pioneer leader in plant pathology," *Annual Review of Phytopathology* 24 (1986), 17–25.

102 H.J. Brodie, "Further observations on the mechanism of germination of the conidia of various species of powdery mildews at low humidity," *Canadian Journal of Research* "c" 23 (1945), 198–211.

103 H.J. Brodie, "Twenty years of nidulariology," *Mycologia* 54 (1962), 713–26.

104 G.J. Green, T. Johnson, and I.L. Conners, "Pioneer leaders in plant pathology: J.H. Craigie," *Annual Review of Phytopathology* 18 (1980), 19–25. See also *Canadian Journal of Plant Pathology* 9 (1988), 416–17, and *Nature* 120 (1927), 116–17, and 765–7.

105 J.H. Craigie, "Epidemiology of stem rust in western Canada," *Scientific Agriculture* 25 (1945), 285–401.

106 W.F. Hanna, "A simple apparatus for isolating single spores," *Phytopathology* 18 (1928), 1017–20. See also *Transactions of the British Mycological Society* 11 (1925), 219–38.

107 W.F. Hanna, "Notes on *Clitocybe illudens*," *Mycologia* 30 (1938), 379–84. See also ibid., 526–36, and ibid. 31 (1939), 250–7; anon., "Obituary notice, William Fielding Hanna," *Tableau* (December 1972), 11, and *Proceedings of the Canadian Phytopathological Society* 40 (1973), 23–4.

108 Irene Mounce, "Homothallism and heterothallism in the genus Coprinus," *Transactions of the British Mycological Society* 7 (1922), 256–69.

109 Dorothy Newton, "The bisexuality of individual strains of *Coprinus Rostrupianus*," *Annals of Botany* 40 (1926), 105–28; also ibid., 891–917.

110 A.H.R. Buller, *Researches on the fungi*, vol. 5 (1934), 207–8.

111 A.H.R. Buller and C.W. Lowe, "Upon the number of spores in the air of Winnipeg," *Transactions of the Royal Society of Canada* 4, section 4 (1911), 41–58.

112 P.H. Gregory, "The first benefactors lecture. The fungal mycelium: An historical perspective," *Transactions of the British Mycological Society* 82 (1984), 3.

113 G.R. Bisby, "Fungi of central Manitoba," *Mycologia* 16 (1924), 122–9. Also, G.R. Bisby, I.L. Conners, and D.L. Bailey, "The parasitic fungi found in Manitoba," in *A summary of the prevalence of plant diseases in the Dominion of Canada 1920–1924*, F.L. Drayton, editor. Dominion of Canada Department of Agriculture bulletin no. 71 (1926), 77–83.

114 W.P. Fraser, "Cultures of herbaceous rusts in 1918," *Mycologia* 11 (1919), 129–33.

115 G.R. Bisby, M.L. Jamieson and M.I. Timonin, "The fungi found in butter," *Canadian Journal of Research* 9 (1933), 97–107. Timonin's paper on soil fungi is in ibid. 13 (1935), 32–46.

116 J.E. Machacek, "Prevalence of *Helminthosporium sativum, Fusarium culmorum* and certain other fungi in experimental plots subjected to various cultural and manurial treatments," *Canadian Journal of Plant Science* 37 (1957), 353–65.

117 W.A.F. Hagborg, "William Laurence Gordon 1901–1963," *Mycologia* 56 (1964), 645. See also *Phytopathology* 54 (1964), 1, and *Proceedings of the World's Grain Exhibition Conference* 2 (Regina 1935) 298–9.

118 W.L. Gordon, "A study of the relation of the environment to the development of the uredineal and telial stages of the physiologic forms of *Puccinia graminis avenae* Erikss. and Henn.," *Scientific Agriculture* 14 (1933), 184–237.

119 R.H. Estey, "Margaret Newton: Distinguished Canadian scientist and first woman member of the Quebec Society for the Protection of Plants," *Phytoprotection* 68 (1987), 79–85.

120 Margaret Newton, "Additions to the fungus flora of the Mackenzie River Basin," in I.L. Conners, compiler, *Twentieth annual report of the Canadian Plant Disease Survey* (Ottawa 1940), 100–2.

121 G.J. Green, "Thorvaldur Johnson, 1897–1979," *Canadian Journal of Plant Pathology* 2 (1980), 3–4. See also T. Johnson, *Rust research in Canada and related plant-disease investigations*, Canada Department of Agriculture publication no. 1098 (Ottawa 1961).

122 G.R. Bisby, A.H.R. Buller, J. Dearness, W.P. Fraser, and R.C. Russell, *The fungi of Manitoba and Saskatchewan* (Ottawa 1938), 63.

123 See the University of Saskatchewan catalogue for 1917.

124 J.A. Stevenson, *An account of fungus exsiccati, containing material from the Americas* (1971), 269.

125 W.P. Fraser, "Culture experiments with heteroecious rusts in 1922, 1923 and 1924," *Mycologia* 17 (1925), 78–80.

126 W.P. Fraser, "Additions to the Uredinales of the prairie provinces of Canada," *Proceedings and Transactions of the Royal Society of Canada* 25, section 5 (1931), 85–92.

127 W.P. Fraser and G.A. Scott, "Smut of western rye grass," *Phytopathology* 16 (1916), 473–7.

128 W.P. Fraser and G.A. Ledingham, "Studies of the sedge rust, *Puccinia caricis-shepherdiae*," *Mycologia* 21 (1929), 86–9.

129 T.C. Vanterpool, "*Asterocystis radicis* in the roots of cereals in Saskatchewan," *Phytopathology* 20 (1930), 677–80.

130 T.C. Vanterpool, "Cultural and inoculation methods with Tilletia species," *Science* 75 (1932), 22–3.

131 T.C. Vanterpool and G.A. Ledingham, "Studies on "browning" root rot of cereals. i The association of *Lagena radicicola* n.gen., n.sp. with root injury in wheat," *Canadian Journal of Research* 2 (1930), 171–94.

132 T.C. Vanterpool, "Homothallism in Pythium," *Mycologia* 31 (1939), 124–7.

133 T.C. Vanterpool and J.H. Truscott, "Studies of browning root rot of cereals. ii Some parasitic species of Pythium and their relation to disease," *Canadian Journal of Research* 6 (1932), 68–93. See also *Mycologia* 25 (1933), 263–5.

134 T.C. Vanterpool, "Flax diseases in Saskatchewan," in *Annual Report of the Saskatchewan Department of Agriculture* (1947), 92.

135 R.A.A. Morrall and M. Shaw, "Thomas Clifford Vanterpool, 1898–1984," *Canadian Journal of Plant Pathology* 6 (1984), 336–7.

136 H.W. Mead, "Studies of methods for the isolation of fungi from wheat roots and kernels," *Scientific Agriculture* 13 (1933), 304–12.

137 See records for these fungi in Bisby et al., *The fungi of Manitoba and Saskatchewan*.

138 D.S. Hone, "Some western Helvellineae," in *Postelsia, The Yearbook of the Minnesota Seaside Station* (St Paul 1906), 238–41.

139 This is mentioned in T. Johnson and M. Newton, "The occurrence of stripe rust in western Canada," *Scientific Agriculture* 8 (1928), 464.

140 *Report of the Board of Governors and President, University of Alberta, 1937–8*, 10.

141 E.H. Moss, "The uredinia of *Cronartium comandre* and *Melampsora medusae*," *Mycologia* 20 (1928), 36–40. See also ibid. 21 (1929), 78–83.

142 E.H. Moss, "Overwintered giant puff-balls in Alberta," *Mycologia* 32 (1940), 271–3.

143 E. Silver Dowding, "*Wallrothiella arceuthobii*, a parasite of Jack pine mistletoe," *Canadian Journal of Research* 5 (1931), 219–31. See also ibid. 6 (1932), 2–20; ibid. 9 (1933), 294–305; *Annals of Botany* 45 (1931), 1–14, and 621–37.

144 A.W. Henry, "Influence of soil temperature and soil sterilization on the reaction of wheat seedlings to *Ophiobolus graminis* Sacc.," *Canadian Journal of Research* 7 (1932), 198–203. Also, ibid. 21 (1943), 343–50.

145 E.C. Stakman, A.W. Henry, G.C. Curran, and W.N. Christopher, "Spores in the upper air," *Journal of Agricultural Research* 24 (1923), 599–605.

146 G.B. Sanford, "Some factors affecting the pathogenicity of *Actinomyces scabies*," *Phytopathology* 16 (1926), 525–47.

147 G.B. Sanford, "A root-rot of sweet clover and related crops caused by *Plenodomus meliloti* Dearness and Sanford, nov. spec.," *Proceedings of the Canadian Phytopathological Society* 1 (1930), 26.

148 G.B. Sanford, "Studies on *Rhizoctonia solani* Kühn. i The effect of potato tuber treatment on stem infection six weeks after planting," *Scientific Agriculture* 17 (1937), 225–34. Number 5 of this series is in *Canadian Journal of Research* 19 (1941), 1–8.

149 G.B. Sanford and W.C. Broadfoot, "On the prevalence of pathogenic forms of *Helminthosporium sativum* and *Fusarium culmorum* in the soil of wheat fields and its relation to the root-rot problem," *Canadian Journal of Research* 10 (1934), 264–74.

150 W.C. Broadfoot and M.W. Cormack, "A low-temperature basidiomycete causing early spring killing of grasses and legumes in Alberta," *Phytopathology* 31 (1941), 1058–9.

151 M.W. Cormack, "*Cylindrocarpon ehrenbergi* Wr., and other species, as root parasites of alfalfa and sweet clover in Alberta," *Canadian Journal of Research* "c" 15 (1937), 403–24. See also ibid., 493–510; *Phytopathology* 30 (1940), 700–1; ibid. 35 (1945), 838–55, and *Scientific Agriculture* 26 (1946), 448–59.

152 J. Dearness, "E.W.D. Holway," *Mycologia* 38 (1946), 231–9.

153 Stevenson *An account of fungus exsiccati*, 33 and 139.

154 Hone, "Some western Helvellineae," 238–41.

155 W.N. Cheesman, "A contribution to the mycologic flora and the Mycetozoa of the Rocky Mountains," *Transactions of the British Mycological Society* 3 (1911), 267–76.

156 W.A. Murrill, "Agaricaceae of the Pacific coast," *Mycologia* 4 (1922), 205–17, and 231–62.

157 *Autobiography of John Macoun, Canadian explorer and naturalist 1830–1920*, second edition (Ottawa 1979), 300. See also W.A. Waiser, *The field naturalist: John Macoun, the Geological Survey and natural science* (Toronto 1989).

158 Margaret Barr, "Pyrenomycetes of British Columbia," *Canadian Journal of Botany* 31 (1953), 810–30. See also J. Ginns, "The Canadians John Macoun and John Dearness, and their contributions to North American mycology," *Mycotaxon* 26 (1986), 47–53.

159 J.W. Eastham, "The Myxomycetes of the Ottawa district – A preliminary list," *Ottawa Naturalist* 25 (1912), 157–63. See also *Annual Report of*

the Quebec Society for the Protection of Plants 5 (1916), 66, and anon., "In Memoriam, John William Eastham (1880–1968)," *Proceedings of the Canadian Phytopathological Society* 37 (1970), 32.

160 I.L. Conners, editor, *Plant pathology in Canada* (Winnipeg 1972), 221, and *Canadian Journal of Research* 6 (1932), 253–4.

161 Jean E. Davidson, "Notes on the Agaricaceae of Vancouver (B.C.) district I," *Mycologia* 22 (1930), 80–93.

162 W. Newton and C. Yarwood, "The downey mildew of the hop in British Columbia," *Scientific Agriculture* 10 (1930), 508–28.

163 W. Newton and R.J. Hastings, "*Botrytis tulipae* (Lib.) Lind.," *Scientific Agriculture* 11 (1931), 820–4.

164 W. Jones, "A new technique for obtaining oospores of the hop downy mildew by inoculating cotyledons," *Science* 75 (1932), 108. See also *Journal of the Institute of Brewing* 29 (1932), 194–6, and ibid. 30 (1933), 126–7.

165 N.L. Cutler, "A contribution to the knowledge of the tree-destroying fungi of the Vancouver forestry district," *Phytopathology* 13 (1923), 294.

166 Irene Mounce, J.E. Bier, and Mildred K. Nobles, "A root-rot of Douglas fir caused by *Poria wierii*," *Canadian Journal of Research* "'c'" 18 (1940), 522–33.

167 J.E. Bier and Mildred K. Nobles, "Brown pocket rot of Sitka spruce," *Canadian Journal of Research* "c" 24 (1946), 115–20.

168 Irene Mounce and J.E. Bosher, "Seedling blight of carrot caused by *Alternaria radicina*," *Scientific Agriculture* 23 (1943), 421–3.

169 J. Ginns, "Irene Mounce, 1894–1987," *Mycologia* 80 (1988), 607–8.

170 G.A. Hardy, "Bracket fungus – The Dryad's Saddle," *Canadian Nature* 5:3 (1944), 80–1.

171 D.B.O. Savile, "Nuclear structure and behavior in species of the Uredinales," *American Journal of Botany* 26 (1939), 585–609. See also *Canadian Journal of Research* "c" 24 (1946), 109–14, and *Mycologia* 38 (1946), 453–4.

Index of Names

General Index